Cities Research Series

Series Editor

Paul Burton, Gold Coast Campus, Griffith Univ, Urban Research Program, Southport, QLD, Australia

This book series brings together researchers, professionals and policy makers in the area of urban development and publishes recent advances in the field. It addresses issues to understand and meet urban challenges and make (future) cities sustainable and better places to live. The series covers the following topics, but not limited to:

- Transport policy and behaviour
- Architecture, architectural science and construction engineering
- Urban planning, urban design and housing
- Infrastructure planning and management
- Complex systems and cities
- Urban and regional governance
- Smart and digital technologies

More information about this series at http://www.springer.com/series/16474

Joerg Baumeister · Edoardo Bertone ·
Paul Burton
Editors

SeaCities

Urban Tactics for Sea-Level Rise

Springer

Editors
Joerg Baumeister
Gold Coast Campus
Cities Research Institute
Southport, QLD, Australia

Edoardo Bertone
Gold Coast Campus
Cities Research Institute
Southport, QLD, Australia

Paul Burton
Gold Coast Campus
Cities Research Institute
Southport, QLD, Australia

ISSN 2662-4842 ISSN 2662-4850 (electronic)
Cities Research Series
ISBN 978-981-15-8747-4 ISBN 978-981-15-8748-1 (eBook)
https://doi.org/10.1007/978-981-15-8748-1

This Springer imprint is published by the registered company Springer Nature Singapore Pte Ltd.
The registered company address is: 152 Beach Road, #21-01/04 Gateway East, Singapore 189721, Singapore

This book is dedicated to all the members of the Griffith SeaCities group for your contributions to this book and to your family and friends who supported you in this project.

Foreword by Prof. Andrew T. Smith

Coastal cities will have to change in the future. The higher the predictions of sea-level rise and its associated impacts, the sooner this change will have to occur. While some see the complexity of cities as a hindrance, we see the opposite: complexity can be helpful in exploring cost-effective opportunities to redesign and rebuild our coastal cities. The book is intended to test this assumption. It is the result of collaboration between like-minded academic and built environment professionals who came together at the first SeaCities Symposium, held at Griffith University's Gold Coast campus from 2–4 March 2020.

We believe the SeaCities initiative provides an excellent platform for urban experiments designed in response to the challenges associated with sea-level rise. As part of the Cities Research Institute, one of the largest groups of urban researchers in Australia, and in partnership with a network of researchers from around the world, including at Columbia and Delft University and with industry

SeaCities Symposium, 2 March 2020, Griffith University

partners like DeFacto, we are well placed to draw on a wide range of practical expertise and critical insight. This enables a truly interdisciplinary approach including the perspectives of architecture, urban planning and governance, systems engineering, as well as cultural ecology, human health, physics, oceanography and indigenous jurisprudence.

This book is a first milestone of the SeaCities research group which we hope will lay the foundation for a continuing critical dialogue about how we might live more securely and sustainably in coastal settings. If you would like to join us in this dialogue, we welcome you contacting any of the authors or editors.

Yours Sincerely

Prof. Andrew T. Smith
PVC Science
Griffith University
Brisbane, Australia

Preface

Griffith University was created to be a different kind of university: challenging conventions through interdisciplinary research and teaching to create bold new visions of the future and pioneering solutions to real-world problems. The SeaCities research group is a new example of this pioneering spirit.

According to the World Bank, three billion of the global population live in coastal communities. Over 400 million people around the world, and over one million people in Australia live less than five metres above sea level. With predicted rises in sea level, people living in these coastal settlements will experience many detrimental effects, including loss of land, the impacts of more intense storm surges, deterioration of urban and natural environments, infrastructure vulnerabilities, human health risks, food insecurity and, as a consequence, profound climate-related existential threats.

Developing and applying new digital and physical models in combination with traditional research methods and adaptable design strategies is a key feature of the methodological innovations pioneered by members of the SeaCities group. This holistic research approach—that spans and includes the disciplines of engineering, architecture and urban design, urban and regional planning and other environmental sciences—enables researchers in the group to develop new approaches to building with and for nature, to create ecosystem-based developments on land and sea that respond to the challenges faced by coastal communities in an adaptive and compatible fashion.

The transition from terrestrial to amphibious to aquatic developments creates exciting design opportunities that can relieve land-based population pressures and foster innovative development solutions. Furthermore, it shifts the focus of research and its application from defining and managing increasing risks for coastal communities towards the exploration and development of novel design, engineering and infrastructure solutions. These can seek to blend and merge the sea and land environments in a productive and sustainable fashion.

This book provides a critical overview of different approaches to dealing with the challenges faced by and opportunities available to contemporary coastal cities. Insightful as well as provocative, this book is to be commended to anyone interested in exploring and developing the idea of SeaCities. As well as reporting on research carried out by members of the SeaCities group, it also includes contributions from some of the leading researchers and practitioners in this field from around the world.

Southport, Australia

Joerg Baumeister
Edoardo Bertone
Paul Burton

Acknowledgements

The editors are thankful to all the chapter authors for their outstanding contributions to this book.

The editors would also like to thank all the chapters reviewers for dedicating their valuable time to improving the quality of this book.

Finally, we would like to thank the City of Gold Coast and Griffith University for their support in helping organise and host the SeaCities Symposium in March 2020 and especially to Diane McDonald of the Cities Research Institute for her patience and logistical expertise. We are also grateful to Professor Carolyn Evans, Vice Chancellor and President of Griffith University; Professor Andrew T. Smith, Pro-Vice Chancellor of Griffith Sciences; and Councillor Tom Tate, Mayor of the Gold Coast for supporting and helping launch the SeaCities Research Initiative and the Symposium which made a significant contribution to the conceptualisation and delivery of this book.

Contents

Editors and Contributors

About the Editors

Prof. Dr.-Ing. Joerg Baumeister has been a practitioner, educator, researcher and consultant for Architecture and Urban Design for more than 20 years throughout Europe, Africa, the Arabian Peninsula, Asia and Australia. As University educator with passion for "his" students, he tries to fuse research, higher education and implementation in order to create feedback loops to evolve the fields of Urban Design and Architecture. Joerg founded in 2019 the "SeaCities" research laboratory (SeaCities.org) at the Cities Research Institute, Griffith University to develop water-adapted cities and floating structures. He is consulting governmental institutions on the federal, state and regional level as well as NGOs and private industry leaders to apply his current research interests which comprise SeaCities, ecological cities, affordable housing in serial building technology and design innovation through creative thinking. He is an award-winning Architect and Urban Designer and continues to be an enthusiastic speaker at international conferences. Joerg is DAAD Ambassador and Scientist for Future.

Dr. Edoardo Bertone is a researcher with a civil engineering background, and he has a current research focus on data-driven modelling, Bayesian network and system dynamics modelling applied to the water resources management, water treatment optimisation, water-energy nexus, sustainable cities, environmental health and climate change adaptation fields. He completed his Ph.D. in water resources engineering in 2015, and he is currently a lecturer at Griffith University. He collaborates with several large Australian water utilities, Government departments and city councils, and he is a member of the SeaCities group under Griffith's Cities Research Institute.

Paul Burton is Professor of Urban Management and Planning and Director of the Cities Research Institute at Griffith University. Having trained and worked in London as a town planner, he joined the School for Advanced Urban Studies at the University of Bristol, UK, to carry out research for his Ph.D. on the redevelopment of London's docklands. Over the next three decades, he led and contributed to research projects on living conditions in European cities, homelessness among young people, the links between housing and labour markets as well as a range of national urban policy evaluations. Since moving to Australia in 2007, Paul's research has focused on multi-level governance in metropolitan settings, new techniques of community engagement and the professional lives of planners. He is the Series Editor for the Springer Cities Series and an active member of the Planning Institute of Australia.

Contributors

Marco Bagliani University of Torino, Turin, Italy

Joerg Baumeister Cities Research Institute, Griffith University, Brisbane, Australia

Edoardo Bertone School of Engineering and Built Environment, Cities Research Institute, Australian Rivers Institute, Griffith University, Brisbane, QLD, Australia

Cecilia Bischeri Architecture and Design, School of Engineering and Built Environment, Gold Coast Campus, Griffith University, Brisbane, QLD, Australia

C. F. Black Griffith Centre for Coastal Management, Cities Research Institute, Griffith University, Brisbane, QLD, Australia

Elena Comino Politecnico di Torino, Turin, Italy

Bruno De Meulder KU Leuven, Leuven, Belgium

Aysin Dedekorkut-Howes Griffith University, Gold Coast, Australia

Ioana Corina Giurgiu SeaCities, Cities Research Institute, Griffith University, Gold Coast, QLD, Australia

Olv Klijn FABRICations, Amsterdam, The Netherlands

Despina Linaraki SeaCities, Cities Research Institute, Griffith University, Gold Coast, Queensland, Australia

Han Meyer Delft University of Technology, Delft, The Netherlands

Thang Viet Nguyen SeaCities, Griffith University, Griffith, Australia

Anne Loes Nillesen Defacto Urbanism, Rotterdam, The Netherlands

Eugen Bogdan Petcu New York Institute of Technology, Old Westbury, NY, USA;
School of Medicine, Menzies Health Institute of Queensland, Griffith University, Brisbane, QLD, Australia

Alessandro Pezzoli Politecnico di Torino, Turin, Italy

Eva Pfannes OOZE Architects & Urbanists, Rotterdam, The Netherlands

Carlotta Quagliolo Politecnico di Torino, Turin, Italy

Avery Robertson Independent Scholar, New York, USA

Oz Sahin School of Engineering and Built Environment, Cities Research Institute, Griffith University, Brisbane, QLD, Australia;
Griffith Climate Change Response Program, Brisbane, QLD, Australia

Kelly Shannon KU Leuven, Leuven, Belgium

Elnaz Torabi Griffith University, Gold Coast, Australia

Julia Watson Graduate School of Design, Harvard University, Cambridge, USA;
School of Architecture, Planning and Preservation, Columbia University, New York, USA

Mona zum Felde Defacto Urbanism, Rotterdam, The Netherlands

Re-Building Coastal Cities: 20 Tactics to Take Advantage of Sea-Level Rise

Check for updates

Joerg Baumeister

Abstract Current urban adaptation methods to Sea-level rise (SLR) derive mostly from technical- and economic-driven risk-management. It does not consider opportunities created by the complexity of cities which include social, productive, cultural, and ecological components. This chapter explores this research gap by reflecting known adaptation methods onto individual urban components. It focuses on the creation of opportunities resulting in a systematics that provides 20 urban adaptation tactics. These different tactics can be either applied individually or combined to strategies to achieve made-to-measure solutions. Compared to current adaptation methods, the tactics are expected to enable solutions which are more adaptable to specific challenges and locations. Thereby the tactics provide the additional potential to offset expenditures of SLR and to be more city, citizen, and environmentally friendly.

1 Sea-Level Rise (SLR)

To better understand the effects and dimensions of global warming, we have to mention these effects briefly at the beginning: in the 1970s, the Club of Rome published with "Limits to Growth" forecasts on global warming and its impact (Meadows et al. 1972). Today, fifty years later, global warming is a fact, and no rational individual is questioning global warming anymore. The data situation and discussion have meanwhile refined, with probabilities of different warming scenarios being assessed (IPCC 2019)

It becomes increasingly evident as to what extent global warming has so far produced unexpectedly high and diverse economic, ecological, social, and geopolitical risks. While "Limits to Growth" still largely addressed experts and interested laypersons, the problems of global warming are now perceived by the general public (World Economic Forum 2020).

J. Baumeister (✉)
Cities Research Institute, Griffith University, Brisbane, Australia
e-mail: j.baumeister@griffith.edu.au

J. Baumeister et al. (eds.), *SeaCities*, Cities Research Series,
https://doi.org/10.1007/978-981-15-8748-1_1

Today, it is common sense that greenhouse gases such as CO_2 or methane, which are produced when fossil fuels are burned, are responsible for global warming. An increase in greenhouse gases in the atmosphere leads to higher temperatures which in turn causes sea-levels to rise. So there is a direct correlation between the proportion of greenhouse gases and SLR. A temperature-related melting or freezing, e.g. of the north and south polar ice, is, however, a slower process than an increase or decrease in temperature, which is why the greenhouse gas concentration and sea-level curve are shifted in time (see Fig. 1.1).

The current greenhouse gas concentration of now over 400 ppm is the highest in millennia, and a corresponding record high of the sea-level will logically follow: in earlier phases of history with a similarly high greenhouse gas concentration, the

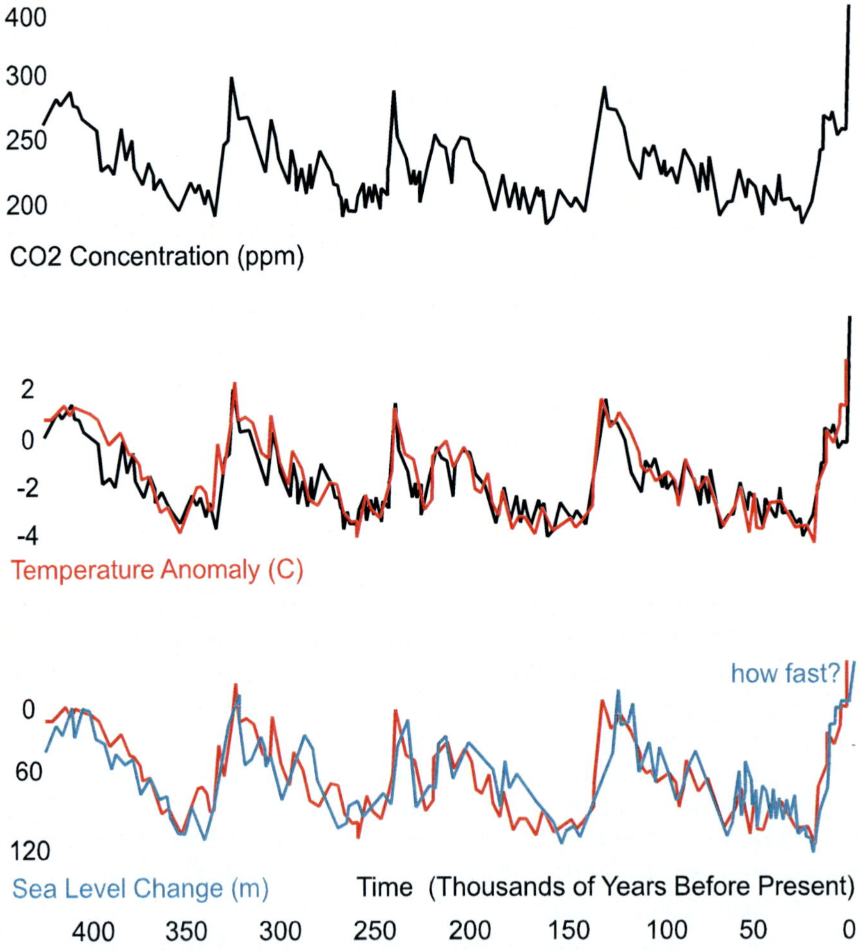

Fig. 1.1 (based on Hansen and Sato 2012): CO_2 concentration, temperature anomaly, and SLR changes for the past 400,000 years

SLR was 15–25 metres higher (Jones 2020). The crucial question is therefore not WHETHER, but HOW FAST the ice will melt in the next decades.

Current intensive research suggests that the more the melting behaviour and its collaterals are understood, the higher the predictions for SLR. New estimates still cannot "explicitly calculate changes in ice mass, partly because of the small scales necessary to resolve ice sheet processes properly, and partly because the relevant physical processes are poorly understood" (Pörtner et al. 2019). But the (always upward developing) predictions estimate a SLR of up to 1.40 m or more by 2100, and a simultaneous up to a 100-fold increase of "extreme sea-level events" (Pörtner et al. 2019).

2 Urban Flooding

Global warming has several other effects in addition to rising sea-levels: the hydrological pressure on coasts changes the condition of groundwater, which can rise and become saline (Ketabchi et al. 2016). And the rising temperatures cause more dynamic movements of air and water, which leads to more extreme weather. Water-related consequences include an increase in the likelihood of heavy rain, which can lead to flooding of land and rivers (King et al. 2015). More water-related impacts of global warming and potential combinations of fatal impacts are highlighted in Fig. 1.2.

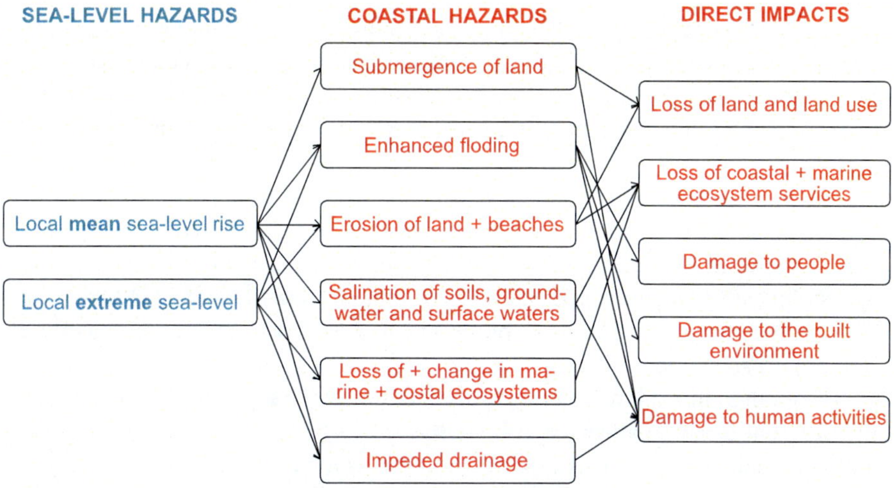

Fig. 1.2 (based on Pörtner et al. 2019): Overview of the main cascading effects of SLR. "Mean SLR" describes thereby the sea-level halfway between high and low water, whereas "extreme sea-level" expresses the maximum level during a selected period like a year

The example of New Orleans impressively demonstrates to what extent flooding events can substantially damage a city: New Orleans is located largely below sea-level and must, therefore, be protected by dykes. The weak point of this so-called polder situation is its susceptibility to flooding as a result of hurricanes, including the associated heavy rain and river flooding (Jonkman et al. 2013). Unfortunately, New Orleans had failed to maintain the levees so that they could not defy hurricanes.

Before hurricane Betsy flooded the city in 1965, New Orleans had 625,000 inhabitants. Afterwards, considerable investments in flood protection subsequently led to a population increase to 500,000. In 2005, hurricane Katrina destroyed the city again. This time 80% of the city was flooded resulting in direct damage of US $ 40 billion and further investments of US $ 15 billion to improve flood protection. Despite the enormous efforts, the population has since halved to 300,000, and a large part of the economic power has moved away (Link 2010).

Globally, the economic impact of flooding in correlation to SLR could average between 1 and 5% of GDP annually (see Fig. 1.3). An exclusive focus on the 400 million population living in the 136 endangered coastal cities with more than 1 million inhabitants will move the annual average costs of 1–5% GDP much higher (Hanson et al. 2010).

3 Re-Building Coastal Cities

We have to Re-Build Coastal Cities: The upcoming rise in mean sea-level and extreme sea-level events will flood coastal cities increasingly (see 1.). Built and natural environment of coastal cities will be impacted dramatically including citizen's health and the urban economy. This poses the question: How can coastal cities adapt best to SLR? One logical approach is to try to risk-manage SLR. Each measure is thereby often considered just as expenditure which tries to minimise damage of SLR (see 2.).

The aim of this publication is to suggest a broadened perspective from risk management to the creation of urban opportunities (see 4.). It will consider thereby cities as specific complex structures consisting of buildings and spaces, economy, community, infrastructure, and natural environment (Baumeister and Ottmann 2015). This creates the opportunity to explore potential SLR adaptation methods for each of these urban elements.

The method to explore potential opportunities will happen in the following way: the different urban elements will be described (5.) and current adaptation models analysed and, if required, optimised (6.). The combination of the urban elements with the adaptation methods in form of a matrix will allow the development of 20 tactics to take advantage of SLR (7.) which will be discussed at the end (8.).

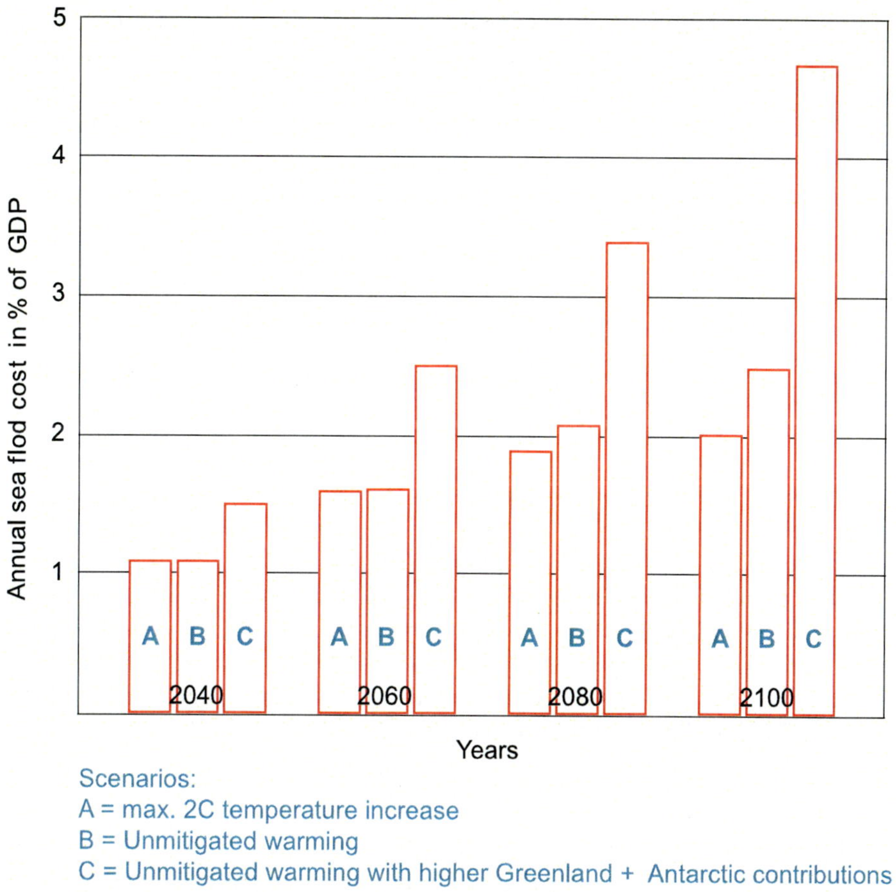

Fig. 1.3 (based on Jevrejeva et al. 2018) showing predicted annual global flood costs (in % of GDP) without additional adaptation for three climate scenarios

4 Risks as Opportunities

The effects of SLR after the last ice age are well documented (Galili et al. 2019). Related myths about coastal flooding developed thousands of years ago are still held today as part of indigenous cultures. An angry deity destroyed, for example, a port city with a flood in the Tamil epic Manimekhalai (Nunn 2014). Physical evidence of SLR thousands of years ago can be found in many coastal locations around the world, e.g. on the Yangtze River, where the Baoden people had to leave areas partially or built seawalls to protect themselves from flooding (Zeng et al. 2016).

Another monument to the effects of SLR was recently discovered on the Israeli coast: During the underwater excavation of an approximately 7000-year-old settlement, a wall made of large stones with a length of more than 100 m was discovered, which stretched parallel to the coastline. The wall is now 90 m offshore about 3 m below the sea-level (Galili et al. 2019). Apparently, the residents had tried to save their city, which makes this sea wall to a memorial that can remind us of the limited nature of protective measures against SLR and its risk of physical and social disintegration.

On the other hand, flooding can stimulate the reorganisation and regeneration of states and their economic resources, e.g. historically in Mesopotamia and Egypt. The relationship between human societies and floods created their advantages for agriculture, transportation, and trade that were decisive for the emergence of early civilisations (Soroush and Mordechai 2018).

The history of other countries like the Netherlands (which is almost a third below sea-level) also shows how SLR can be viewed not only as a risk but also as an opportunity: as early as the ninth century, the collective reaction to flooding events played a community-building role in the Netherlands, which led to the success of elected boards as one of the first democratic institutions. The Dutch society had to be always open to new technologies for water management, construction, and transportation, and their joint development and financing models were the reasons for Holland's rise to become a world power. Nowadays, in the period of increasing SLR, the Netherlands plays a central role in areas such as water management, land reclamation, and land defence (Van Kongingsfeld et al. 2008).

Or, as announced in the New York Times, "The Dutch Have Solutions to Rising Seas. The World Is Watching. In the waterlogged Netherlands, climate change is considered neither as hypothetical nor as a drag on the economy. Instead, it's an opportunity "entirely in the sense of "live with the water rather than struggle to defeat it" (Kimmelmann in New York Times 2017).

5 Urban Elements

To achieve a far-reaching change from "defeat the water" to "living with the water", the complete urban system has to be considered. Therefore, cities should not only be viewed from, e.g., purely economic but from a holistic perspective.

The publication "Urban Ecolution" (Baumeister and Ottmann 2015) offers this holistic description of the city system which arranges the urban elements as a taxonomy. This reference system connects the physical, economic, social, and technical elements of a city. To achieve this, "Urban Ecolution" first analyses the history of urban configurations and tries to understand them as an urban metabolism. Then, ecology is presented as a system to remind us that we can do things differently to live with nature rather than the struggle to defeat it. Finally, it applies the principles of ecological systems to urban systems and describes the following system elements:

- "Buildings and Space" are thereby residential, commercial, industrial, cultural, and special building typologies in combination with the surrounding urban space.

- "Production" includes industry, commerce, retail, aquaculture, agriculture, and forestry.
- "Community" comprises culture, spirituality, education, research, recreation, governance, and politics.
- "Infrastructure" describes the combination of transport, water, sewage, waste, power, and communication.
- "Natural Environment" is the amalgamation of sun, air, biomass (plants, animals), minerals, and water.

6 Urban Adaptation Models for SLR

After the description of the urban elements above, current adaptation models have to be analysed now. The aim is to combine afterwards the urban elements with the adaptation methods to develop a systematics of different opportunities for SLR. Therefore, the question is now how far there are already models (e.g. from coastal protection, ecology, and economics) that describe urban adaptation methods to SLR. The goal is either to adopt an existing model or, if necessary, to develop a new one.

(a) A first adaptation model focuses on the adaptation method "Protect" to defend urban "assets" as suggested, for example, by C40: 94 cities and Ramboll Foundation thereby developed a framework, which measures the progress of urban climate-change adaptation. To combat storm surges and SLR, the framework includes the installation of flood gates to reduce storm surge flooding, a permanent coastline protection by building dykes and seawalls, a stabilisation of river banks with vegetation or concrete reinforcement as well as a relocation of assets which are at risk or an adaptation of assets by elevating or hardening them (C40 2019).

(b) Other experts have proposed to organise a model which is based on the history of coastal engineering. For example, K. Hill suggests a model which builds on the physical form of walls and landforms (Hill 2015). Walls are thereby subdivided into static walls (seawalls, floodwalls) and dynamic walls (tide gates, movable surge barriers). The same distinction happens between landforms which are either static (levees, dykes, mounds, breakwaters) or dynamic (beaches, sand dunes and bars, marshes). These tools allow for the application of different adaptation methods called "Robust" (= to prevent negative consequences), "Resilient" (= able to recover), or "Adaptive" (not greatly affected by the event).

(c) The World Bank suggests a model which considers three adaptation methods which react more divers on urban coastal areas: "Protect" reduces the likelihood of hazards, "Accommodate" reduces the impact of the hazard event, and "Retreat" reduces exposure by moving away from the source of the hazard.

Technological and policy options are for

- "Protect": Dykes, levees, floodwalls, seawalls, bulkheads, flood gates and tidal barriers, detached breakwaters, periodic beach nourishment, wetland restoration, afforestation, wooden and stone walls.
- "Accommodate": Emergency planning, insurance, modification of buildings to cope with floods (strengthen and lift), improved drainage, strict regulation in hazard zones, modification of land use planning.
- "Retreat": Increase or establish retreat zones, relocate threatened buildings, phase out or ban development in areas susceptible to flooding, rolling easements, erosion control easements, upland buffers (World Bank 2017).

(d) The United Nations' Environment Programme created together with the World Meteorological Organisation in 1988 the IPCC (Intergovernmental Panel on Climate Change) which describes in Chapter 4 of its updated 2019 sea-level report one of the most comprehensive models. It follows thereby mostly the World Bank's suggestions but adds subcategories and a new adaptation method:

- **Protection** "reduces coastal risk and impacts by blocking the inland propagation and other effects of mean or extreme sea-levels" (Oppenheimer). Elements of the so-called hard protection are dykes, seawalls, breakwaters, barriers, and barrages to protect against flooding, erosion, and salt-water intrusion. Elements of the "sediment-based protection" are beach and shore nourishment and dunes.
- **Accommodation** "includes diverse biophysical and institutional responses that mitigate coastal risk and impacts" (Oppenheimer et al. 2019). Accommodation measures include to lift buildings on stilts and podiums or to move water-sensitive functions of buildings onto higher floors or "floating houses and gardens" (Trang 2016).
- **Retreat** "reduces coastal risk by moving exposed people, assets, and human activities out of the coastal hazard zone" (Oppenheimer et al. 2019). This includes "migration" which is voluntary, "displacement" as involuntary and unforeseen movement, and "relocation" which is a managed retreat typically implemented by governments.
- **Advance** "creates new land by building seaward, reducing coastal risks for the hinterland and the newly elevated land" (Oppenheimer et al. 2019) comprising land reclamation by land filling, natural accretion by planting specific vegetation.

Comparing the four SLR adaptation models, model (a) mainly considers minimising damage by preserving assets and is therefore excluded. The second model (b) suggests a variability between static and dynamic measures, but the small number of components allows only limited use in an urban context. In contrast to that, model (c) considers and sorts a much larger number of components. Model (d) is the most comprehensive and has a lot in common with proposal number (c). However, there is sometimes ambiguity of adaptation methods which swing between human, technical, and physical aspects.

Consequently, in the following, a new model is developed to describe urban adaptation methods to SLR. It is based on model (d) and integrates components of (c) but also takes the following aspects into account:

- The IPCC report describes "ecosystem-based protection" by plants as a separate adaptation method. Its main purpose is still to protect; therefore, we keep it as part of this response type.
- Due to increasing attention to floating structures as urban development measures, we propose to include them in the adaptation method "Advance".

This results in the following four adaptation methods:

Protect:

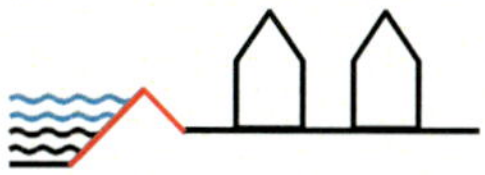

Either there is a hard respond as dykes, seawalls, breakwaters, barriers, and barrages, or it is sediment-based as beach and shore nourishment and dunes, or it is ecosystem-based including plants, wetland restoration, afforestation.

Accommodate:

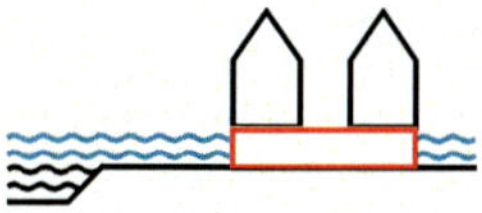

Accommodation measures include to lift buildings on stilts and podiums or to move water-sensitive functions of buildings onto podiums.

Retreat:

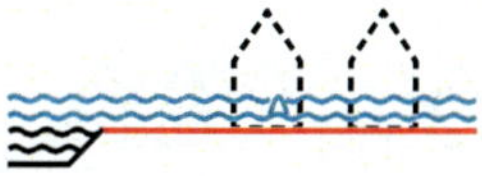

Retreat zones, rolling easements, erosion control easements, upland buffers are increased or established, and threatened buildings relocated.

Advance

Comprising land reclamation by landfilling as well as natural accretion by planting specific vegetation and floating structures are introduced.

7 20 Tactics

The combination of both the defined adaptation methods and the described urban elements together create a systematics of opportunities responding to SLR. The individual elements will be called "tactics" which, in normal usage, are often equated with the term "strategy". However, there are differences: While tactics reflect a single activity or technique, strategy rather refers to the "large, overarching plan in which parallel or successive tactics are coordinated" (Strategy in Encyclopaedia Britannica 2020). Therefore, the individual parts of the systematics are the tactics which will be created and described now. Afterwards, corresponding strategies which use different tactics as part of an overarching plan will be discussed.

The tactics will describe opportunities for the different adaptation methods projected onto the various elements of the city, both as described above. This results in a matrix where adaptation methods can be applied in the vertical direction and the urban system elements in the horizontal.

	Buildings/Space	Production	Community	Infrastructure	Natural environment
Protect	1.	2.	3.	4.	5.
Accommodate	6.	7.	8.	9.	10.
Retreat	11.	12.	13.	14.	15.
Advance	16.	17.	18.	19.	20.

Each combination results in a tactic describing an opportunity for SLR. This is executed in rows for 1. – 5. (Protect), 6.– 10. (Accommodate), 11. – 15. (Retreat), and 16. – 20. (Advance).

Tactics for **"Protect"** (1 .– 5.)

The construction of dykes and other hard protection elements such as walls, barriers, and barrages align themselves against SLR. However, their height must be adapted to increasing SLR, whereby in the case of "Protect" the measures are becoming more and more complex. The base of dykes and the foundations of protective walls have physical limits to growth, which is the reason why it is more interesting in short to medium term.

Exceptions are sediment—and eco-based nourishments, which ideally adapt to SLR automatically.

Tactic 1 (Protect × Buildings):

The logic of this opportunity consists in the functional development from a pure protective structure to the creation of additional building volumes for residential, commercial, industrial, cultural, and special buildings. This can reduce the construction costs of the purely protective structure, and at the same time, the city and urban spaces can expand.

Tactic 2 (Protect × Production):

In this case, the surface of the protective structure is used to enable additional functions resulting in commercial opportunities to produce wind or solar energy. Extensive as well as intensive agriculture is an option, also because the slightly salty note of the products (e.g. lamb meat or strawberries) is very popular for gourmets. In case of soft solutions, beaches and dunes contribute to the attractiveness for the tourism industry.

Tactic 3 (Protect × Community):

The sense of community strengthens with the common interest in the maintenance of the protective measures and the shared responsibility in extreme water events. The created landscape spaces such as dykes or dune landscapes also offer leisure areas to be enjoyed together.

Tactic 4 (Protect × Infrastructure):

The top of dykes as well as dune trails are attractive bike and footpaths. Additional infrastructure for extra energy supply (e.g. due to the wind or solar energy generated in tactic 2.) or for decentralised wastewater disposal can also be accommodated there.

Tactic 5 (Protect × Ecosystems):

Hard protective measures such as dykes or protective walls can be ecologically upgraded with plantings and thus contribute to the diversity of flora and fauna. The natural environment of sediment-based and ecosystem-based environments will undergo a constant process of change that can still be controlled by human interventions (such as planting indigenous trees).

Tactics for **"Accommodate"** (6. – 10.)

While "Protect" asserts itself horizontally against SLR, "accommodate" chooses the vertical. If possible, buildings can be raised, or water-sensitive functions can be moved upwards. Instead of these short-term limitations to flooding damages, the vertical "escape" from SLR can be achieved more sustainably in the longer term by building podiums either from piles or from sediments. If the podium is stable enough, it can be raised further in the future.

While individual property owners invest in "Protect" in a common protective measure, the "Accommodate" measures take place directly on the individual properties, so that SLR can be reacted to both jointly as community and as private individual.

Tactic 6 (Accommodate × Buildings):

The volume required to create the podiums produces additional interior and exterior space. This allows the expansion of usable building space, and the protection of outdoor space against SLR. In a transition period in which the surroundings are only flooded during extreme events, the lower level and the outside space in front of it can be used almost without restrictions and, if necessary, protected by bulkheads.

Tactic 7 (Accommodate × Production):

Additional areas created in the podium can be used for industry, commerce, and retail. The podium level first produces added value as an additional access level. With increasing flooding, the relationship to water and the associated waterfront is strengthened, which can have a positive effect on the commercial value of uses ("Blue Economy").

Tactic 8 (Accommodate × Community):

A slow move of public space from the natural surface to the podiums creates a restriction of freedom of movement on one side. On the other hand, a community feeling is strengthened, and public places are used more intensively which is comparable to the "sense of community" of island residents. The increasing proximity to water will also enhance "Blue Health" (describing the impact of water-based environments on human health and wellbeing) and will create new leisure opportunities.

Tactic 9 (Accommodate × Infrastructure):

Infrastructure in flooded areas must adapt to SLR to ensure the functionality of buildings. The podiums can stay connected to the centralised infrastructure, which requires joint coordination and financing. However, SLR can also drive the opportunity to increasingly consider decentralised solutions (e.g. decentralised energy and water supplies or wastewater disposal) for the podiums.

Tactic 10 (Accommodate × Ecosystems):

The edges of podiums and pile foundations are amphibious transition zones between land and water, which, depending on the climate zone, can be ecologically upgraded with plants such as reeds or mangroves.

Tactics for "**Retreat**" (11. – 15.):

Retreat might appear as the least obvious opportunity to SLR since it abandons human-cultivated areas either temporary or permanent: Existing buildings and infrastructure become unusable, and therefore, values are destroyed and communities are exiled. However, a change of perspective can help to recognise positive aspects, because there are amphibious transition zones created which can be discovered and used by both humans and nature.

Tactic 11 (Retreat × Buildings):

While existing buildings and urban spaces are increasingly giving way to SLR, the edges of the flooding areas are becoming more and more attractive due to their increasing water frontage which can result in a denser development with urban spaces. At the same time, the floodplains can be equipped with observation or research stations and buildings can also be adapted to the amphibious way of life, which is described under Tactic 16.

Tactic 12 (Retreat × Production):

Depending on the respective climatic zone, aquaculture can grow products in flood areas such as rice or shrimp. Additionally, the cultivation of high-efficiency micro-algae can encourage a change of production towards bio-economy to substitute fossil carbon and to grow bioplastics, livestock food, and nutrients for food and health industry.

Tactic 13 (Retreat × Communities):

SLR displaces existing communities that have to be located elsewhere. At the same time, a change from an urban to more natural water landscape creates additional leisure opportunities, for example, for water sports, which makes these areas more attractive for communities.

Tactic 14 (Retreat × Infrastructure):

Amphibious transition zones can serve as flexible flooding areas for extreme events until these areas are flooded permanently. For example, what used usually as a parking lot can be flooded in an emergency. If surfaces are permanently flooded, they are ideal for boat traffic as a supplement to road traffic on the non-flooded surfaces.

Tactic 15 (Retreat × Ecosystems):

Increasingly flooded areas will be taken over by the natural environment wherever there is no countermeasure. Ecologically effective and indigenous vegetation should be explicitly supported (such as mangroves in tropical and subtropical climates with the associated ecosystems).

Tactics for "**Advance**" (16. – 20.):

"Protect" strengthens and protects the boundary between water and land, "Accommodate" moves it vertically, and "Retreat" draws the interface back horizontally. "Advance" also acts horizontally but in the opposite direction by moving the border from land to water. This land reclamation can take place either by expanding the mainland, by planting specific vegetation, or by placing floating structures in front of the land. Especially in the latter case, the elegance of the solution lies in its unique automatic adaptation to SLR, which is why we will concentrate on it.

In all cases, legal boundaries shift from land to water so that the legal planning principles and new property relationships often have to be redefined.

Tactic 16 (Advance × Buildings):

Floating buildings not only extend urban uses and spaces but reduce the swell and thus protect the coast behind them. This construction method and the attractive water position can compensate for higher construction costs.

Tactic 17 (Advance × Production):

Proximity to water is desirable for commerce and retail. Its extension and the mobility of individual floating industrial plants also increases the flexibility of work processes and supply chains. Additionally, potential aquaculture farming enables efficient production of biomass either as a food base or to strengthen the secondary industrial sector (e.g. production of pharmaceutical products, energy and bioplastics).

Tactic 18 (Advance × Community):

Recreation areas and cultural institutions can expand from the shore to the water. What was preserved previously for boat owners can now be enjoyed by the entire population. The increasing density creates additional aquatic benefits to a larger community like facilities for "Blue Culture" (e.g. floating exhibitions) or "Blue Recreation" (e.g. floating parks).

Tactic 19 (Advance × Infrastructure):

Transport by boat, which is much more efficient than land transport is the most obvious opportunity. But in addition to that, there is the possibility of becoming more independent and at the same time more ecological from the existing infrastructure: wastewater can be processed, and freshwater and energy can be produced on site.

Tactic 20 (Advance × Ecosystems):

The massive positive impact of coastal protection by coastal and marine ecosystems is well understood. The creation of artificial reefs as well of floating islands will add value as wave-breakers and will create even more amphibious and aquatic ecological diversity.

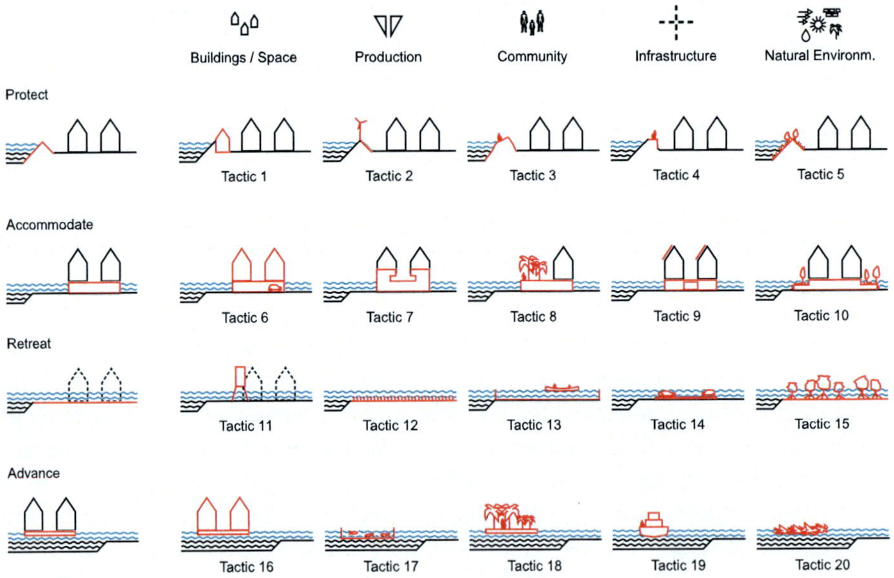

In summary, examples of the different tactics exist (1. – 15.) or are yet to be fully explored (16. – 20.): Tactics 1. – 5. have been practised since the creation of polder landscapes (De Boer 2019) or even before, Tactics 6.–10. can be studied in storm surge areas such as in HafenCity Hamburg (Ge et al. 2013), Tactics 11. – 15. are currently used, e.g. in projects in New York (Howarth, 2014) and in the Delta Plan in the Netherlands (Van Hagen et al. 2015). And Tactics 16. – 20. can be seen for example in projects from SeaCities' (SeaCities.org) or UN's Floating Cities approach (UN-Habitat 2019).

8 Discussion

The tactics have been created systematically by projecting different adaptation methods onto different urban elements. Therefore, all are different, but there are also relevant similarities for further applications:

PHASING: Tactics using the "Protect" and "Accommodate" adaptation method will only adapt to SLR to the point where the reinforcement of the security measures becomes more costly than the value of the areas to be protected. With rising sea-levels, the production and maintenance of protective measures becomes more and more complex and consumes more and more space that was used previously for urban or ecological purposes. High costs of the protective measures in more impoverished cities will lead to the withdrawal of urban areas earlier (King et al. 2015).

If areas cannot be protected or accommodated any more, they can follow tactics of the "Retreat" or "Advance" method. The latter can develop further offshore as proposed or it can also be applied on flooded areas inland to compensate the "Retreat".

OWNERSHIP: Tactics which follow the adaptation methods "Protect" and "Accommodate" try to preserve the current ownership structure and are therefore often the more preferred option by politicians and the community. The former ones want to protect the ownerships as a community, whereas the latter offers the possibility to develop the protection individually. Tactics following "Retreat" can only hope for compensation or an alternative property (like the following), whereas "Advance" tactics are opening up new development opportunities.

GOVERNANCE: To maintain trust among the population and prepare for additional expenses, good governance must develop (and also communicate) a specific long-term plan for the radical measures that will be required to adapt to the SLR (King et al. 2015).

COMBINATIONS: The coordination of above-mentioned components phasing, ownership, and governance can happen during the preparation and application of individual tactics. Due to the dimension of the SLR challenge and the complexity of cities, it can be more advisable to embed tactics in a strategy as "a large, overarching plan in which parallel or successive tactics are coordinated" (Strategy in Encyclopaedia Britannica 2020):

STRATEGIES can be thereby developed out of the systematics in different directions, either as a combination of tactics

- horizontally (= same adaptation methods) like Tactics 1. – 5. or 6. – 1 0. etc.
- vertically (= same urban elements) like Tactics 11. + 6. + 11. + 16 or 2. + 7. + 12. + 17. etc.
- selectively following a synergistic approach for an integrated and sequenced response to SLR (Oppenheimer et al. 2019).

One example of a strategy of selectively combined tactics is the Deltaplan project in the Netherlands, which takes an SLR of 4 m in 2020 (Deltacommissie 2008) into account. It uses a combination of protection-tactics like Tactics 1.– 4. as well as retreat tactics like Tactics 11.– 14. and Tactics 5 .+ 10 . + 15. + 20. for the (re-) production of a diversity of ecological flooding areas (Van Koningsveld et al. 2008).

The approach of a strategic combination of selected tactics goes hand in hand with the World Bank strategy, which declares, for example, tactics related to hard structures as not an economically viable solution. “In some cases, protective structures will still be necessary. However, the consensus is that the climate-change challenge should be used as an opportunity to adopt a long-term strategic approach” (World Bank 2017).

According to a global survey, “failure of climate-change mitigation and adaptation is this year’s number one long-term risk by impact” (World Economic Forum 2020). The developed 20 tactics should be helpful to generate a variety of made-to-measure strategies for a successful climate-change mitigation and adaptation. They can be applied in a next step to any coastal city under special consideration of its specific bioclimatic, socio-cultural, economic requirements. Regardless of whether the reader is an academic, government representative, entrepreneur or interested layperson, these 20 tactics resemble a tool set for everyone to develop and apply tactics and strategies as innovative opportunities for SLR which is one of the most urban challenges in the near future.

References

Baumeister J, Ottmann DA (2015) Urban evolution—a pocket generator to explore future solutions for healthy and ecologically integrated cities. UWAP, Crawley. ISBN 9781742589985

C40 (2019) In: Measuring progress in urban climate change adaptation, monitoring-evaluating-reporting framework. https://www.c40.org/blog_posts/mer-framework-launch. Accessed 11 May 2020

De Boer HP (2019) Europolders a European program on polder landscape, heritage, and innovation. In: Adaptive strategies for water heritage, pp 230–249. https://doi.org/10.1007/978-3-030-00268-8-12

Deltacommissie (2008) Working together with water. a living land builds for its future. findings of the Deltacommissie. http://www.deltacommissie.com/doc/deltareport_full.pdf. Accessed 11 May 2020

Galili E, Benjamin J, Eshed V, Rosen B, McCarthy J, Kolska Horwitz L (2019) A submerged 7000-year-old village and seawall demonstrate earliest known coastal defence against sea-level rise. PLOS ONE 14(12): e0222560. https://doi.org/10.1371/journal.pone.0222560

Ge J, Much D, Kappenberg J, Nino O, Ding P, Chen Z (2013) Simulating storm flooding maps over HafenCity under present and sea level rise scenarios. J Flood Risk Manage 7(4):319–331. https://doi.org/10.1111/jfr3.12054

Hansen JE, Sato M (2012) Paleoclimate implications for human-made climate change. Clim Change 21–47. https://doi.org/10.1007/978-3-7091-0973-1-2

Hanson S, Nicholls R, Ranger N, Hallegatte S, Corfee-Morlot J, Herweijer C, Chateau J (2010) A global ranking of port cities with high exposure to climate extremes. Clim Change 104(1):89–111. https://doi.org/10.1007/s10584-010-9977-4

Howarth D (2014) BIG awarded $335 million to improve lower Manhattan storm defences. https://www.dezeen.com/2014/06/03/big-lower-manhattan-storm-defences-rebuild-by-design/. Accessed 11 May 2020

Hill K (2015) Coastal infrastructure: a typology for the next century of adaptation to sea-level rise. Front Ecol Environ 13(9):468–476. https://doi.org/10.1890/150088

IPCC (2019) Summary for policymakers. In: Shukla PR, Skea J, Calvo Buendia E, Masson-Delmotte V, Pörtner H-O, Roberts DC, Zhai P, Slade R, Connors S, van Diemen R, Ferrat M, Haughey E, Luz S, Neogi S, Pathak M, Petzold J, Portugal Pereira J, Vyas P, Huntley E, Kissick K, Belkacemi M, Malley J (eds) Climate change and land: an IPCC special report on climate change, desertification, land degradation, sustainable land management, food security, and greenhouse gas fluxes in terrestrial ecosystems (in press)

Jevrejeva S, Jackson LP, Grinsted A, Lincke D, Marzeion B (2018) Flood damage costs under the sea level rise with warming of 1.5 °C and 2 °C. Environ Res Lett 13(7):074014. https://doi.org/10.1088/1748-9326/aacc76

Jones N (2020) How the world passed a carbon threshold and why it matters. Yale E360. https://e360.yale.edu/features/how-the-world-passed-a-carbon-threshold-400ppm-and-why-it-matters

Jonkman SN et al (2013) Costs of adapting coastal defences to sea-level rise—new estimates and their implications. J Coast Res 290:1212–1226

Ketabchi H, Mahmoodzadeh D, Ataie-Ashtiani B, Simmons CT (2016) Sea-level rise impacts on seawater intrusion in coastal aquifers: review and integration. J Hydrol 535:235–255. https://doi.org/10.1016/j.jhydrol.2016.01.083

King D, Schrag D, Dadi Z, Ye Q, Ghosh A (2015) Climate change, a risk assessment. University of Cambridge, Centre for Science and Policy (CSaP). http://www.csap.cam.ac.uk/projects/climate-change-risk-assessment/ Accessed 20 April 2020

Kimmelmann M (2017) The Dutch have solutions to rising Seas. The World is watching. in the waterlogged Netherlands, climate change is considered neither a hypothetical nor a drag on the economy. Instead, it's an opportunity. In: The New York times 15 June 2017. https://www.nytimes.com/interactive/2017/06/15/world/europe/climate-change-rotterdam.html?mtrref=www.google.com&gwh=BFD3681FD8413E67B2F85FCBD82B27EC&gwt=pay&assetType=REGIWALL. Accessed 20 April 2020

Link LE (2010) The anatomy of a disaster, an overview of Hurricane Katrina and New Orleans. Ocean Eng 37(1):4–12. https://doi.org/10.1016/j.oceaneng.2009.09.002

Meadows DH, Meadows DL, Randers J, Behrens WW (1972) The limits to growth: a report for the club of rome's project on the predicament of mankind. Universe Books, New York

Nunn PD (2014) Geohazards and myths: ancient memories of rapid coastal change in the Asia-Pacific region and their value to future adaptation. Geosci Lett 1(1):3. https://doi.org/10.1186/2196-4092-1-3

Oppenheimer et al (2019) Chapter 4: Sea level rise and implications for low lying islands, coasts and communities. In: Pörtner H-O et al (eds) IPCC special report on the ocean and cryosphere in a changing climate, Cambridge University Press, Cambridge, UK. https://www.ipcc.ch/srocc/download-report/

Pörtner H- O, Roberts DC, Masson-Delmotte V, Zhai P, Tignor M, Poloczanska E, Mintenbeck K, Nicolai M, Okem A, Petzold J, Rama B, Weyer N (2019) IPCC 2019: summary for policymakers. In: IPCC special report on the ocean and cryosphere in a changing climate (in press). https://www.ipcc.ch/site/assets/uploads/sites/3/2019/11/SROCC_SOD_Ch04_FINAL.pdf

Soroush M, Mordechai L (2018) Adaptation to short-term cataclysmic events: flooding in premodern riverine societies. Human Ecol 46(3):349–361

Strategy (n.d.). Encyclopedia Britannica. https://www.britannica.com/topic/strategy-military. Accessed 11 May 2020

Trang NTT (2016) Architectural approaches to a sustainable community with floating housing units adapting to climate change and sea-level rise in Vietnam world academy of science, engineering and technology. Int J Civil Environ Struct Constr Architectural Eng 10(2):168–179

UN-Habitat (2019) Convened a roundtable discussion about floating cities https://unhabitat.org/roundtable-on-floating-cities-at-unhq-calls-for-innovation-to-benefit-all. Accessed 30 April 2020

Van Haegen M, Wieriks K (2015) The Deltaplan revisited: changing perspectives in the Netherlands' flood risk reduction philosophy. Water Policy 17(S1):41–57

Van Koningsveld M, Mulder JP, Stive MJ, VanDerValk L, VanDerWeck AW (2008) Living with sea-level rise and climate change: a case study of the Netherlands. J Coastal Res 242:367–379. https://doi.org/10.2112/07a-0010.1

World Bank (2017) Adaptation to climate change in coastal areas of the ECA region: a contribution to the umbrella report on adaptation to climate change in ECA (English). Washington, D.C., World Bank Group. http://documents.worldbank.org/curated/en/377981484811872690/Adaptation-to-climate-change-in-coastal-areas-of-the-ECA-Region-a-contribution-to-the-umbrella-report-on-adaptation-to-climate-change-in-ECA. Accessed 20 April 2020

World Economic Forum (2020) In: The Global Risks Report 2020, World Economic Forum. http://www3.weforum.org/docs/WEF_Global_Risk_Report_2020.pdf. Accessed 11 May 2020

Zeng M, Ma C, Zhu C, Song Y, Zhu T, He K, Chen J, Huang M, Jia T, Guo T (2016) Influence of climate change on the evolution of ancient culture from 4500 to 3700 cal. yr BP in the Chengdu plain, upper reaches of the Yangtze river China. Catena 147:742–754

Joerg Baumeister Prof. Dr.-Ing. Joerg Baumeister has been a practitioner, educator, researcher and consultant for Architecture and Urban Design for more than 20 years throughout Europe, Africa, the Arabian Peninsula, Asia, and Australia. As University educator with passion for "his" students, he tries to fuse research, higher education, and implementation in order to create feed-back loops to evolve the fields of Urban Design and Architecture. Joerg founded in 2019 the "SeaCities" research lab (SeaCities.org) at the Cities Research Institute, Griffith University to develop water-adapted cities and floating structures. He is consulting governmental institutions on the federal, state and regional level as well as NGOs and private industry leaders to apply his current research interests which comprises SeaCities, Ecological Cities, Affordable Housing in serial building technology, and Design Innovation through creative thinking. He is an award-winning Architect and Urban Designer and continues to be an enthusiastic speaker at international conferences. Joerg is DAAD Ambassador and Scientist For Future.

Design Strategies for Coastal Adaptation Urban Speculation in Palm Beach, Gold Coast—Australia

Cecilia Bischeri

Abstract Sea-level rise has put under the spotlight the foreshore of our cities. The architecture of the foreshore has become subject to considerable investigation to understand how our cities can cope and thrive in a progressively mutable context. This slow-onset event and the physical transformation provoked present a fascinating scenario and an incredible opportunity to challenge the way we conceive, organise and design coastal cities. At the base of this opportunity lies a paradigmatic shift on the belief that the land we occupy controls water. As da Cunha (2019) states, "the act of separation [between land and water] is a land-centric idea conceived to contain and control wetness." From this perspective, any change to this "state of control" is deemed as a disruption or risk which promotes uncertainties in our social systems and infrastructures. This study investigates design strategies for urban adaptation of coastal developments. Considering the architectural perspective, the precedents and approaches presented combine landform infrastructures with interventions at the architectural scale. The preferred context of the application is Palm Beach in the City of Gold Coast, Queensland (Australia). The primary goal of this study is to present a speculative scenario for a neighbourhood within Palm Beach. Considered as a speculative project, the proposal aims to spur a constructive conversation on innovative design solutions able to influence the local architectural practice. The design proposal tests two different, yet related options. The first attempt is based on the Fingers of High Ground project by Mathur et al. (2014). The project's key aspect is the manipulation of contours with the intent to mitigate and contain water while providing a territorial infrastructure. The second iteration, labelled Hybridised Canal Estate, couples the moulding of contours with the provision of sought-after real-estate waterfront properties. This iteration pivots around two main characters of the target community and context: the leisure driven tourism and real-estate market. Ultimately, this study aspires to promote a broader conversation on how to envision an innovative architecture of the foreshore able to transform the threat of sea-level rise into an opportunity to rethink our cities.

C. Bischeri (✉)
Architecture and Design, School of Engineering and Built Environment, Gold Coast Campus, Griffith University, Brisbane, QLD 4222, Australia
e-mail: c.bischeri@griffith.edu.au

J. Baumeister et al. (eds.), *SeaCities*, Cities Research Series,
https://doi.org/10.1007/978-981-15-8748-1_2

1 Introduction: Study Drivers

Sea-level rise falls within the slow onset events provoked by climate change. Sea-level rise is here investigated in its relationship with the architecture of the foreshore and the development of design strategies required for coastal cities to adapt in this ever-changing context. The global average sea-level rose at a rate of 3.3 mm per year from 1993 to 2009 (Cazenave and Llovel 2010; Nicholls and Cazenave 2010). This figure is destined to grow and, in fact, the dramatic change and accelerated melting of the polar ice sheet mass support the depiction of a future where sea-level will reach, in comparison to the present-day situation, an estimated +0.8 m by 2100 (IPCC 2012). This last value is the most cited in the scientific arena, and the projection is based on intermediate carbon emissions and relatively stable Antarctic ice sheets (Kopp et al. 2014).

In October 2019, Scott Kulp and Benjamin Strauss published a study on a more accurate way to measure land elevation from satellite images—called CoastalDEM.[1] This new method has cast light on the actual number of people who live on low-lying coastal areas and, therefore, are going to be affected by sea-level rise. Kulp and Strauss (2019) state that "190 million people [...] currently occupy global land below projected high tide lines for 2100 under low carbon emissions, up from 110 million today". "Under high emissions, CoastalDEM indicates up to 630 million people live on land below projected annual flood levels for 2100." Considering the national and global calibre of many of the cities involved in the projections, such as Shanghai, Bangkok and Amsterdam, it is reasonable to argue that relocation will not be an option and sophisticated strategies need to be implemented to cope with this mutable context. Urban planner, Lynch (1990, p109) stated, "A city is hard to kill, in part because of its strategic geographic location, its concentrated, persisting stock of physical capital, and even more because of the memories, motives, and skills of its inhabitants." The question that awaits to be answered is, how are our cities and community going to adapt?

This study emerged from the desire of testing design solutions for coastal resilience at the neighbourhood scale in the City of Gold Coast, Queensland—Australia. Three drivers need to be unfolded to clarify the research aims. The first driver is constituted by understanding the role of the architectural discipline in the provision of resilient solutions for cities and communities against, in this particular case, sea-level rise. The role of architecture in supporting resilient communities has almost solely been circumscribed to restorative actions; in other words, architects have been consulted and deployed with the main scope of providing solution during the reconstruction phase in the aftermath of a disaster. The shift required aims to invest the discipline with a more proactive role in the enhancement of resilience for the targeted communities. In addition, it is here argued that the distinctive role of architecture, in comparison to sister disciplines such as the engineering disciplines, lies in picturing a vision for the city. In fact, while engineering focuses on technical solutions and their implementation are primarily delivered through an additional

[1] See: https://go.climatecentral.org/coastaldem/.

apparatus to the city forms, architecture enables the integration of those technical strategies, with programs and places for the community to live in and experience. Within this scenario, architectural innovation can proliferate.

Secondly, this study relies upon previous work dedicated to understanding how to enhance community resilience and how architecture can facilitate the process. The main lesson learned is the value of architecture in supporting the construction, maintenance and intensification of social networks. What differs the current study from the previous is the change in scale. This investigation, in fact, moves from the architectural artefact to the neighbourhood scale. Before going any further, the role of social networks within this study has to be outlined. Notwithstanding the vast and multidisciplinary literature acknowledging the part played by social networks (Cutter et al. 2008, Sherrieb et al. 2010, Vale and Campanella 2005), also defined as social capital, in building community resilience and driving recovery, government guidelines and investments have often overlooked this factor. This study, on the contrary, acknowledges the role of social networks and foreground it as one of the design principles of the presented proposal. Focusing on tangible costs and aiming to minimise expenditure while providing safety to communities, the government response to risk has focused on investing resources in hard infrastructures notwithstanding the evidence that social infrastructures are at the core of resilient communities (Aldrich and Meyer 2015).

A fundamental aspect of a healthy, functioning and resilient community is constituted by the three types of networks that social capital embodies. Those are bonding, bridging and linking. The primary reference for a detailed description of the three types of social capital remains the work of Daniel Aldrich (2012). However, *in nuce*, bonding is the network an individual established with their family members, friends and neighbours. Bridging involves the connection with external networks, different for demographic, religion or ethnicity. Aldrich (2012) labels bonding and bridging as horizontal, connecting groups of the same status.[2] Linking, instead, is a vertical connection where groups interact with formal or institutional power or authorities. Mileti and Gailus (2005, p 498) have recognised the disenfranchisement "from power and influence" as an important factor in affecting negatively the community's capacity to recover from natural disasters. The three levels are equally essential. In conclusion, social capital is more than an ideological stand and informs *how* and *what* can be designed to enhance community resilience by providing a framework of the community's needs. Architecture, therefore, cannot limit its relevant space of manoeuvre to the single building and its technical aspects; on the contrary, the discipline needs to engage with the city and focus on the systems of networks which can fundamentally enhance community resilience. Foregrounding social relationship has to be the driver for an architecture of resilience, and understanding the character and needs of the target community is of paramount importance to be successful in this process. Needless to say, the promoted position does not intend to discredit the role

[2]See: Woolcock (2002). Social Capital in Theory and Practice: Reducing Poverty by building Partnership between states, Markets and Civil Society. In Social Capital and Poverty reduction: Which Role for Civil Society Organizations and the State? 20-44. France: UNESCO.

of hard infrastructures and disaster-ready artefacts. Yet, the emphasis posed on social relationships aspires to encourage a shift, where the built environment is conceived in relation to its larger context. Infrastructures with doubling-up programs able to address the need to adapt while providing assets to the community should be the norm.

Lastly, the financial return cannot be emphasised enough. Looking into the Australian panorama, in fact, the damages on community connectedness together with health and wellbeing impacts are considered intangible costs provoked by a natural disaster (Australian Business Roundtable for Disaster Resilience and Safer Communities 2016). While generally, the economic costs for natural disasters have been estimated around tangible costs (direct and indirect), recent literature has demonstrated that intangible costs may be underestimated by at least 50% (Australian Business Roundtable for Disaster Resilience and Safer Communities 2016). Therefore, a vision for the architecture of the foreshore constitutes an incredible opportunity to provide alternative visions for our cities combining urban forms with resilience. In addition, the architecture discipline could take a leading role in supplying solutions which engage with tangible (built environment) and intangible (wellbeing and community connectedness) domains.

The last driver is constituted by a design intuition originated by observing how past civilisations adapted to the phenomenon of sea-level rise. Among the results presented by Kulp and Strauss in a divulgatory article published by The New York Times, the city of Basra is showcased to demonstrate the impact of sea-level rise and the extent of inundation to be expected (Lu and Flavelle 2019). Apparent is the need to equip the city and its territory with design solutions enabling communities to adapt to the profound reshaping of their context. The case of Basra becomes instrumental to this discussion. Basra, today located in southern Iraq, was part of the region known as Mesopotamia. Jennifer Pournelle, archaeologist specialised in the genesis of urban settlements, is here cited for her studies on the ancient Mesopotamian landscape during the dramatic sea-level rise occurred in 4000BCE. What is relevant to the present discussion is the similarity between the reconstruction of the Gulf's inundation and the projections depicted by the future sea-level rise (Fig. 2.1).

Hence, it appears pertinent to report how the then threaten population coped and adapted. To respond to the then rising water and vast inundation, the nomadic populations occupied finger-shaped ridges, called turtlebacks, which were elevated from the flood overlay, and surrounded by marshes and wetlands. Those locations of high-ground did not just offer a dry spot but supported by the richness of flora and fauna of the marshes became the sites for the proto-urban developments which morphed the basis for the rise of urban centres like Uruk, where writing was first invented (Pournelle 2003, 2007). This digression aims to spur a constructive conversation on how the threaten cities should undertake the journey for adaptation; and reflect on the opportunities offered by a high-impact phenomenon to revise and ameliorate current urban practices; acknowledging the need for a transdisciplinary approach able to surpass palliative and punctual solutions; and finally the urgency of innovative urban forms to enable adaptation.

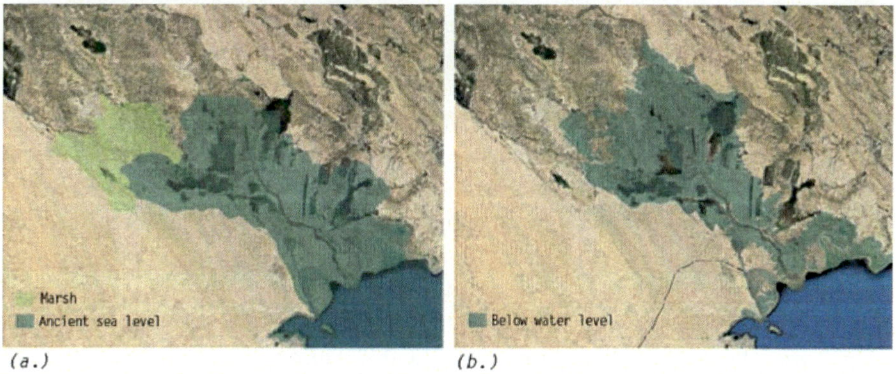

Fig. 2.1 Comparison of inundations in Basra (Southern Iraq) in 4000BCE and 2050CE **a** Basra in 4000BCE. Visualisation based on Pournelle's reconstruction and visualisation. *Source* Lawler (2011 p.141) **b** Basra in 2050CE. Visualisation based on Kulp and Strauss' projection and visualisation. *Source* https://coastal.climatecentral.org/map/7/49.4576/30.7947/?theme=sea_level_rise&map_type=coastal_dem_comparison&contiguous=true&elevation_model=best_available&forecast_year=2050&pathway=rcp45&percentile=p50&return_level=return_level_1&slr_model=kopp_2014

2 Design Strategies

In two previous studies (Dupre and Bischeri 2019, Bischeri 2015) focused on the architecture of resilience, a set of transferable guidelines on design criteria to support the planning and realisations of resilient communities have been produced. The guidelines have been drawn from comparative studies and good-practice examples of architectural precedents that pose the intensification of social networks as their driving concept. Four criteria, namely, (1) location, (2) function, (3) compositional strategies and formal expression, and (4) community involvement, have been identified as aspects which have a clear impact on providing an architecture of resilience. In addition, a definition of contextual factors related to the moulding of community identity has been taken into consideration. This framework lies on the background of the present study and has guided the selection and review of relevant factors to form an initial understanding of the targeted community identity. Seeking to provide a design response to the neighbourhood level, rather than the single building, further investigations are required.

Four aspects have been central in shaping the proposal: (a) adopting a system to support the choice and application of coastal adaptation strategies; (b) investigating the neighbourhood scale which provides the opportunity to engage with the urban fabric while experimenting with the urban forms at a controllable scale; (c) investigating the identity of the targeted community via its context, physical and social characters; (d) identify good-practice examples which address similar issues and provide solutions through architectural strategies. A brief summary is here introduced to outline the main findings of this investigative phase.

a. *Design strategies for coastal adaptation: embracing a fluid ordering system*

The recent literature on design approaches to cope with sea-level rise pivots around a clear message: adaptation can be achieved when coastal solutions work with nature rather than against it. Most authors have organised the available coastal engineering practices into categories with the intention of providing an interpretative key to the understanding of options and type of approaches for designers, governments or other stakeholders. This study has no ambition of proposing a new system of categorisation, but rather to reflect on two specific aspects which accompany conventional engineering strategies and support their translation into architectural artefacts for the foreshore. The two aspects are introduced with the aim of advancing a more holistic approach for the urban foreshore via the surpass of palliative solutions and the push forth of revised urban forms to support adaptation.

This study relies on the system of classification presented in the work of Kristina Hill, to promote a shared understanding of the existing infrastructures for coastal resilience. Particular interest is laid on her simple, yet sophisticated, approach which bridges the gap between listing the options available within the infrastructures for coastal resilience and the provision of valuable instruments to guide the selection and application of those very options (Hill 2015). In fact, rather than pigeonholing strategies according to the most broadly adopted two main criteria—1. Are those infrastructures natural and nature-based or conventional?[3] 2. What is the level of control those infrastructures apply to water?—Hill proposes a more fluid way to proceed.

Hill's method puts forth the use of typologies which are defined following Doty and Glick (1994, p 230) definition of "conceptually derived interrelated sets of ideal types." The adoption of typologies is deemed as an important tool during the decision-making process (Hill 2015). The infrastructures are grouped under four ideal types. Those are walls, static (i.e. seawalls and floodwalls) or dynamic (i.e. tide gates and movable surge barriers); and landforms, static (i.e. mounds, dikes and canals) or dynamic (i.e. beaches, dunes and wetlands). The available options have clear pros and cons which need to be considered in the design process. Any coastal infrastructure, in fact, has an impact on natural systems and has to be accounted for during the feasibility study. For a more thorough analysis, the work of Hill is recommended. However, a brief overview is here offered via the Table 2.1.

Hill's classification does not provide a definitive matrix of solutions but rather a field, open to testing. And, Hill (2015, p 472) states, "the intent of the typology

---

[3]See: Bridges et al. (2015) Use of natural and nature-based features for coastal resilience. Washington, DC: US Army Corps of Engineers. http://www.nad.usace.army.mil/Portals/40/docs/NACCS/NNBF%20FINAL.pdf.

The adoption of wetlands, marshes, beaches, dunes, corals and other coastal ecosystem have been proved to be not only effective in supporting coastal resilience but also beneficial to maintain balanced and healthy ecosystems. The US Army Corps of Engineers has labelled those components natural and, nature-based features, which can be considered as an upgrade of conventional engineering strategies, they might occur naturally in the observed landscape or are engineered and built to mimic natural conditions.

Table 2.1 Comparing advantages and disadvantages in Hill's typology approach

	Advantages	Disadvantages
Walls	-well suited for punctual interventions	-incapability to adapt to new scenarios -impossibility to provide doubling up functions -potential of mechanical or power failure (limited to dynamic walls)
Landforms	-well suited for addressing city-wide or regional issues -can be built in phases -can be adjusted to new circumstances -opportunity of doubling-up functions (this expands the benefits that landform interventions provide to the community)	-need for constant maintenance and monitoring

is to represent the range of choices that *could* be applied, but the selection of the specific types of infrastructure that *should* be applied requires a review of the specific context." Hill's approach removes the safe feeling of having a list of possibilities and engages with a more inquisitive approach where the specificity of the context becomes of paramount importance. The context put forth in Hill's argument refers primarily to the ecological dimension and built environment (use, density and urban morphology) with a broader consideration to the adjacent territory which is likely to impact and be impacted by the design and implementation of coastal infrastructures. The present study adopts solutions based on the landform type. Notably, Hill's cited study targets the city or region; while, the current research aims to produce speculative solutions for the neighbourhood scale. In this scenario, therefore, the context has to assimilate and reflect on the fine grain of the socio-economic conditions and level of social capital present in the area.

b. *The Neighbourhood scale*

Dealing with the complexity of an entire city is beyond the capability of the present study; however, testing design solutions at the neighbourhood level offers the opportunity to engage with the intricacy of the urban fabric and its formal and functional requirements while maintaining a manageable size. This scale, in fact, encompasses the building dimension without alienating the city system and its infrastructures. In addition, it provides the prospect to reflect on the neighbourhood's components and their adoption for facilitating the introduction of structures of coastal resilience at the urban scale. This approach draws inspiration from a speculative matrix proposed by Paul Lewis in the recent publication (Nordenson et al. 2018) titled "Structures of Coastal Resilience." Yet, the presented proposal, instead of limiting the outcome to the hybridisation of a single element (i.e. a road, with a resilient infrastructure, such as a bulkhead), encompasses the entire urban block. The reason behind this approach lies on the opportunity to surpass punctual interventions to promote a more holistic approach where all the main elements considered are involved. A series of diagrams support the visualisation of the process which has been undertaken.

The pictorial apparatus provides a storyboard for the advanced proposals. The first step consists of identifying the most basic components which form a neighbourhood notwithstanding its primary driver, i.e. residential, mixed-use, privately owned, etc. The three elements identified are green, buildings and circulation. It is here argued that we can achieve any type of neighbourhood just by varying the connotation of each of those elements. In other words, we can consider the component *green* as an element able to vary between private, the backyard garden of a house, for instance, to recreational, a park equipped with facilities. Similarly, *buildings* vary between private and public while *circulation* between vehicular and non-vehicular. Those parameters can be adjusted with the intent of reflecting the character of a current or envisioned neighbourhood. Having set the connotations for the three components, the following step consists of implementing strategies of coastal resilience in the urban form of the neighbourhood with the intent of embedding, rather than overlaying, resilience and adaptation. To maintain the speculation valuable, two premises need to be acknowledged. Firstly, it is assumed that the examined neighbourhood is equipped with a safe connection to the primary arterial system. Secondly, the speculative design work has focused on a neighbourhood where a water frontage is present (Fig. 2.2).

c. *Architecture for the bigger picture: Two Exemplars*

With the intent of presenting a speculative project for a neighbourhood in Palm Beach, the proposal has required some groundwork and the selection of two good-practice examples. The first project is the Big "U" by the BIG Team.[4] The Big "U" aims to improve flooding resilience for the southern side of the island of Manhattan, New York. Yet, it also provides an opportunity to demonstrate the role of social capital and its impact on *how* and *what* it can be designed to enhance community resilience. The low-lying terrain of the area contains, on the one hand, Wall Street and on the other 35,000 affordable housing units. The project was initiated in the aftermath of Hurricane Sandy (2012). The project constitutes a protective system which stretches for approximately sixteen kms. Labelled by the designer team as a social infrastructure, the proposal interacts with a very dense, highly vibrant and socio-economically diverse community. The Big "U" is here presented as an exemplar in highlighting the existent inequality among the pockets of communities living in the area and investigating design solutions to mitigate the discrepancy. The mapping of ethnicities, annual income, public transport system, outline a strident fracture between the wealthy, white and well served West Side and the economically disadvantaged, not white and not well-served Lower East Side (LES). The architectural response focuses primarily on understanding how the Big "U" could facilitate change through the provision of resilient infrastructures and amenities able to enrich, connect and address with specificity the LES community's needs while providing access to the river.

The second project is titled Fingers of High Ground (FHG) by Anuradha Mathur and Dilip da Cunha. The project seems to emulate the aforementioned Mesopotamian

[4]See: http://www.rebuildbydesign.org/our-work/all-proposals/winning-projects/big-u https://big.dk/#projects-hud.

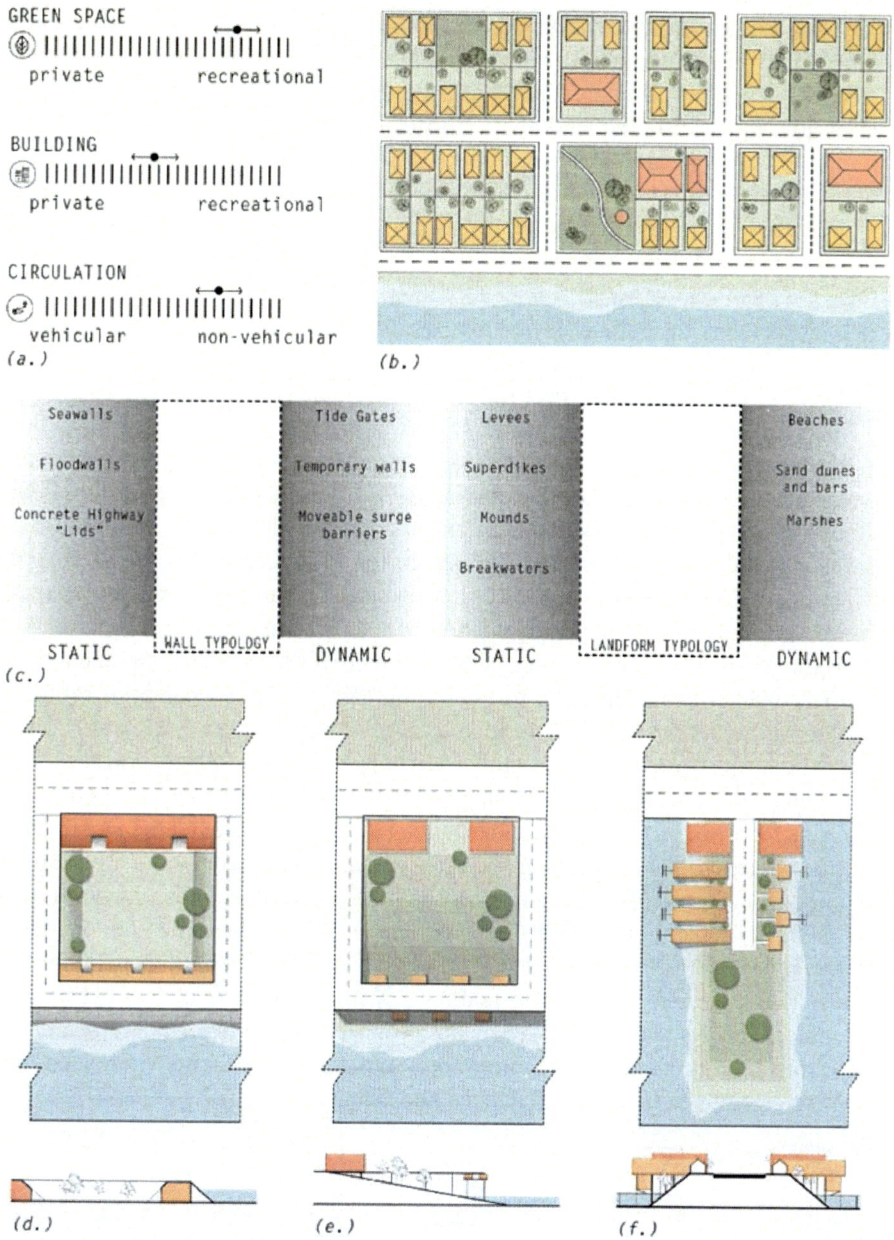

Fig. 2.2 Structures of coastal resilience at the neighbourhood scale **a** Neighbourhood components **b** Typical residential neighbourhood layout **c** Coastal resilience strategies *Source* Hill (2015) **d, e, f** Embedding coastal resilience strategies in the neighbourhood urban forms: Dike crossed with neighbourhood components

turtlebacks via a highly constructed and anthropised proposition. The project is part of a larger venture supported by the Rockefeller Foundation which was dedicated to studying and proposing resilient infrastructure for the North Atlantic coast of the USA after the devastating landfall of Hurricane Sandy and the consequent flooding events. Mathur and da Cunha focused on Norfolk in Virginia. The potent message the projects puts forth for the site is "Norfolk does not call for barriers and protection; it calls for design interventions on the basis of a new visualisation of the coast as a place where land meets sea, not across a line, but in a field of points." And, "land does not meet sea across a 'front', but rather through a number of discrete fingers of high grounds. Each finger is a unique gradient or a unique gathering of gradients between land and sea, working to structure a coast that is more fractured, cumulative and diverse than it is continuous, linear and absolute" (Mathur et al. 2014, p147).

The FHG project is here presented for two reasons. Firstly, the proposal deems to provide fertile material for the discussion and formalisation of a hybrid canal estate able to combine innovative, resilient urban forms with the extension of water frontage for real estate development and facilities. This scope is informed by the investigated context and further explored in the following section, *d. The context: Palm Beach*. Yet, differently from the canal estate, FHG are a territorial infrastructure morphed around an attentive observation of the existing geology, morphology and topography of the region subject of the study. FHG have a tectonic anchorage to the land which lies behind the foreshore and "can serve as islands of refuge and protective barriers in times of storm and surge" (Mathur et al. 2014, p149). In fact, they do not just provide a safe higher territory to occupy; FHG protect and limit the extent of inundation while reshaping the coast. The fundamental belief which informs the work of Amathur and da Cunha (2019) pivots around the fact that "land and water are products of an intentional act of separation carried out with the help of a uniquely endowed line drawn in a chosen moment of time. Water, however, seems to always be lesser in this difference, confined as it is by the same line to a place in waterbodies that are made to serve land—draining it, irrigating it, providing it with a waterfront for real estate, even making it with depositions of silt, but primarily keeping it dry for settlement." The slant proposed by the two landscape architects offers a shift of perspective, enabling for a reimagined foreshore where land, water and infrastructures find a new balanced coexistence.

d. *The context: Palm Beach*

The City of Gold Coast, in Queensland—Australia, is located within a greater urbanised region, known as South-east Queensland. Characterised by subtropical climate and a low-lying, flood-prone coast, the region is nestled between the Great Escarpment, on the west, and the Pacific Ocean on the east side.[5] The slow-moving and bend-rich water streams are the ideal habitat for wetlands and marshes. Within

[5] See: Oilier C. D. (1982) The Great Escarpment of eastern Australia: Tectonic and geomorphic significance, Journal of the Geological Society of Australia, 29(1-2): 13-23, https://doi.org/10.1080/00167618208729190.

this broader context, the City is punctuated by canal estates which are broadly recognised as part of the local character. Their ubiquitous presence is the testimonial of the transformation process the City has experienced. Starting from the beginning of the last century, the ocean, its sandy beaches and tourism turned a series of frontier townships in the Gold Coast phenomenon. The City is tightly connected with water, not just for its physical presence but for its exploitation as one of the elements of a larger apparatus concentrated around leisure. In 1997, architecture and design theorist Goad (1997, p36) in preparing an urban heritage and character study for the City Council wrote: "In Australia, the Gold Coast represents a peculiar and unique urbanism. Shunned by the tastemakers, it is nevertheless a vital part of our hybrid culture with its artistic, social and economic allegiances drawn from all over the globe." And, "It is almost certain, that the current growth and continuing vibrancy of a culture sustained almost solely by leisure will presage a new cultural and urban condition for the twenty-first century."

The appearance of the canal estates in Gold Coast marked the turn of the city from an agricultural and sparsely inhabited region to a tourist destination. The first canal estate subdivision was approved in Gold Coast in 1956. This new form of development seemed to match the expectation of the developers, who aspired to extend the coastline to provide direct water access to the holiday-house owners while reclaiming cheap land from swamps and lowland forests. In the beginning, the process was driven by speculation and lacked regulation. However, the opportunity of exponentially extend the waterfront available for real estate demand favoured their adoption. Notwithstanding the environmental toll on the native wetlands and the increase of property exposure to flooding risks, the canal estates have been a significant instrument in the modern urban development for the City of Gold Coast.[6]

The suburb of Palm Beach, in Gold Coast, has been selected as a site for speculative propositions for four reasons. Firstly, due to its low-lying terrain, Palm Beach is exposed to seasonal flooding having the Tellebudgera Creek and Currumbin Creek to define, respectively, its northern and southern boundaries. Secondly, due to the proximity with the estuaries of the two aforementioned creeks and presence of canals, sea-level rise projections forecast that large part of the suburb will be inundated during king tide events. Those projections consider future changes to climate, incorporating the projected increase in sea-level of 0.8 metres by 2100. Additionally, a projected 10% storm tide intensity and 10% rainfall intensity increase have been factored in the flood overlay map. Lastly, Palm Beach is characterised by a low socio-economic demographic profile, which is generally associated with lesser community resilience. The Index of Relative Socio-economic Disadvantage (IRSD), one of the four indexes which form the Socio-Economic Indexes for Areas (SEIFA), is adopted to underpin the previous statement (Australian Bureau of Statistics 2018). The Index considers

[6]See: Bosman et al. (2016) Off the Plan : The Urbanisation of the Gold Coast. Clayton, Vic: CSIRO PUBLISHING. https://search-ebscohost-com.libraryproxy.griffith.edu.au/login.aspx?direct = true&db = nlebk&AN = 1164333&site = ehost-live&scope = site In particular the following two chapters: *All that glitters: an environmental history 'sketch' of Gold Coast City* by Bosman and Houston and *City with/out a Plan* by Dedekorkut-Howes and Mayere.

Leach (2018) Gold Coast: City and Architecture. London: Lund Humphries.

criteria such as income, educational attainment, unemployment and jobs in relatively unskilled occupations.[7] Ultimately, the site analysis has highlighted the scarce and scattered presence of socio-cultural facilities in the area. This last factor showcases the lack of structures supporting the flourishing of social capital. Those considerations have played a fundamental role in driving the choice for the project site (Fig. 2.3).

3 Project Exegesis

The previous analyses have provided the design strategies and criteria to inform the development of the proposal for Palm Beach. Considering all the listed factors—from the lack of a system of social and cultural facilities; the topography and character of the site; exposure to flooding and the essential role played by leisure within the framework of the city—a landform strategy is preferred for the benefits that can provide to the community. Therefore, thinking on the opportunities to yield on the integration of strategies of coastal resilience with the three identified neighbourhood's components and local urban form, the proposition for a static landform (dyke) is put forth. The ambition lies in presenting a dyke hybridised with a canal estate. This aims to embed the desirable aspects of the canal estate developments while bypassing the detrimental effects of inappropriate development and the re-establishment of marshes.

Additionally, the Hybridised Canal Estate (HCE) model accommodates the predicted population growth, which sees the Gold Coast as the most populous region in Queensland (Queensland Statistician's Office Queensland Government projections 2018); provides a more democratised access to the water; and, double-ups and maximises the financial return for the investment in resilience strategies, via the enhancement of quality and quantity of community assets. In this context, community assets refer to those associated with place, namely built and natural assets (Flora et al. 2015). The attention to these specific components is linked to the importance of social capital in building community resilience. The proposal of an HCE might provide a plausible option in the design of a resilient future where the built environment has a proactive, rather than defensive, role in equipping communities with resilience strategies.

The manipulation of a static landform, a dyke, with the three components of the neighbourhood has progressed via a process of formal analogy in the testing of the concept of Fingers of High Ground introduced by Mathur and da Cunha for the selected site. The FHG are identified with dykes which lay perpendicular to the shoreline and extend inland. The proposal has incorporated and valued aspects of

[7]The SEIFA Index is developed by the Australian Bureau of Statistics and ranks areas in Australia according to relative socio-economic advantage and disadvantage. The indexes are based on information from the five-yearly Census. The last data available refer to the 2016 census. For this study, the data have been gathered focusing on SA1 (Statistical Areas Level 1). The definition of SA1 can be found here: https://www.abs.gov.au/ausstats/abs@.nsf/Lookup/by%20Subject/1270.0.55.001 ~ July%202016 ~ Main%20Features ~ Statistical%20Area%20Level%201%20(SA1) ~ 10013.

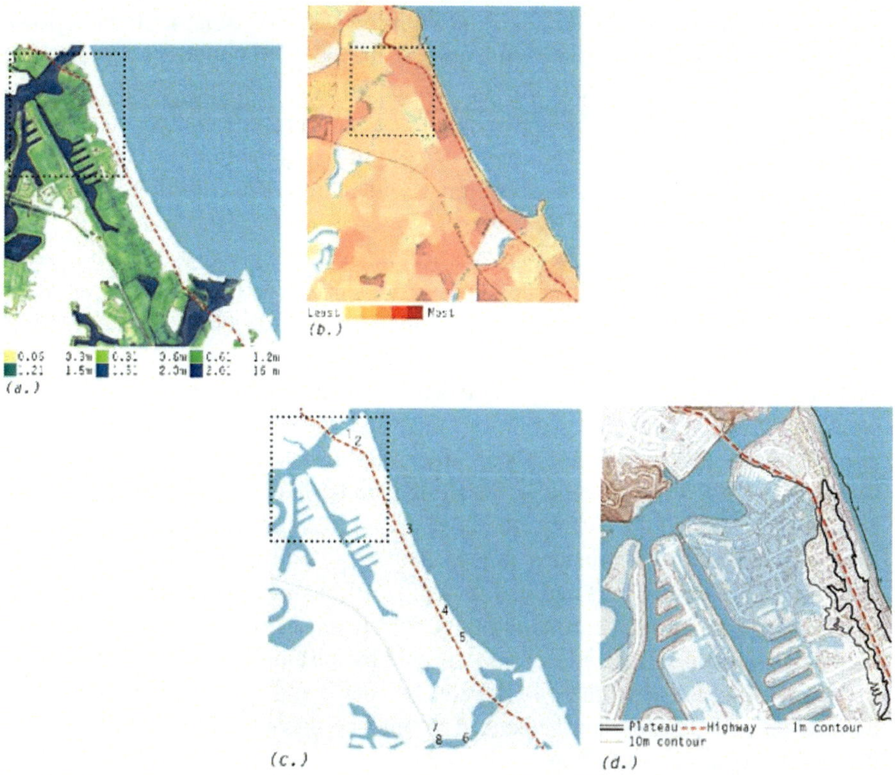

Fig. 2.3 Palm Beach. Framing the Context **a** The updated mapping considers future changes to climate, incorporating the projected increase in sea level of 0.8 metres above present-day levels by 2100 established by the State Government in 2015. The updated mapping also includes the State Governments projected 10 per cent storm tide intensity and 10 per cent rainfall intensity, based on advice from industry representatives. *Source*https://www.goldcoast.qld.gov.au/council/flood-heights-maps-2222.html
Source **b** Index of Relative Socio-economic Disadvantage (IRSD). *Source* https://www.abs.gov.au/ausstats/abs@.nsf/Lookup/by%20Subject/2033.0.55.001~2016~Main%20Features~SOCIO-ECONOMIC%20INDEXES%20FOR%20AREAS%20(SEIFA)%202016~1 **c** Existing social and cultural facilities: 1. Gold Coast Recreation Centre, 2. Tallebudgera Leisure Centre, 3. Beach, 4. Palm Beach Surf Club, 5. Palm Beach Neighbourhood Centre, 6. Thrower House Youth & Community Hub, 7. Palm Beach Scout Den, 8. Palm Beach & Currumbin Sports Club **d** The selected neighbourhood for the testing design propositions

the physical (i.e. topography, flooding points of access) and societal (i.e. cultural and social facilities) context, guiding the selection of the anchorage points for the insertion of this territorial infrastructure. The Gold Coast Highway runs parallel to the ocean and constitutes the highest point in the area. This artery provides one of the anchorage points for the FHG which branches towards the existing canals. The positioning of the secondary branches provides connections with the surrounding neighbourhoods and their primary social infrastructure, a tourist park, school and sports facilities. Concurrently, the FHG limit the impacts of flooding events. In fact, the branches are positioned strategically with the scope of isolating the two lowest elevation points of the area and circumscribing the extent of flooding water accessing the neighbourhood and inundating the residential developments. The architectural solution for the FHG is presented via architectural precedents and diagrammatic sections which demonstrate the intent of coupling architectural artefacts with the consistent presence of green space equipped with recreational activities and designed to adapt to the ever-changing water levels (Fig. 2.4).

For the second test, the lessons learned from the application of the FHG are applied to the canal estate. The intent is to provide a resilient urban form paired with the favourable aspect of the canal estate, an extension of water access. Furthermore, the provision of additional community facilities, which work in synergy with the existing one, enriches community assets while facilitating the strengthening of social networks. To achieve such an ambitious outcome, the neighbourhood is profoundly reshaped in its urban form and topographic conformation. Networks are morphed through new and existing facilities. The system underpinning the network of facilities becomes the spine of the new residential development and arms the proposition with stepped parks and infrastructure able to adapt to flooding events. The master plan of the HCE presents a balanced diversity of functions—commercial-, medium- and low-density residential—and community-driven infrastructures. Their positioning takes advantage of calculated locations offered by the Gold Coast Highway (i.e. high traffic and visibility) for the commercial business, and prime water front site for the low-density development enhanced by intense greenery, amenities and a more democratised access to the water (Fig. 2.5).

4 Conclusion

This study has investigated design strategies for urban adaptation of coastal developments in Palm Beach. The precedents and approaches presented combine landform infrastructures with interventions at the architectural scale. The study has presented two speculative scenarios for a neighbourhood. The proposals have moved through stages and testing which were recorded in the presented diagrams. Considered as speculative projects, the proposal aims to spur a constructive conversation on innovative design solutions able to influence the local architectural practice.

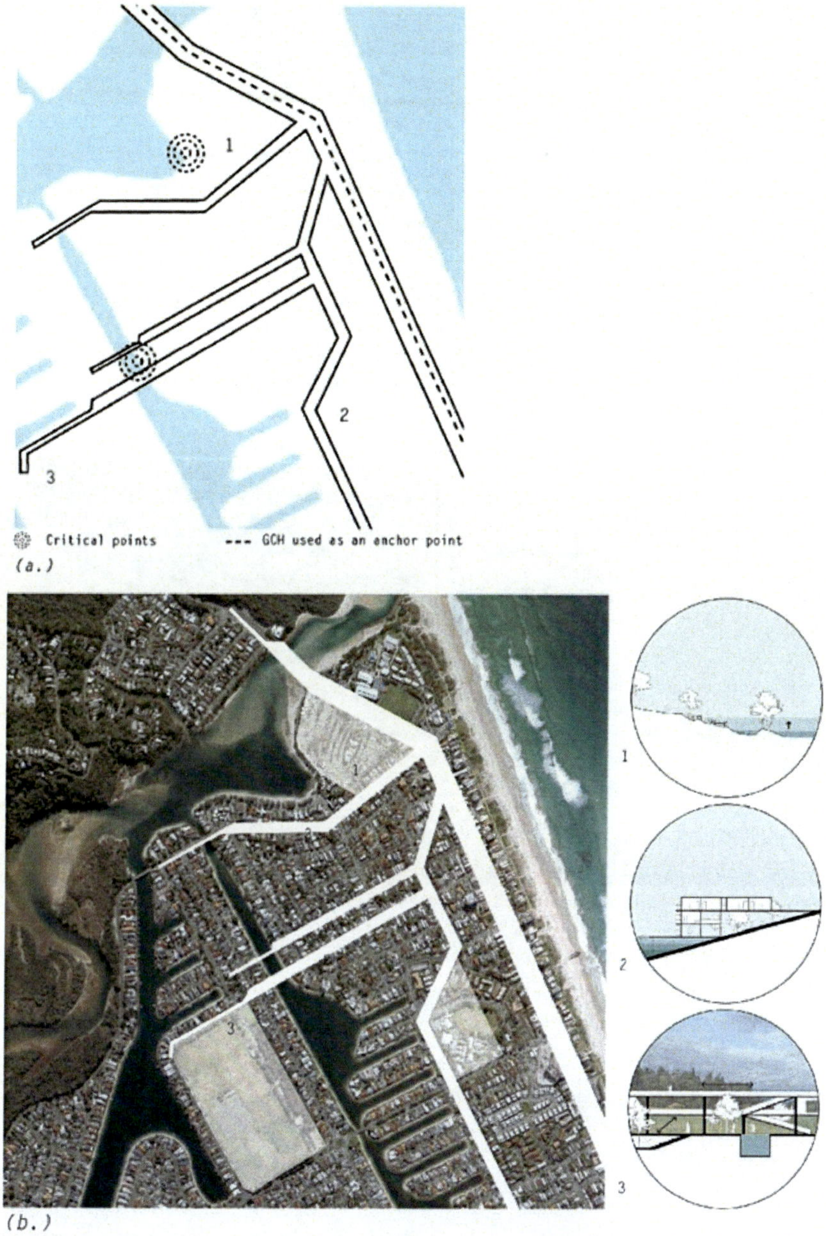

Fig. 2.4 Testing fingers of high ground in Palm beach (with Carina Soleckhan) **a** The proposal responds to the context. Existing activity anchorage points: 1. Tourist park, 2. Education facility, 3. Sport/recreation facility **b** Schematic master plan and design propositions via architectural examples. 1. Adoption of Minghu Wetland Park, China by Turenscape (2012) to promote adaptation in the tourist park; 2. Student housing, Italy by Aldo Rossi (1974), to promote housing solutions to occupy the slope and maximise the use of floodable green areas; 3. Swimming pools, Switzerland by Aurelio Galfetti (1967), inhabited urban connector to support and enrich the network of community facilities while providing adaptative solutions. **c, d** Housing solution on fingers of high ground

(c.)

(d.)

Fig. 2.4 (continued)

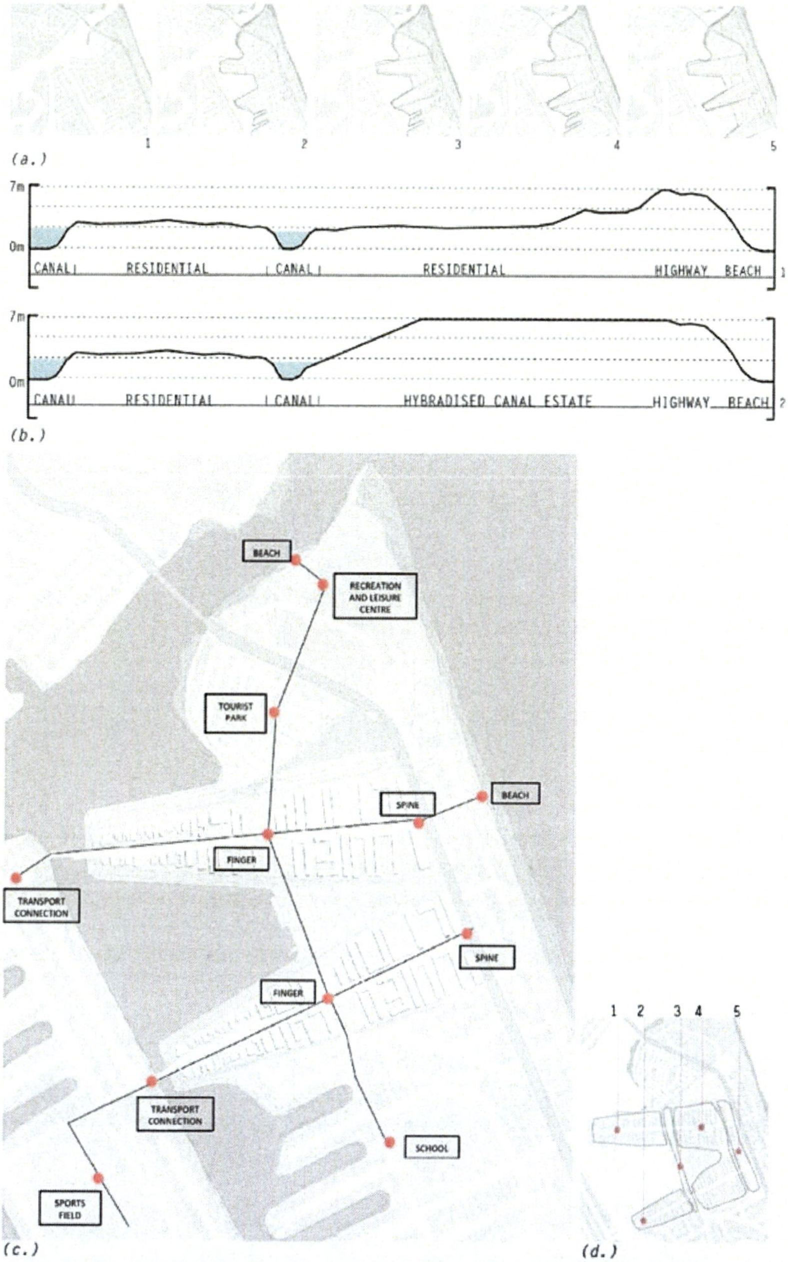

Fig. 2.5 Testing hybridised canal estate in palm beach (with Tahlia Wright) **a** Stages of implementation: 1. Present day, 2. Stage 1: 10 years, 3. Stage 2: 30 years, 4. Stage 3: 40 years, 5. Future extensions. **b** 1. Existing site section, 2. Proposed site section. **c** Network of existing and proposed community facilities **d** Program zones: 1. Low-density residential, 2. low-density residential, 3. community, 4. medium-density residential, 5. commercial. **e** master plan, **f**, **g** community facilities

Fig. 2.5 (continued)

References

Aldrich DP (2012) In: Building resilience: social capital in post-disaster recovery, University of Chicago Press

Aldrich DP, Meyer MA (2015) Social capital and community resilience. Am Behav Sci 59(2):254–269. https://doi.org/10.1177/0002764214550299

Australian Bureau of Statistics (2018) Technical Paper Socio-Economic Indexes for Areas (SEIFA) 2016 Catalogue No. 2033.0.55.001 https://www.ausstats.abs.gov.au/Ausstats/subscriber.nsf/0/756EE3DBEFA869EFCA258259000BA746/$File/SEIFA%202016%20Technical%20Paper.pdf

Australian Business Roundtable for Disaster Resilience and Safer Communities (2016) The economic cost of the social impact of natural disasters. http://australianbusinessroundtable.com.au/assets/documents/Report%20-%20Social%20costs/Report%20-%20The%20economic%20cost%20of%20the%20social%20impact%20of%20natural%20disasters.pdf

Bischeri C (2015) A cyclone-proof community centre for Atherton: tropical monumentality to enable resilience in far North Queensland communities. MPhil Thesis, School of Architecture, The University of Queensland. https://doi.org/10.14264/uql.2015.839

Bosman C, Dedekorkut-Howes A, Leach A (2016) Off the plan : the urbanisation of the gold coast. CSIRO Publishing, Clayton, Vic https://search-ebscohost-com.libraryproxy.griffith.edu.au/login.aspx?direct=true&db=nlebk&AN=1164333&site=ehost-live&scope=site

Bridges TS, Wagner PA, Burks-Copes KA, et al. (2015) Use of natural and nature-based features for coastal resilience. US Army Corps of Engineers, Washington, DC. www.nad.usace.army.mil/Portals/40/docs/NACCS/NNBF%20FINAL.pdf

Cazenave A, Llovel W (2010) Contemporary sea level rise. Annu Rev Marine Sci 2:45–173

Cutter SL et al (2008) A place-based model for understanding community resilience to natural disaster, Global Environ Change 18

da Cunha D (2019) The Jungle's Call Harvard Design Magazine No. 45/ Into the Woods http://www.harvarddesignmagazine.org/issues/45/the-jungles-call

Doty DH, Glick WH (1994) Typologies as a unique form of the-ory building: toward improved understanding and modeling. Acad Manage Rev 19:230

Dupre K, Bischeri C (2019) The architecture of resilience in rural towns. Archnet-IJAR. https://doi.org/10.1108/ARCH-07-2019-0178

Flora CB, Flora JL, Gasteyer SP (2015) In: Rural communities: legacy and change, Routledge

Goad P (1997) "4.4 The gold coast: architecture and planning. In: Marquis-Kyle AL (ed) Henshall Hansen Associates, Context, HJM, and Staddon Consulting."Gold coast urban heritage and character study," prepared for Gold Coast City Council

Hill K (2015) Coastal infrastructure: a typology for the next century of adaptation to sea-level rise. Front Ecol Environ 13(9):468–476. https://doi.org/10.1890/150088

Intergovernmental Panel on Climate Change (2012) Managing the risks of extreme events and disasters to advance climate change adaptation A special report of the intergovernmental panel on climate change. Cambridge University Press, Cambridge, UK, New York, NY, USA, p 582

Kopp RE, Horton RM, Little CM, Mitrovica JX, Oppenheimer M, Rasmussen DJ, Strauss BH, Tebaldi C (2014) Probabilistic 21st and 22nd century sea-level projections at a global network of tide-gauge sites. Earth's Future 2(8):383–406

Kulp SA, Strauss BH (2019) New elevation data triple estimates of global vulnerability to sea-level rise and coastal flooding. Nature Commun 10:4844. https://doi.org/10.1038/s41467-019-12808-z

Lawler A (2011) Did the first cities grow from marshes? Science 331:2011. https://doi.org/10.1126/science.331.6014.141

Lynch K (1990) In: Michael S (ed) Wasting away, San Francisco, Sierra Club Books

Lu D, Flavelle C (2019) Rising seas will erase more cities by 2050, New Research Shows, The New York Times https://nyti.ms/2NkOqOQ

Leach A (2018) Gold coast: city and architecture. Lund Humphries, London

Mathur A, da Cunha D, Squier-Roper C, Laird Prentice G, Weiner MJ (2014) Turning the frontier Norfolk and Hapton roads, Virginia. In: Structure of coastal resilience, Phase 1 context, site, and vulnerability analysis

Mileti DS, Gailus JL (2005) Sustainable development and hazards mitigation in the United States: disasters by design revisited. Mitigation and adaptation strategies for global change (3)

Nicholls RJ, Cazenave A (2010) Sea-level rise and its impact on coastal zones. Science 328:517–1520

Nordenson CS, Nordenson G, Chapman J (2018) Structures of coastal resilience. Island Press, Washington

Oilier CD (1982) The Great escarpment of eastern Australia: tectonic and geomorphic significance. J Geol Soc Aust 29(1–2):13–23. https://doi.org/10.1080/00167618208729190

Pournelle JT (2003) The littoral foundations of the Uruk State: using satellite photography toward a new understanding Of 5th/4th millenium BCE landscapes in the Warka survey area, Iraq. In: Gheorghiu D (ed) Chalcolithic and early bronze age hydrostrategies. Archaeopress, Oxford, pp 5–23

Pournelle JT (2007) KLM to CORONA: a bird's-eye view of cultural ecology and early mesopotamian urbanization. In: Stone EC (ed) Settlement and society: essays dedicated to Robert McCormick Adams. Cotsen Institute of Archaeology Press at UCLA, Los Angeles, pp 29–62

Queensland Statistician's Office Queensland Government projections edition (2018). https://www.qgso.qld.gov.au/statistics/theme/population/population-projections/regions#current-release-qld-population-projections-region-reports-2018-edn

Sherrieb K, Norris FH, Galea S (2010) Measuring capacities for community resilience. Soc Indic Res 99:2

Vale J, Campanella TJ (2005) Axioms of resilience. In: Lawrence J V, Thomas JC (ed) The resilient city. how modern cities recover from disaster, Oxford University Press, Oxford, New York

Woolcock M (2002). Social capital in theory and practice: reducing poverty by building partnership between states, markets and civil society. In: Social capital and poverty reduction: which role for civil society organizations and the State? UNESCO, France, pp 20–44

Cecilia Bischeri Cecilia Bischeri has been specialising in architectural design and its integration with the urban forms since 2006. She has been interested in developing this subject coupling theoretical and applied research. Concurrently, she has been involved in architectural education and has lectured design in Italy and Australia. Cecilia's main areas of interest regard architectural projects that ground their strengths in connecting the technical requirements of large-scale projects with the provision of a societal dimension for the targeted community. Through the integrative approach of urban and architectural design, the rehabilitation of socially disadvantaged communities and the provision of ameliorative conditions for accelerating the recovery of communities threatened by natural disasters constitute the main body of her work. Cecilia is the co-founder and Head at the Architecture Et cetera Lab. The AEc Lab provides a productive forum for the overlap of academia and architectural practice via the production of practice-based research. Cecilia's design projects have been shortlisted for international architectural prizes, published and part of expositions in Italy and China. Cecilia is a registered architect.

When It's Time to Let Go: Re-Imagining Coastal Urban Living in the Face of Rising Seas

Elnaz Torabi and Aysin Dedekorkut-Howes

Abstract Sea cities are at the forefront of climate change. Globally, the developed coastline of many cities is at high risk of sea level rise, coastal flooding, and storm surge. Such risks, however, can turn into important opportunities for re-imagining the future of cities and their resilience and sustainability. Despite being controversial, unsettlement, re-settlement, retreat, temporary and permanent relocation, and climigration are concepts that are rapidly becoming an inevitable urban policy and planning consideration. Yet, the idea of a nomadic city, a city that moves, is not new. In fact, the oldest forms of human adaptation to coastal hazards have been through relocation to higher grounds. While temporary forms of relocation have long been prominent in disaster risk reduction efforts, permanent retreat and relocation of urban communities are increasingly becoming critical discussions in climate change adaptation. This chapter focuses on urban resilience to coastal hazards and explores strategies ranging from temporary retreat and managed retreat to migration as a land use policy that can create transformative opportunities for adaptation and re-imagines the future of sea cities in a changing climate. Case studies from around the world are presented, exploring the potential of policy responses as well as the key barriers and drivers for their implementation.

1 Introduction

The increasing impacts of climate change put pressure on many cities around the world that are already dealing with urbanisation and rapid population growth issues. Today, nearly half (44%) of the world's population lives near the coast (United Nations n.d.). Coastal hazards including flooding triggered by sea level rise and increased frequency/intensity of storm surge and cyclones will impact many cities

E. Torabi (✉) · A. Dedekorkut-Howes
Griffith University, Gold Coast, Australia
e-mail: e.torabi@griffith.edu.au

A. Dedekorkut-Howes
e-mail: a.dedekorkut@griffith.edu.au

J. Baumeister et al. (eds.), *SeaCities*, Cities Research Series,
https://doi.org/10.1007/978-981-15-8748-1_3

(IPCC 2018). In response, cities will need to build their resilience and adapt to the impacts of climate change. Adaptation is defined as "the process of adjustment to actual or expected climate and its effects" in order to moderate harm and exploit beneficial opportunities (IPCC 2014: 1785), while resilience is the capacity of a city to not only return back and restore its previous functions, but also adjust to change (for example by building back better) and/or transform the way it deals with the impacts of hazards (Rockefeller Foundation n.d.; Torabi 2017).

Traditionally, cities have built their resilience and adapted to the impacts of disasters by *protecting* the vulnerable areas where population, economic activity, and natural resources are located, *accommodating* the impacts by enhancing the capacity of the natural and built environment and people to deal with the impacts while continuing to use vulnerable areas, and/or *retreating* people and structures from vulnerable areas via temporary or permanent relocation to safer areas (Dronkers and Mulder 1990). There are arguments for or against each strategy. Protection and defence via engineering structures have been the typical planning responses for many major cities around the world as they are easier to implement and do not disrupt the existing institutional and social contexts (Dedekorkut-Howes et al. 2020b). However, engineering solutions are only as good as their design benchmarks and protect for the short-term. Peterson (2019: 203) warns against planning for individual temporary hazards "rather than the combined effect of more extensive storm surges followed by permanent inundation of increasing amounts of coastal land" and deems attempts at defence a "losing battle" which will only incur further expense of removal in the long run. Dedekorkut-Howes et al. (2020b) agree that retreat will eventually become inevitable for some cities as they cannot be defended indefinitely. According to Siders et al. (2019: 761) "the question is no longer whether some communities will retreat—moving people and assets out of harm's way—but why, where, when, and how they will retreat." Cities increasingly need to recognise the need to consider other options including a shift from fighting water to living with it (Rijke et al. 2012).

Considering the increasing cost of investment in protection and accommodation strategies (see Donner and Webber 2014), many governments around the world are considering retreat as a viable strategy for the long-term (Dedekorkut-Howes et al. 2020b; Hanna et al. 2020; Hino et al. 2017; Peterson 2019). The retreat strategy applies to a wide range of risks including those associated with sea level rise and coastal flooding, inland flooding, bushfires, and landslides. Retreat also spans across different spatio-temporal scales, involving planned or managed relocation of entire communities, strategic relocation of critical infrastructure, and/or more asset-based responses (Dedekorkut-Howes et al. 2020b; Hanna et al. 2019, 2020; Tadgell et al. 2018). Accordingly, different terminology has been used in framing retreat in urban practice, policy, and research including relocation or realignment; planned or managed retreat; unsettlement, resettlement, migration, abandonment (Dedekorkut-Howes et al. 2020b; Hanna et al. 2019), or as Matthews and Potts (2018) term it climigration (climate-related immigration and displacement). King (2017: 66–67) distinguishes between relocation and resettlement in that the former is "the ad hoc migration of people" whereas the latter denotes "the permanent or long-term movement of a community from one site to another", and in this sense, resettlement is often the preferred option.

The various forms of retreat can be categorised as: (1) *vertical retreat* which involves elevating structures, (2) *engineered retreat* which involves land reclamation and filling, (3) *temporary relocation* during hazards, (4) *horizontal planned retreat* or managed realignment involving moving permanently to higher land, and (5) *migration* which involves permanent abandonment of an area (Dedekorkut-Howes et al. 2020b). Peterson (2019: 204) distinguishes between tactical retreat which may involve "relocating buildings landward on the same property" from larger scale strategic relocation which relocates "coastal neighborhoods, infrastructure, communities, or ecosystems to new, safer sites".

While retreat can enhance the natural coastal processes and is less costly in some cases compared to hard protection (Abel et al. 2011; Harman et al. 2013), it is still a controversial "dirty word" or a "last resort" in planning and decision making, sometimes considered to be surrender or defeat in the fight against climate change. For example, the negative connotations of the term retreat have led to the use of "managed realignment" as a more politically acceptable term in countries such as the UK (Harman et al. 2013; Pethick 2002).

The implications of retreat can be very complex, involving difficult decisions. It will require significant resources, communities will lose their homes and neighbourhoods, and there will be interruptions to business and the economy (Lonsdale et al. 2008; Scott et al. 2012). Some view retreat as an unsuitable option for developed urban areas and argue that it can lead to inequality by disproportionately affecting the community, and cause loss of heritage in areas of historical significance (Harman et al. 2013; Lonsdale et al. 2008). There are several barriers to implementation of managed retreat. For instance, existing settlement patterns and lack of suitable land for relocation can be a physical limitation (Harman et al. 2013; Munji et al. 2013; Primo 1997; Scott et al. 2012). Retreat can lead to the clearing of more mangroves and trigger environmental problems (Munji et al. 2013). It is also very likely to result in major legal disputes and have considerable impact on property values (O'Donnell 2019; Scott et al. 2012). Land use controls that prohibit or limit new development in high risk areas can be challenged in the courts as the "taking" of private land and create land tenure and liability issues for local governments (Dedekorkut-Howes et al. 2020b; Primo 1997).

The socio-political cost of retreat is high, making it a publicly unacceptable option (Harman et al. 2013; Lonsdale et al. 2008; Sahin et al. 2013). This is mainly due to communities' desire to live close to water (Torabi et al. 2018) and attachment to their existing lifestyles and cultural heritage (Donner and Webber 2014; Douglas et al. 2012; Primo 1997). The attachment of a community to a place is closely linked to sense of pride and belonging to a community and considered to be the most powerful contributor to a final decision regarding migration and retreat (Jamero et al. 2017). It is important to consider the socioeconomic impacts of population dispersal and mixing as part of retreat due to cultural perception of land loss and community preference for in situ (structural) adaptation options (Harman et al. 2013;

Lonsdale et al. 2008). There are also negative political implications for land use planning, as local governments are perceived to be withholding tenable land from the market (Alexander et al. 2012). Wealthy property owners want to build along the coasts and politicians waver between managing community demand and responding to increasing coastal risks. This translates into significant financial burden on the community (Harman et al. 2013).

Yet, retreat also provides unique opportunities for re-imagining the future of cities in the age of climate change and resource depletion that are so far unexplored (Siders et al. 2019; Black et al. 2011). Many cities around the world have to deal with the legacy development and infrastructure that have been laid out without any consideration of future hazards and climate change. Retreat can provide a second chance for good planning and undoing past mistakes. This chapter focuses on urban resilience to coastal hazards and explores strategies ranging from temporary retreat and managed retreat to migration as a land use policy that can create transformative opportunities for adaptation and re-imagines the future of sea cities in a changing climate.

2 Living Harmoniously with Water

Relocation to higher grounds is an old form of human adaptation to coastal hazards. This involves relocation to temporary houses or use of relocatable structures for housing that can be easily transferred to higher grounds, as traditionally practiced in many island and delta communities in the Philippines, Federated States of Micronesia, and Indigenous communities in Australia (Primo 1997; Munji et al. 2013; Jamero et al. 2017). Historically, some sea nomad communities such as the Bajo in Southeast Sulawesi, Indonesia have also spent their entire lives on houseboats (Kusuma et al. 2017).

Temporary relocation allows communities to retain temporary houses on the low-lying areas for their livelihood, while using permanent houses on the relocated areas during natural hazards. The underground cities of Cappadocia in Turkey dating back to 1200 BC provide an interesting example of temporary resettlement. They were designed to serve as temporary shelter to tens of thousands of people during enemy invasions (Çiner and Aydar 2019; Bertini 2010). While temporary relocation has long been prominent in disaster risk reduction, permanent retreat and relocation of urban communities are increasingly becoming critical discussions in climate change adaptation (Nalau and Handmer 2018).

In the context of western societies, the idea of a nomadic city is also getting traction in the literature. Fry (2011) proposed the concept of the *urmadic* city to connote urban nomadism and describe non-permanent, moving cities. The idea of an urmadic city, a city that moves, is not new. As opposed to moving existing cities, the urmadic city is designed to move if and when it becomes necessary. These examples throughout history across the globe provide us with opportunities to take lessons from the past to apply them to the future. But this requires the transformation of traditional assumptions that underpin urban planning. Floating houses and neighbourhoods are no longer part of a utopian dream, but a possibility of a new form of sustainable urban living, as supported by the United Nations (2019). Yet, a recent systematic review of the literature (Dedekorkut-Howes et al. 2020b) shows a gap in research on the complexities of decision making and planning in this context.

3 Managed Retreat and Resettlement

There are several examples of managed retreat and resettlement projects across the world, both successful and unsuccessful. Here we provide a brief overview of some of them focusing on barriers and drivers of their implementation and opportunities they provide.

3.1 Australia

With 85% of its population living in coastal areas Australia is extremely exposed to the impacts of climate change, yet despite the urgency of the matter, it does not have a national coastal policy. The lack of leadership from the national level had made coastal adaptation a responsibility of the states and territories, leading to a variety of inconsistent responses (Howes and Dedekorkut-Howes 2016). A review of the coastal policies and plans of all Australian states and territories shows that while some states such as South and Western Australia provide more policy guidance on retreat at higher strategic level there is little on implementation (Dedekorkut-Howes et al. 2020a).

Managed retreat has been implemented in Australia in several cases, in less populated areas (towns) and usually as an ad hoc response to a disaster. One well-known case of failure to retreat is the town of Byron Bay in New South Wales, Australia. The significance of this case relates to its demonstration of the critical role of the community for implementation of managed retreat policies. In response to coastal storms in 1970, the local council adopted a managed retreat policy in 1988, which received significant community backlash (England 2013; Harman et al. 2013). In 2009, coastal storms caused severe damage to private property along the beach. Despite the Council's plans for retreat of property and refusal to reinstate existing sandbag walls, in a legal challenge the community demanded action from the Council to reinstate the sandbags to continue to live in the same area (Leitch 2009). Ironically, a few years later, the continued exposure of the community led the Council to consider more engineering protection measures such as seawalls to defend existing settlements, much like the Gold Coast across the state border (Torabi et al. 2017a). This decision faced significant backlash from the local community that demanded the public amenity of the beach (Lovejoy 2015; O'Donnell 2016). Similar resistance by the community to government initiatives of retreat was also experienced at Smith Island in the Chesapeake Bay in the United States of America (US) after hurricane

Fig. 3.1 The central pacific island of Tarawa in Kiribati (Government of Kiribati 2005, creative commons attribution 3.0 unported (https://creativecommons.org/licenses/by/3.0/deed.en))

Sandy causing tens of millions of dollars to be committed to protection (Peterson 2019). The island nation of Kiribati (Fig. 3.1) provides another example of backing away from proactive relocation despite previously purchasing land in Fiji for planned resettlement (Walker 2017).

On the opposite side of the coin, managed retreat has proved to be successful where appropriate institutional setting is in place and the affected community is properly engaged in the process. The town of Grantham in Queensland was hit hard by flash flooding in 2011 (Fig. 3.2). After the floods the government implemented a land swap strategy to facilitate the relocation of the community. The local government in Grantham purchased a parcel of land on higher ground near the existing flood affected town. Through communication of its land swap strategy by using maps and engaging with the community, the local council facilitated the relocation of the entire community. The proximity of the new area to the existing flood affected community proved helpful as it facilitated engagement: people could see what the new estate looked like and were more interested in the program (Simmonds 2020). Grantham's experience highlights the importance of existing planning regulations which can hinder action, local political leadership, collaboration between all sectors involved, community participation, and ongoing assessments for the success of resettlement process (Okada et al. 2014; Sipe and Vella 2014). However, the more common approach in Australia is relocation through voluntary buy-back schemes particularly after disasters (King et al. 2014).

Fig. 3.2 Motor vehicles and motorboat colliding with the railway bridge during the floods in Grantham, Queensland, 2011 (Geoff Purton, John Oxley Library, State Library of Queensland, Image No: 28510-0145-0018)

3.2 The United States

Much like Australia, the US is highly exposed to coastal impacts of climate change. In the US flood management has historically focused on enabling people and infrastructure to remain in the same area. Unlike Australia, however, the US has the longest-running programs of managed retreat globally, driven by its national government (Hino et al. 2017). Voluntary buyouts of flood prone properties funded by US federal agencies, especially the Federal Emergency Management Agency (FEMA), underpins the country's approach to retreat (Dyckman et al. 2014). There has, however, been significant equity concerns around managed retreat in the US. (Siders et al. 2019). A recent review of FEMA-funded voluntary buyouts by Mach et al. (2019) showed that local governments in counties with higher population and income were more likely to administer buyouts, whereas within cities the buyouts were concentrated in more socially vulnerable neighbourhoods. There are examples such as Soldiers Grove, Wisconsin where managed retreat created opportunities. The deliberate relocation of the business centre of the city away from the river and closer to the highway in 1978 created economic opportunities for a city that is now known as a Solar Town (Siders et al. 2019; David and Mayer 1984). In San Francisco's Ocean Beach, efforts are under way via the *Ocean Beach Master Plan* (2012) to address sea level rise, protect existing infrastructure, restore coastal processes, and improve public access to the beach. An important part of the masterplan is the rerouting of

Fig. 3.3 Artist's sketch of Ocean view master plan's vision (Courtesy of SPUR and AECOM)

parts of the Great Highway. Retreat of the existing road in this case, combined with other measures such as asphalt overlays to redirect water runoff and sewer/drainage improvements, is used as an opportunity to improve public access (pedestrian and bicycle) and connections to the beach (SPUR 2012) as shown in Fig. 3.3.

The fourth *National Climate Assessment* (USGCRP 2018: 1329) remarked that "in all but the very lowest sea level rise projections, retreat will become an unavoidable option in some areas along the U.S. coastline" but acknowledged that "the potential need for millions of people and billions of dollars of coastal infrastructure to be relocated in the future creates challenging legal, financial, and equity issues that have not yet been addressed." So far, the examples of managed retreat are small scale such as the relocation of a bike path and parking lot inland in Surfers Point, Ventura, CA (Kochnower et al. 2015). An example of community resettlement is the Biloxi-Chitimacha-Choctaw Tribe who reside in Isle de Jean Charles in southeastern Louisiana. The ideal resettlement scenario produced by the community highlight the opportunities provided by resettlement: "The residents envision a sustainable community that utilizes ground-breaking technology and resilience measures while integrating the history, traditions, and culture of the Biloxi-Chitimacha-Choctaw tribe" (King 2017: 313). The features of the new community include stormwater detention in community parks, treating water as a resource through family gardens, water management technology, and strategic tree planting and forest and water features that allow for crawfish ponds and migratory bird sanctuaries. The success of the resettlement depends on it being community-led and voluntary. So far, the state of Louisiana has named the master planner of the project and selected a site approximately 40 miles northwest of the Isle de Jean Charles (Crepelle 2018). However, the project is not without its critics. Crepelle (2018) questions why the resettlement funding cannot be used for restoration and erosion prevention and Marino (2018) suggests that Isle de Jean Charles' exclusion from the levee project due to the cost-benefit analysis may have been a politically sanctioned sacrifice. Jessee (2020) reports that the State's recent redefinition of the scope of the resettlement and lack of commitment to the existing tribal vision has confirmed the fears of the community that it was turning the tribal resettlement into a subdivision.

On a larger scale, Hawaii is in the early stages of exploring the feasibility and implications of managed retreat (OP-CZM 2019). The Office of Planning's project report found that managed retreat is more likely after a catastrophic event (as in the case of Australia's Grantham) when resources and the resulting political and social will to relocate are more plentiful and urgent. Yet, it is currently not possible to develop a step-by-step plan to implement managed retreat given the various unknowns and competing priorities (OP-CZM 2019).

3.3 The Netherlands

The Netherlands is the most famous example of a country with a long history of living with water, implementing multiple managed retreat strategies including the realignment of the shoreline to create space for coastal habitat development and a natural buffer zone for flood protection (Stronkhurst and Mulder 2014). The country's *Room for the River* program focused on strategic retreat and removed human-built barriers, reshaping landscapes at more than 30 locations around the country. As part of this program and through extensive consultation with the community, a managed retreat strategy was pursued to lower the existing dikes in De Noordwaard while 75 households were resettled elsewhere in elevated mounds or outside the area (Rijke et al. 2012; Schut et al. 2010; Hino et al. 2017). In this case, residents who initially opposed retreat came around after several repeated inundation events. The project created several opportunities for the landscape, housing, and the agricultural businesses in the area including nature, economic activity, and recreation. The retreat strategy in this case was combined with measures to accommodate the impacts of flooding including safe in situ housing opportunities, evacuation guidelines, construction of a high-water channel to drain large volumes of water to the sea, building higher quays for more sustainable agriculture practices, as noted by the Directorate-General for Public Works and Water Management (Rijkswaterstaat n.d.).

A more urban example from the *Room for the River* program is Nijmegen, one of the Netherlands' oldest cities. The city's high-density centre is located in close proximity to the Waal River, extremely exposed to flooding. Here the development along the river did not leave any space to set back from the river or bulk up seawalls. In close collaboration with the residents, the government successfully bought 50 homes, using the space to move the defence structures inland and create a recreational space (Baurick 2020).

3.4 The United Kingdom

One retreat approach is to provide compensation (or support) for private owners whose homes are threatened. Through the UK *Coastal Change Pathfinder Programme*, five local councils relocated or bought out households at high risk from coastal hazards. Between 2009 and 2011, the programme funded pilot projects and bought out properties, relocating around ten households in different communities (OECD 2019). The North Norfolk District Council was one of the five councils in the UK to receive major funding (£3 million) in 2009–10 under the programme (DEFRA 2012). Owners of the nine houses at immediate risk of coastal hazards were paid a market value. The project initially planned to test the strategy of "buy-and-lease back", but ultimately this was not pursued due to lack of sufficient funding and perceived financial risk to the Council. This project is reported to have increased the Council's understanding of coastal management strategies and created a sense of community and cohesion (DEFRA 2012).

Currently the Environment Agency and local councils are developing *Shoreline Management Plans* (SMPs) in England and Wales to manage the threat of coastal change (Environment Agency 2019). These include "holding the line" policies which protect settlements and infrastructure with hard defences either for part of this century or all the way to the end of the century with an estimated implementation cost of £18–30 billion. However, a recent report by the UK's Committee on Climate Change (2018) suggests that this policy is for the most part not cost-beneficial, thus unlikely to get funded raising the need for realistic plans to adapt. The *Draft National Flood and Coastal Erosion Risk Management Strategy for England* accepts that some areas will flood and promotes managed realignment in areas of high flood risk from rivers. The 2012 *Shoreline Management Plan* for the area encompassing 600-household village of Fairbourne in Wales (Fig. 3.4) has recommended realignment of the coast and eventual decommissioning of Fairbourne eventually prompting lawsuits from residents who argued that the value of their properties decreased (Buser 2020, Peterson 2019). This case of the first and largest UK community to be abandoned also highlights the importance of involving affected residents in the planning process early and in meaningful ways and the need for a robust communication plan that involves the media (Buser 2020).

3.5 New Zealand

In line with the IPCC recommendations, adaptation strategies were integrated in the *New Zealand Coastal Policy Statement,* which provides a national policy direction for coastal management. The policy focuses on managed retreat "by relocation or removal of existing structures or their abandonment in extreme circumstances, and designing for relocatability or recoverability from hazard events" (Department of Conservation 2010: 24). Despite being controversial and facing opposition from the

Fig. 3.4 Beach at Fairbourne showing the dragon's teeth sea defences, 2011 (cc-by-sa/2.0 - © SMJ - geograph.org.uk/p/1459592)

affected communities, managed retreat has been implemented by some local councils in New Zealand (Bloomfield 2018). Yet the path towards retreat and relocation of communities has not always been clear. A case in point is Matatā, a town in the Bay of Plenty in the North Island of New Zealand hit by debris flows in 2005 after exceptionally heavy rainfall (Fig. 3.5). In the absence of a guiding framework, regulations and the necessary resources, the local council faced significant challenges in managed retreat of affected properties. This resulted in disputes between residents and local government as well as trauma and stress on the community and the officers involved (Hanna et al. 2020). After 14 years, and while some residents continued to be exposed to risk, the Whakatāne District Council, the Bay of Plenty Regional Council, and the Department of Internal Affairs agreed on a joint funding for property acquisition in this area. Yet the ad hoc approach to managed retreat has been criticised as being "disruptive and inequitable" resulting in a fragmented risk management approach (Hanna et al. 2020).

A successful case of retreat, on the other hand, was carried out by the Waitakere City Council in Auckland. In 2002, the Council developed the *Project Twin Streams*, running for 10 years to reduce flood risk and improve ecological functioning of the waterways (Bell et al. 2017). Rather than using compulsory acquisition of land through the *Public Works Act 1981*, the council adopted a voluntary buy back approach through an inclusive participatory process engaging property owners and community representatives such as politicians as well as the Māori. As a result, 78 full and 78 partial purchases of properties were successfully negotiated

Fig. 3.5 Debris flow in Matatā in May 2005 (Courtesy of *Whakatāne Beacon*)

to allow for floodplain redesign, linking managed retreat objectives with broader environmental, social, economic, and cultural goals (Bell et al. 2017). In this case, managed retreat created an important opportunity to strengthen the social fabric of the community by providing new public infrastructure (parks, cycleways, and walkways) and accommodating the needs of those who moved to other areas (Vandenbeld and Macdonald 2013). This was in the main achieved by the effective and efficient approach of the Council to engagement by building on the local knowledge of the affected people, having extended periods of engagement to accommodate people's needs, and negotiating equitable and individualised solutions (Bell et al. 2017).

4 Opportunities: Re-Imagining the Future of Sea Cities in a Changing Climate

Fighting with the ocean is a losing battle. Managed retreat is a promising but controversial approach to adaptation. In theory, the idea of removing people and property out of the harm's way is a no regrets best practice approach to address sea level rise and coastal flooding. However, in practice it is politically sensitive and contentious. To make the technically sound solutions politically acceptable, managed retreat policy and practice needs to be better informed.

Managed retreat can create many opportunities for the community. For instance, it can provide funding opportunities for the people who want to leave but cannot afford to do so (Siders et al. 2019). The abandoned areas can then be used to restore the natural amenity of the coastline, create opportunities for absorbing flood waters, or can be used as public amenity spaces (Dedekorkut-Howes et al. 2020b; Siders et al. 2019). Above all, managed retreat is an opportunity for social, economic, and ecological transformation and new beginnings (Freudenberg et al. 2016; Siders et al. 2019). Realising the opportunities provided by managed retreat requires a transformational approach to adaptation, adding long-term innovative approaches to change the existing ways of life and practices (Matyas and Pelling 2015, Torabi et al. 2018). This approach is underpinned by a change to embed long-term planning, proactive social learning, critical re-appraisal of the economy, fostering creativity and openness, and incorporating future uncertainty in decision making (Restemeyer et al. 2015; Tchakert and Dietrich 2010).

There is an increasing need for embedding managed retreat as a viable option in long-term strategic planning. Siders et al. (2019) emphasise the importance of strategic, managed retreat rather than ad hoc reactive actions to ensure opportunities to contribute to broader societal goals (such as sustainability) and economic development are not missed. Key dimensions of such strategic retreat would include decision-making and planning at larger geographic and temporal scales; involvement of multiple agencies and jurisdictions; addressing of multiple hazards; and integration into planning for economic, social, and environmental goals. Ryan et al. (2015) develop five key principles to govern the large-scale relocation of jobs and housing: (1) out of harm's way, (2) minimise stress, (3) receiving capacity, (4) build it back better, and (5) feasibility of implementation. Two of these principles illustrate the opportunities provided by relocation. Careful selection of relocation sites would not only protect them from risk of flooding but also other risk factors such as steep slopes, contaminated brownfields, polluted areas, and areas adjacent to heavy industry. Relocation also provides the opportunity to build back better where a holistic plan can be prepared to consider all aspects of sustainable development from energy and water footprints of settlements to opportunities for more active and public transport. Managed retreat can be a catalyst for a resilient and sustainable future, one that includes water sensitive and smart cities, supported by blue-green infrastructure, and providing equitable opportunities for everyone in the city.

To be effective managed retreat needs to be implemented on a larger scale, as the retreat of an isolated section of the coast can increase erosion in the remainder of the system, especially if constrained by coastal defence structures (French 2008; Pethick 1993). Often the focus on logistics of moving people and settlements as an ad hoc response to a disaster can undermine the strategic direction and benefit from these opportunities in the longer-term (Siders et al. 2019). Effective governance and management are critical for administering retreat. In Peru, for example, due to limited institutional capacity to ensure the sustainability of the changes, residents of flood prone areas returned to their original housing (Lavell 2016). Such long-term strategic planning needs to be complemented by a comprehensive community engagement for the success of managed retreat (Abel et al. 2011; Alexander et al. 2012). This creates opportunities to not only inform and engage the community about the change process, but also shifts the negative connotation of retreat to a positive experience by strengthening the sense of community and creating a common purpose (Siders et al. 2019). The availability of financial resources also plays a big role. For example, retreat of communities can depend on the possibility of obtaining low priced housing in new areas (Douglas et al. 2012; Jamero et al. 2017) and availability of jobs can directly impact the willingness of a community to migrate (Donner and Webber 2014). As the examples in this chapter show, there is also a need for a tailored approach to every context. For instance, relocation of inland communities (in most cases to similar neighbourhoods) is considered to be less contentious compared to the relocation of coastal communities that might lose their immediate connection with the coast (Bloomfield 2018).

Several financial and regulatory tools can facilitate the process of managed retreat and exploiting opportunities. These include land use planning, market-based incentives/disincentives, and community engagement (Dedekorkut-Howes et al. 2020b; Peterson 2019). Land use planning includes zoning regulations and overlays requiring increased buffers, setback, and density conditions that can limit development of high-risk areas in cities (Susskind 2010; Harman et al. 2013). Planning policies can facilitate phasing out existing development and/or promote more conditional development options (Frazier et al. 2010). Temporary development (such as ephemeral buildings) can be built in high-risk areas with conditions to abandon properties on specific timeframes or when the risk is too high (Harman et al. 2013). Incentives include facilitating land swaps that trade ownership of developed areas near the coast via transferrable development rights for land owned by the government located outside risk zones (Frazier et al. 2010; Susskind 2010), and providing legal foundations for government buy-outs such as establishing and funding a buy-out program as discussed in the case studies. Rolling conservation easement, transfer of development rights, mandatory disclosure of risk in real estate transactions are important policy tools in this context (Alexander et al. 2012; Peterson 2019; Siders et al. 2019; Susskind 2010).

5 Conclusions

To be effective, a managed retreat policy needs to be supported by government leadership at all levels (Taylor et al. 2013; Peterson 2019). Higher levels of government play an important role as local governments do not necessarily have the capability to manage legal, political, and financial risks associated with managed retreat (Harman et al. 2013). Recently there has been a global call for national level leadership on retreat as an opportunity to redesign underlying norms and infrastrcture of our cities (Siders et al. 2019). In fact, Peterson (2019) believes that a national relocation framework that illustrates the financial infeasibility of structural protection everywhere will improve the chances of relocation to be considered as a policy option. Research also demonstrates that local governments are more likely to act when there are guidance, benchmarks, and requirements from higher level governments (Dedekorkut-Howes et al. 2020b; Dedekorkut Howes and Vickers 2017; Torabi et al. 2017b).

Yet, such an approach will require significant shifts in institutions and behavioural change to support the transformation. There is a need for a collaboratory and integrated approach to planning, underpinned by reliable and large scale hazard data to better inform decision making. This is an important consideration for sea cities globally, as climate change may cause hazard information, such as local council flood maps, to be dangerously outdated (Meyer 2020). In conclusion, managed retreat will require many difficult decisions to be made, implementation of which will require reconceptualising our relationship with our properties and neighbourhoods (Siders et al. 2019), imagining urmadic cities and communities moving across land and the ocean that are not bounded by land, but on sociocultural values that create communities (Fry 2014).

References

Abel N, Gorddard R, Harman B, Leitch A, Langridge J, Ryan A, Heyenga S (2011) Sea level rise, coastal development and planned retreat: analytical framework, governance principles and an australian case study. Environ Sci Policy 14(3):279–288. https://doi.org/10.1016/j.envsci.2010.12.002

Alexander KS, Ryan A, Measham TG (2012) Managed retreat of coastal communities: understanding responses to projected sea level rise. J Environ Planning Manage 55(4):409–433. https://doi.org/10.1080/09640568.2011.604193

Baurick T (2020) *Water Ways: The Dutch Are Giving Rising Rivers More Room. Should We Follow Suit?* New Orleans Public Radio. https://www.wwno.org/post/water-ways-dutch-are-giving-rising-rivers-more-room-should-we-follow-suit

Bell R, Lawrence J, Allan S, Blackett P, Stephens S (2017) Coastal hazards and climate change: appendices. New Zealand Government. https://www.mfe.govt.nz/sites/default/files/media/coastal-hazards-appendices-final_0.pdf

Bertini A (2010) Underground cities, cave dwelling, cave homes: yesterday, to day, to-morrow. In: Bucci A, Mollo L (eds) Regional architecture in the Mediterranean area. Alinea Editrice, Firenze

Black R, Bennett SR, Thomas SM, Beddington JR (2011) Climate change: migration as adaptation. Nature 478(7370):447–449

Bloomfield SEM (2018) Stepping back: a look at managed retreat in NZ. In: Rajagopalan P, Andamon MM (eds) Engaging architectural science: meeting the challenges of higher density, The Architectural Science Association and RMIT University, Australia, pp 553–559

Buser M (2020) Coastal adaptation planning in fairbourne, wales: lessons for climate change adaptation. Planning Practice Res 1–21. https://doi.org/10.1080/02697459.2019.1696145

Çiner A, Aydar E (2019) A fascinating gift from volcanoes: the fairy chimneys and underground cities of Cappadocia. In: Kuzucoğlu C, Çiner A, Kazancı N (eds) Landscapes and landforms of Turkey. Springer, Cham, pp 535–549

Committee on Climate Change (2018) In: Managing the coast in a changing climate. London, UK. https://www.theccc.org.uk/publication/managing-the-coast-in-a-changing-climate/

Crepelle A (2018) The United States first climate relocation: recognition, relocation, and indigenous rights at the Isle de Jean Charles. Belmont Law Rev 6:1–40. Available at SSRN: https://ssrn.com/abstract=3326691

David E, Mayer J (1984) Comparing costs of alternative flood hazard mitigation plans: the case of soldiers Grove, Wisconsin. J Am Planning Assoc 50(1):22–35

Douglas EM, Kirshen PH, Paolisso M, Watson C, Wiggin J, Enrici A, Ruth M (2012) Coastal flooding, climate change and environmental justice: identifying obstacles and incentives for adaptation in two metropolitan Boston Massachusetts communities. Mitig Adapt Strat Glob Change 17(5):537–562. https://doi.org/10.1007/s11027-011-9340-8

Dedekorkut-Howes A, Torabi E, Howes M (2020a) Planning for a different kind of sea change: lessons from Australia for sea level rise and coastal flooding. Clim Policy. https://doi.org/10.1080/14693062.2020.1819766

Dedekorkut-Howes A, Torabi E, Howes M (2020b) When the tide gets high: a review of adaptive responses to sea level rise and coastal flooding. J Environ Planning Manage 63(12):2102–2143. https://doi.org/10.1080/09640568.2019.1708709

Dedekorkut-Howes A, Vickers J (2017) Coastal climate adaptation at the local level: a policy analysis of the gold coast. In: Filho W (ed) Climate change adaptation in Pacific Countries: fostering resilience and improving the quality of life. Springer, Cham, pp 401–415

DEFRA [Department of Environment, Food and Rural Affairs]. (2012) Coastal Change Pathfinder Review Final Report. DEFRA, London, UK. https://assets.publishing.service.gov.uk/government/uploads/system/uploads/attachment_data/file/69508/pb13720-coastal-pathfinder-review.pdf

Department of Conservation (2010) New Zealand coastal policy statement. New Zealand Government: Wellington. https://www.doc.govt.nz/globalassets/documents/conservation/marine-and-coastal/coastal-management/nz-coastal-policy-statement-2010.pdf

Donner SD, Webber S (2014) Obstacles to climate change adaptation decisions: a case study of Sea-level rise and coastal protection measures in Kiribati. Sustain Sci 9(3):331–345. https://doi.org/10.1007/s11625-014-0242-z

Dronkers J, Mulder J (1990) Strategies for adaptation to Sea level rise. report of the coastal zone management subgroup. Geneva, IPCC

Dyckman CS, St. John C, London JB (2014) Realizing managed retreat and innovation in state-level coastal management planning. Ocean Coastal Manage 102(PartA):212–223. https://doi.org/10.1016/j.ocecoaman.2014.09.010

England P (2013) Too much too soon? on the rise and fall of Australia's coastal climate change law. Environ Planning Law J 30(5):390. https://research-repository.griffith.edu.au/bitstream/handle/10072/57341/90527_1.pdf?sequence=1&isAllowed=y

Environment Agency (2019) Shoreline management plans. https://www.gov.uk/government/publications/shoreline-management-plans-smps/shoreline-management-plans-smps

Frazier TG, Wood N, Yarnal B (2010) Stakeholder perspectives on land-use strategies for adapting to climate-change-enhanced coastal hazards: sarasota Florida. Appl Geogr 30(4):506–517. https://doi.org/10.1016/j.apgeog.2010.05.007

French JR (2008) Hydrodynamic modelling of estuarine flood defence realignment as an adaptive management response to Sea-level rise. J Coastal Res 24(2B):1–12. https://doi.org/10.2112/05-0534.1

Freudenberg BR, Calvin E, Tolkoff L, Brawley D (2016) Buy-in for buyouts: the case for managed retreat from flood zones. Lincoln Institute of Land Policy, Cambridge, MA, US

Fry T (2014) City futures in the age of a changing climate. Routledge, London

Fry T (2011) The urmadic city: the idea and the image. Griffith University Queensland College of Art, Brisbane

Hanna C, White I, Glavovic B (2019) Managed retreat in practice: mechanisms and challenges for implementation. Oxford Research Encyclopedia of Natural Hazard Science, Oxford University Press, New York, USA. https://doi.org/10.1093/acrefore/9780199389407.013.350

Hanna C, White I, Glavovic B (2020) The uncertainty contagion: revealing the interrelated cascading uncertainties of managed retreat. Sustainability 12(2):736. https://doi.org/10.3390/su12020736

Harman BP, Heyenga S, Taylor BM, Fletcher CS (2013) Global lessons for adapting coastal communities to protect against storm surge inundation. J Coastal Res 31(4):790–801. https://doi.org/10.2112/JCOASTRES-D-13-00095.1

Hino M, Field CB, Mach KJ (2017) Managed retreat as a response to natural hazard risk. Nature Clim Change 7(5):364–370

Howes M, Dedekorkut-Howes A (2016) The rise and fall of climate adaptation governance on the Gold Coast, Australia. In: Knieling J (ed) Climate adaptation governance in cities and regions: Theoretical fundamentals and practical evidence. Wiley Inc, Hamburg, pp 237–250

IPCC [Intergovernmental Panel on Climate Change] (2018) Global Warming of 1.5C. Special Report. Cambridge University Press, Cambridge

IPCC (2014) Climate change 2014: Impacts, adaptation, and vulnerability. Part A: global and sectoral aspects. Contribution of working group II to the fifth assessment report of the intergovernmental panel on climate change. Cambridge University Press, Cambridge, United Kingdom and New York, NY, USA

Jamero ML, Onuki M, Esteban M, Billones-Sensano XK, Tan N, Nellas A, Takagi H, Thao ND, Valenzuela VP (2017) Small-island communities in the philippines prefer local measures to relocation in response to sea-level rise. Nature Climate Change 7(8):581–586. https://doi.org/10.1038/nclimate3344

Jessee N (2020) Community resettlement in louisiana: learning from histories of horror and hope. In: Laska S (ed) Louisiana's response to extreme weather. Springer, Cham, pp 147–184

King D, Bird D, Haynes K, Boon H, Cottrell A, Millar J, Okada T, Box P, Keogh D, Thomas M (2014) Voluntary relocation as an adaptation strategy to extreme weather events. Int J Disaster Risk Reduc 8:83–90

King M (2017) A tribe faces rising tides: the resettlement of Isle de Jean Charles. LSU J Energy Law Res 6:295–317

Kochnower D, Reddy SM, Flick RE (2015) Factors influencing local decisions to use habitats to protect coastal communities from hazards. Ocean Coast Manag 116:277–290

Kusuma P, Brucato N, Cox M, Letellier T, Manan A, Nuraini C, Grange P, Sudoyo H, Ricaut FX (2017) The last Sea nomads of the Indonesian archipelago: genomic origins and dispersal. Eur J Hum Genet 25:1004–1010. https://doi.org/10.1038/ejhg.2017.88

Lavell A (2016) Colombia, Peru and Mexico Closure Report 4/4, Reducing relocation risk in urban areas. The Bartlett Development Planning Unit, UK. https://www.ucl.ac.uk/bartlett/development/sites/bartlett/files/wp4_closure_rep_flacso_eng.pdf

Leitch AM (2009) Community response to sea level rise policy of planned retreat: Localised debates in community newspapers. In: Paper presented at the 18th NSW Coastal conference. 3–6 November 2009, Ballina, Australia

Lonsdale KG, Downing TE, Nicholls RJ, Parker D, Vafeidis AT, Dawson R, Hall J (2008) Plausible responses to the threat of rapid Sea-level rise in the thames estuary. Clim Change 91(1–2):145–169. https://doi.org/10.1007/s10584-008-9483-0

Lovejoy H (2015) Protest planned against Byron councillors' rock walls. Echo Netdaily. 19 May 2015. https://www.echo.net.au/2015/05/protest-planned-against-byron-councillors-rock-wall/

Mach KJ, Kraan CM, Hino M, Siders AR, Johnston EM, Field CB (2019) Managed retreat through voluntary buyouts of flood-prone properties. Sci Adv 5(10):eaax8995

Marino E (2018) Sea level rise and social justice: the social construction of climate change driven migrations. In: Murphy C, Gardoni P, McKim R (eds) Climate change and its impacts: risks and inequalities. Springer, Cham, pp 181–193

Matthews T, Potts R (2018) Planning for climigration: a framework for effective action. Clim Change 148(4):607–621

Matyas D, Pelling M (2015) Positioning resilience for 2015: the role of resistance, incremental adjustment and transformation in disaster risk management policy. Disasters 39(s1):1–18. https://doi.org/10.1111/disa.12107

Meyer C (2020) What are flood maps your property value depends on them and they may be dangerously outdated. Canada's National Observer. 10 Feb 2020. https://www.nationalobserver.com/2020/02/10/news/what-are-flood-maps-your-property-value-depends-them-and-they-may-be-dangerously

Munji CA, Bele MY, Nkwatoh AF, Idinoba ME, Somorin OA, Sonwa DJ (2013) Vulnerability to coastal flooding and response strategies: the case of settlements in cameroon mangrove forests. Environ Develop 5(1):54–72. https://doi.org/10.1016/j.envdev.2012.10.002

Nalau J, Handmer J (2018) Improving development outcomes and reducing disaster risk through planned community relocation. Sustainability 10(10):3545

O'Donnell T (2016) Coastal law shift from property rights to climate adaptation is a landmark reform. The Conversation. 23 May 2016. https://theconversation.com/coastal-law-shift-from-property-rights-to-climate-adaptation-is-a-landmark-reform-59083

O'Donnell T (2019) Contrasting land use policies for climate change adaptation: A case study of political and geo-legal realities for Australian coastal locations. Land Use Policy 88:104145. https://doi.org/10.1016/j.landusepol.2019.104145

OECD [Organisation for Economic Co-operation and Development] (2019) In: Responding to rising Seas, OECD Country approaches to tackling coastal risks, OECD. https://www.oecd.org/env/responding-to-rising-seas-9789264312487-en.htm

Okada T, Haynes K, Bird D, van den Honert R, King D (2014) Recovery and resettlement following the 2011 flash flooding in the Lockyer Valley. Int J Disaster Risk Reduc 8:20–31

OP-CZM [Office of Planning, Coastal Zone Management Program]. (2019) Assessing the Feasibility and Implications of Managed Retreat Strategies for Vulnerable Coastal Areas in Hawaii. Final Report. https://planning.hawaii.gov/czm/ormp/ormp-action-team-project-on-the-feasibility-of-managed-retreat-for-hawaii/

Peterson J (2019) A new coast: strategies for responding to devastating storms and rising seas. Island Press, Washington

Pethick J (1993) Shoreline adjustments and coastal management: physical and biological processes under accelerated sea-level rise. Geogr J 159(2):162–168. https://doi.org/10.2307/3451406

Pethick J (2002) Estuarine and tidal wetland restoration in the United Kingdom: policy versus practice. Restor Ecol 10(3):431–437. https://doi.org/10.1046/j.1526-100X.2002.01033.x

Primo LH (1997) Sea-level rise vulnerability of kosrae, federated states of micronesia and strategies for adaptation. J Coastal Res 24:107–115

Restemeyer B, Woltjer J, van den Brink M (2015) A strategy-based framework for assessing the flood resilience of cities–A Hamburg case study. Planning Theory Pract 16(1):45–62

Rijke J, van Herk S, Zevenbergen C, Ashley R (2012) Room for the river: delivering integrated river basin management in the Netherlands. Int J River Basin Manage 10(4):369–382. https://doi.org/10.1080/15715124.2012.739173

Rijkswaterstaat (n.d.) Room for the River. https://www.rijkswaterstaat.nl/english/about-us/gems-of-rijkswaterstaat/room-for-the-river/index.aspx

Rockefeller Foundation (n.d.) 100 resilient cities: what is urban resilience? http://www.100resilientcities.org/resilience#/-_/

Ryan BD, Vega-Barachowitz D, Perkins-High L (2015) Rising tides: relocation and sea level rise in metropolitan boston. Project Primer, MIT Norman B. Leventhal Center for Advanced Urbanism, Cambridge. http://lcau.mit.edu/sites/lcau.mit.edu/files/attachments/project/RisingTides_ProjectPrimer.pdf

Sahin O, Mohamed S, Warnken J, Rahman A (2013) Assessment of sea-level rise adaptation options: multiple-criteria decision-making approach involving stakeholders. Struct Surv 31(4):283–300. https://doi.org/10.1108/SS-01-2013-0006

Schut M, Leeuwis C, van Paassen A (2010) Room for the river: room for research? the case of depoldering De Noordwaard, the Netherlands. Sci Public Policy 37(8):611–627

Scott D, Simpson MC, Sim R (2012) The vulnerability of caribbean coastal tourism to scenarios of climate change related sea level rise. J Sustain Tourism 20(6):883–898. https://doi.org/10.1080/09669582.2012.699063

Siders AR, Hino M, Mach KJ (2019) The case for strategic and managed climate retreat Why, where, when, and how should communities relocate? Science 365(6455):761–763

Simmonds J (2020) Rising from the flood: moving the town of Grantham. Bad Apple Press, Arcadia, New South Wales

Sipe N, Vella K (2014) Relocating a flood-affected community: good planning or good politics? J Am Planning Assoc 80(4):400–412. https://doi.org/10.1080/01944363.2014.976586

SPUR (2012) Ocean beach master plan. SPUR, San Francisco, CA. https://www.spur.org/sites/default/files/migrated/anchors/Ocean_Beach_Master_Plan052012.pdf

Stronkhorst J, Mulder J (2014) Considerations on Managed Realignment in The Netherlands. In: Esteves L (ed) Managed realignment: a viable long-term coastal management strategy? springer briefs in environmental science. Springer, Dordrecht

Susskind L (2010) Responding to the risks posed by climate change: cities have no choice but to adapt. Town Planning Rev 81(3):217–235. https://doi.org/10.3828/tpr.2010.5

Tadgell A, Doberstein B, Mortsch L (2018) Principles for climate-related resettlement of informal settlements in less developed nations: a review of resettlement literature and institutional guidelines. Climate Dev 10(2):102–115. https://doi.org/10.1080/17565529.2017.1291401

Taylor BM, Harman BP, Inman M (2013) Scaling-up, scaling-down, and scaling-out: local planning strategies for sea-level rise in New South Wales, Australia. Geographical Res 51(3):292–303. https://doi.org/10.1111/1745-5871.12011

Torabi E (2017) To be beside the seaside: urban resilience to climate related disasters in coastal cities. Thesis (PhD Doctorate). Griffith University, Gold Coast Campus. https://experts.griffith.edu.au/publication/n2e1110fa6fb4e00928a4d743348b9d9b

Torabi E, Dedekorkut-Howes A, Howes M (2017a) Urban resilience to climate-related disasters: emerging lessons from resilience policy and practice in coastal tourism cities. In: Filho W (ed) Climate change adaptation in pacific countries: fostering resilience and improving the quality of life. Springer, Cham, pp 241–254

Torabi E, Dedekorkut-Howes A, Howes M (2017b) Not waving, drowning: comparing local government initiatives on climate change adaptation and disaster risk management. Urban Policy Res 35(3):312–332. https://doi.org/10.1080/08111146.2017.1294538

Torabi E, Dedekorkut-Howes A, Howes M (2018) Adapting or maladapting: building resilience to climate-related disasters in coastal cities. Cities 72(Part B):295–309. https://doi.org/10.1016/j.cities.2017.09.008

Tschakert P, Dietrich KA (2010) Anticipatory learning for climate change adaptation and resilience. Ecol Soc 15(2):11. http://www.ecologyandsociety.org/vol15/iss2/art11/

United Nations (2019) Sustainable floating cities can offer solutions to climate change threats facing urban areas, deputy secretary-general tells first high level meeting. 22 Feb 2020. https://www.un.org/press/en/2019/dsgsm1269.doc.htm

United Nations. (n.d.) In: Atlas of the Oceans. http://www.oceansatlas.org

USGCRP [U.S. Global Change Research Program] (2018) In: Impacts, risks, and adaptation in the united states: fourth national climate assessment, vol 2. U.S. Global Change Research Program, Washington, DC, USA. https://doi.org/10.7930/nca4.2018

Vandenbeld A, MacDonald J (2013) Fostering community acceptance of managed retreat in New Zealand. In: Palutikof J, Boulter SL, Ash AJ, Smith MS, Parry M, Waschka M, Guitart D (eds) Climate adaptation futures, Wiley-Blackwell, pp 161–167. https://doi.org/10.1002/9781118529577.ch15

Walker B (2017) An Island nation turns away from climate migration, despite rising seas. Inside Climate News. 20 Nov https://insideclimatenews.org/news/20112017/kiribati-climate-change-refugees-migration-pacific-islands-sea-level-rise-coconuts-tourism

Elnaz Torabi Dr. Elnaz Torabi is a Lecturer in Environmental Policy/Planning at Griffith University, School of Environment and Science. She is a city planner/architect and has a background in urban and environmental planning research and practice. Her research explores urban resilience and adaptation to climate-related disasters in coastal cities. This research is topical at a time when governments and communities are struggling to effectively respond to future uncertainty and change. Elnaz is affiliated with Griffith's Cities Research Institute and its Adaptation Science Research theme focusing on planning and decision-making for building resilience. Her research interests include water resilient cities, strategic planning, and decision-making processes under uncertainty.

Aysin Dedekorkut-Howes Dr. Aysin Dedekorkut-Howes is Program Director of Urban and Environmental Planning and member of the Cities Research Institute at Griffith University. She has a PhD from Florida State University, a Master's degree from Clemson University, and a Bachelor's degree from Middle East Technical University in urban and regional planning. She has taught and conducted research in the United States, Turkey, and Australia. Aysin has expertise in environmental planning, natural resource management, governance and collaborative planning, regional planning, and growth management. Her research focuses on climate change adaptation and disaster resilience, water resource management, and urbanisation in subtropical areas and coastal cities. She is the co-editor of the book *Off the Plan: The Urbanisation of the Gold Coast.*

Lo-TEK: Underwater and Intertidal Nature-Based Technologies

Julia Watson, Despina Linaraki, and Avery Robertson

Abstract This chapter considers the underwater and intertidal nature-based technologies of indigenous cultures and explores their innovations as solutions for the impacts of climate change to low-lying coastal areas. Indigenous people have been living with and developing water-responsive infrastructures for generations that engage and support the complex ecosystems they inhabit. Rooted in traditional ecological knowledge, or TEK, these technologies work symbiotically with, rather than against nature, ushering in a more comprehensive approach to underwater and intertidal design. Indigenous peoples' responses to sea level rise and storm events improve coastal resiliency, yet remain undocumented and unexplored in the evolution of contemporary solutions. This chapter places these technologies in the modern scientific framework, cross-referencing indigenous people's local nature-based technologies using the five sea level rise response strategies outlined in the Intergovernmental Panel on Climate Change's 2019 Special Report on the Ocean and Cryosphere in a Changing Climate: protect, accommodate, retreat, advance, and ecosystem-based adaptation. Reframed through an architectural and scientific rather than anthropological lens, the challenges cultures were facing and the resources that were available to them are explored to inform us in designing for global climate resiliency today.

Keywords Lo-TEK · TEK · Indigenous · Nature-based solutions · Hybrid technologies · Resilience · Climate change · Coastal resilience · Ecosystem protection · Symbiotic design

---

J. Watson (✉)
Graduate School of Design, Harvard University, Cambridge, USA
e-mail: info@juliawatson.com

School of Architecture, Planning and Preservation, Columbia University, New York, USA

D. Linaraki
SeaCities, Cities Research Institute, Griffith University, Gold Coast, Queensland, Australia
e-mail: d.linaraki@gmail.com

A. Robertson
Independent Scholar, New York, USA
e-mail: robertson.av1@gmail.com

J. Baumeister et al. (eds.), *SeaCities*, Cities Research Series,
https://doi.org/10.1007/978-981-15-8748-1_4

1 Changing Grounds

Coastlines have been occupied since the time of the earliest settlements for access to trade and resources.[1] However, this proximity to water now poses a threat; climate change is forcing us to rethink our relationship with water—as the seas rise, we must adapt. Currently, 11% of the world population lives in coastal areas which are less than ten meters above sea level.[2] After unprecedented disasters, the fear of flooding motivates responses to coastal resilience, preferring fortification to accommodation, elevation, or migration. Overwhelmingly homogeneous, high-tech infrastructures are deployed irrespective of local communities and environmental conditions. Designed primarily by and for affluent cities, these single-purpose solutions which are adopted globally as "best practice" are counterintuitively incompatible with local skill sets and ecosystems.[3] This chapter challenges the "best practice" narrative by documenting traditional indigenous responses to coastal resilience that instead amplify cultural, ecological, economic, and agricultural resilience.

The impacts of climate change to coastal communities, including sea level rise, storm surges, and salination will uniquely, not universally, affect ecosystems, ecosystem services, natural resources, and human systems.[4] Avoiding catastrophe will depend largely on the individual responses by communities and local governments. As stated in the United Nations' Intergovernmental Panel on Climate Change's Special Report on the Ocean and Cryosphere in a Changing Climate of 2019, "Before technical limits are reached, economic and social limits will be reached because societies are neither economically able nor socially willing to invest in coastal protection."[5] Climate change is predominantly addressed in the aftermath of an environmental disaster, through recovery and resilience efforts by government agencies. In these times of crisis, adopted technologies become market driven as quick solutions are required. In major global cities like New York and Amsterdam, hard infrastructural and technical solutions to coastal resilience are deployed. For other less affluent cities, resilient technologies must take a different approach.

The most vulnerable climate crisis communities are often those in impoverished areas or undeveloped nations, which lack the capital and resources needed for high-tech, costly, contemporary-resilient infrastructures. These communities are sold a belief that the high-tech solutions of developed cities are superior to their own innovations, even though the latter embody the intelligence of the environments and cultures that have evolved them. This research explores how hybridizing and scaling indigenous coastal, nature-based technologies is both an effective and pragmatic solution to be included in today's toolkit for resilient design.

This chapter examines how systems evolved from traditional ecological knowledge can work as multi-dimensional ecosystem approaches to reducing the impact

[1]Nunn (2007).

[2]Oppenheimer et al. (2019).

[3]Nunn et al. (2017).

[4]See Footnote 2.

[5]Ibid.

of climate change. Some examples explored include the fish ponds of the Hawaiian people, the Langalanga Islands of the saltwater people on the Solomon Islands, and the taro islets of the Yapese people in Micronesia, all of which function as integral pieces of larger multi-ecosystem scale mitigating interventions. While indigenous peoples' responses to coastal resilience remain largely unexplored in the search for design solutions, these studies at the material, module, structural, and system scale, which are reframed through an architectural rather than anthropological lens, will inform the future of design for climate resilience.

2 Lo-TEK

This chapter retells an ancient mythology—that humankind can and must live symbiotically with nature. Lo-TEK, a term coined by Harvard and Columbia University Lecturer of Urban Design Julia Watson, is identified as resilient infrastructures developed by indigenous people through traditional ecological knowledge (TEK). The movement to bring these innovations to the forefront of the design field counters the idea that Lo-TEK indigenous innovation is low tech: unsophisticated, uncomplicated, and primitive. Instead, Lo-TEK aligns to today's sustainable values of low energy, low impact, and low cost, while producing complex nature-based innovations that are inherently sustainable. Forming the foundation of indigenous technologies, TEK is a field of study in anthropology defined as a cumulative body of knowledge, practice, and belief, handed down through generations by traditional songs, origin stories, and everyday life. By using TEK, humans have been able to harness the energy of ecosystems and adapt to environmental obstacles using soft and symbiotic living systems. Developed through direct contact with nature, TEK is engineered to sustain, rather than exploit resources. It fosters symbiosis between species, while making biodiversity the building block used to construct sustainable technologies.

Lo-TEK is how humans have been dealing with the extremes of the climate for millennia, by harnessing the energy and intelligence of complex ecosystems. It suggests that it is eminently possible to weave ancient knowledge on how to live symbiotically with nature into how we shape the cities of the future, before this wisdom is lost forever. We can rewild our urban landscapes and apply Lo-TEK ecological solutions to sanitation, storm surge, sea level rise, drought, deluge, wildfire, food supply, and fresh drinking water, that have worked for indigenous peoples for thousands of years.

Many contemporary green technologies, like green roofs and floating wetlands, have been around for thousands of years, being rediscovered only when packaged as new. The vision for this research is something of the same. By gathering a compendium of indigenous design and innovation, a framework for adaptation and innovation is posed that reroots our very relationship with nature from superior to symbiotic, exploring the intersection of design and radical indigenism. This concept takes its name from the Latin derivation of the word "radical": radix, meaning "root."[6] Radical indigenism, as defined by Princeton Professor and citizen of the Cherokee Nation Eva Marie Garroutte, argues for a rebuilding of knowledge and explores indigenous philosophies capable of generating new knowledge.[7] Lo-TEK orients us toward a different mythology of technology found across the globe. It expands the definition of contemporary technology by rebuilding our understanding of climate-resilient design using indigenous knowledge and practices that are sustainable, adaptable, and borne out of necessity.

3 Adaptation Pathways

Sea level rise is accelerating, prompting designers, scientists, economists, and politicians to ask "how will we survive? and where can we look for solutions?" in a continually shrinking time frame.[8] The Netherlands is seen as a nation of leading world experts in keeping the ocean at bay. Since 2010, the country has developed the Delta Programme as a water management and flood mitigation strategy with the goal of keeping ahead of the floods. But different cultures across the globe have formed vastly varied relationships with water that may not be synonymous with the Dutch approach. Nor do they have the resources required to replicate some of the proposed solutions, when local conditions might necessitate unique adaptations. Universally, what works for the Netherlands is taken as the best practice benchmark, but in reality this is only true in specific geopolitical locations.

In the book *The Collapse of Western Civilisation*, American science historian Naomi Oreskes, of Harvard University, and Erik Conway outline a dystopian but scientifically plausible prophecy about the impact of climate change after humanity has failed to effectively curb global warming.[9] In this essay, a dystopic future history is imagined in which the Netherlands is the main character, as the Dutch population migrates northward into Germany as the landscape of polders and dikes slowly succumbs to the sea. The essay is not a map of fighting back from an unforeseen environmental catastrophe as is the current climate change approach, but instead shows how the Netherlands could change as a result of gradual developments, rather than major disasters. The essay elucidates an alternative approach to a climate change response where in the face of deep uncertainty a "wait and see" approach to adaptation is taken, until uncertainty is reduced.

Protocols around decision making in the face of great uncertainty, especially in the era of climate change, are changing. To support this shift, Marjolijn Haasnoot, an environmental scientist and the founder of the Dynamic Adaptive Policy Pathways approach, advocates an adaptation pathway approach, in which sequences of

---

[6]Hoad (2003).

[7]Garroutte and Defining (2020).

[8]Church et al. (2013).

[9]Oreskes and Conway (2013).

linked actions can be implemented as conditions change.[10] These adaptation pathways in response to sea level rise are described as four distinct scenarios: defend, surrender, offend, and retreat. Synonymously, the United Nations Intergovernmental Panel on Climate Change's 2019 Report identifies similar categories: protect, accommodate, advance, and retreat, and a fifth ecosystem-based approach.[11] The alternative scenarios described as adaptation pathways expand the climate change conversation beyond simply fighting back.

In this chapter, we use these four adaptive scenarios to categorize the underwater and intertidal technologies of indigenous communities. While the use of TEK in empirical science is not new, applying TEK to a scientific framework is. Traditional ecological knowledge is present in the foundation of scientific knowledge today, only remaining uncredited by an inability to ascribe this knowledge to institutionally based criteria.[12] Categorizing the following indigenous innovations in accordance with the *IPCC's 2019 Special Report on the Ocean and Cryosphere in a Changing Climate* definition of sea level rise responses allows our contemporary-resilient design approaches to embrace TEK. In each scenario, two indigenous innovations found in existence across the globe are described in detail. Lo-TEK: Underwater and Intertidal Nature-based Technologies is a deeper investigation of specifically coastal indigenous technologies.

3.1 Defend/Protect

The Netherlands was settled by the Frisians in 400 BCE by building terpenes or mounded villages that were protected from flooding, and long dikes or embankments that prevented inundation. Since then, the Dutch and their ancestors have been working to hold back and reclaim land from the North Sea for over two thousand years.[13] According to 2019 IPCC Special Report on the Ocean and Cryosphere in a Changing Climate mentioned above, **protection** refers to all the strategies that aim to protect the coastal areas, by eliminating the impact of natural hazards.[14] Effective protection is provided by coastal ecosystems and sediment-based protection, but the most widely used protection strategies involve hard infrastructures. Contemporary protection structures include breakwaters, sea walls, dikes, and surge barriers. While in the short term these solutions are cost-effective in dense urban areas, hard protection techniques can be expensive to maintain and rebuild, and they can lead to flooding, erosion, and the destruction of alternatively protective ecosystems. For example, if the Netherlands were to continue using current systems for keeping the sea at bay, dikes will need to be raised making the polders behind them deeper, more

---

[10]Haasnoot et al. (2013).

[11]See Footnote 2.

[12]Ellen et al. (2000).

[13]Borger et al. (1998).

[14]See Footnote 2.

Fig. 1 This world map is indicating the location of the two indigenous technologies that are analyzed below. The first is the *Loko i'a kalo, loko wai, loko pu'uone, and loko kuapa* Fish Ponds in the *Ahupua'a* of the Hawaiians, Hawaii, and the second is the *Derdernin* and *Ngurruwarr* Fish Traps of the Kaiadilt and Lardil Saltwater People of the Wellesley Islands, Australia

vulnerable, and more expensive to maintain. Higher dikes pose their own problem by preventing natural silting, thereby stopping the delta from being able to grow along with the advancing sea. Even these measures combined in the long term may prove insufficient in preserving the low-lying parts of the Netherlands.[15]

In the following section, the ***Loko i'a kalo*, *loko wai*, *loko pu'uone*, and *loko kuapa*[16] Fish Ponds in the Ahupua'a of the Hawaiians, Hawaii,** and the ***Derdernin*and *Ngurruwarr*Fish Traps of the Kaiadilt and Lardil Saltwater People of the Wellesley Islands, Australia**, are explored as nature-based technologies that can inform the development of protective resilient infrastructures today (Fig. 1). Both of these examples contain the base architecture component of a wall combined with a sustainable aquaculture seascape. Unlike contemporary infrastructures and materials, such as reinforced concrete used to construct sea walls, these indigenous structures utilize local materials, implement existing ecosystem offerings such as mangrove roots, oysters, and algae, and opportunistically offer habitat to species that increase the resilience of the system.

[15]Schuttenhelm (2019).

[16]Kikuchi (1976).

Fig. 2 The longest fish pond that currently exists in Hawaii is located at He'eia, Honolulu on the Island of Oahu (*Source* Google Earth©2020 Maxar Technologies)

3.1.1 *Loko i'a Kalo*, *Loko Wai*, *Loko Pu'uone*, and *Loko Kuapa* Fish Ponds in the *Ahupua'a* of the Hawaiians, Hawaii

The landscapes of the Hawaiian Islands range from barren volcanic mountain tops to highland forests to river valleys ending at the land–sea interface containing farmed land and fish ponds. The Hawaiian people developed a complex land division system, held in trust and controlled by the highest chief or king, dividing up the island, a *mokupuni*, into *ahupua'as* along the natural boundaries of the watershed. Each *ahupua'a* contained the resources for self-subsistence and encouraged trading from fertile lowlands for farming taro or sweet potato, to *koa* and other upland trees for timber, to fish ponds located either offshore (Fig. 2) or inland providing fish and salt. Inland villagers exchanged agricultural foods and wood for canoes for fish with the coastal communities.[17]

Hawaiian fish ponds or *loko* are enclosed bodies of inland or ocean water. These ponds were an important resource in the *ahupua'a* system. They belonged to the local chief, an *ali'i,* who ruled an *ahupua'a.* A well-maintained fish pond and water system was a signifier of a healthy community in the *ahupua'a.*[18] The fish ponds are thought to have originated in the fourteenth century, and were mostly built four to six hundred years ago.[19] Fish ponds are found both inland and in large expanses of shallow sea extending from the coasts of the islands (Fig. 3). There are four basic

[17] Levy and Chernisky (2005).

[18] Joana (2005).

[19] See Footnote 16.

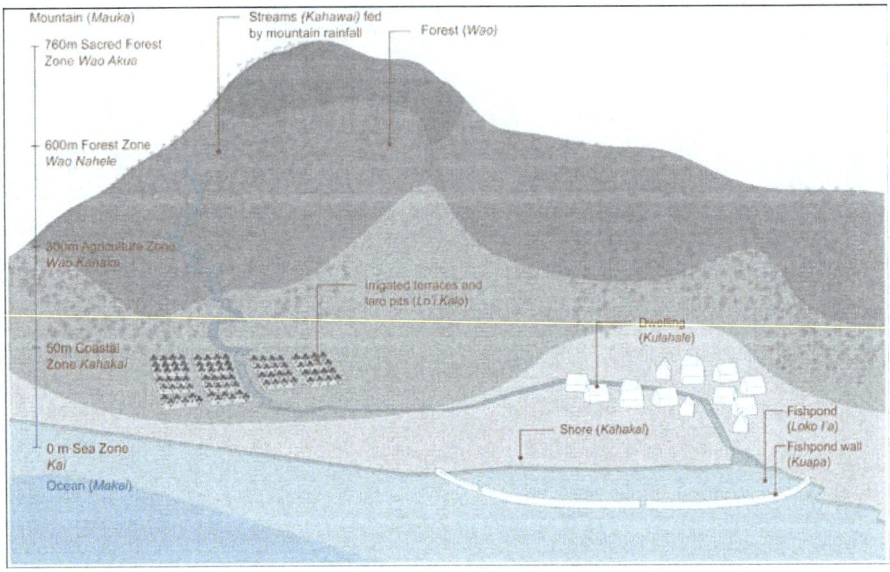

Fig. 3 The divisions in the *ahupua'a* system are based on the watershed, radiating out to the coast from the mountain and containing multiple ecological zones at different elevations. The fish pond, here of the *loko kuapa* type, is located at the coast

types of ponds: *loko i'a kalo*, *loko wai*, *loko kuapa*, and *loko pu'uone*,[20] as well as one fish trap: *loko 'umeiki*.[21]

Loko i'a kalo were freshwater inland irrigated agricultural ponds for the farming of fish and of taro (*Colocasia esculenta*). These ponds, or *lo'i*, began as shallow depressions that were fed with running water from the mountains, and stocked with fish and food to create the system of aquaculture. The pond's shallow depth provided excellent conditions for the growth of taro, algae, seaweed, and plankton. Fish then entered directly from the sea through artificial estuaries or were carried in overland by the farmer.[22] The fish thrived due to the abundance of food in the pond—and the fish, in turn, provided a plentiful supply of food for the farmers.

The *loko wai* was much closer to the shoreline and would often become brackish from tidal waters, and naturally attract fish from the sea. These were excavated by hand from natural depressions and filled from groundwater springs, aquifers, or diverted streams and rivers. Fish were harvested in these ponds by woven reed nets *hala*, which were placed across a channel as they migrated back to the sea for spawning.[23]

The *loko kuapa* fish ponds were located in shallow coastal areas and had semicircular walls built of lava stone or coral (Fig. 4), *kuapa*; canals, *auwai*; and one or more

[20]Ibid.

[21]Costa-Pierce (1987).

[22]Ibid.

[23]Ibid.

Fig. 4 What remains of the Huilua Fish pond: a rock wall extended into the sea. (*Source* Bradshaw 2010)

sluice gates, *makaha* (Fig. 5). These *kuapa* walls ranged from 46 to 1920 m long, and some were built using coralline algae as a natural cement to strengthen them.[24] The *auwai* were built into the walls, connecting the inner pond with the sea, with a *makaha* placed in between.[25] The *makaha* had narrow grates which allowed water and young fish to move in where they would feed on seaweed placed in the pond.[26] The grates were made of timber, or *lama*, that were lashed together vertically with crossbeams, separating the vertical beams by 0.5–2.0 cm spaces. When the mature fish attempt to return to the sea to spawn, they are no longer able to pass through the narrow openings and are harvested. A later addition to the canal was a second *makaha*, which was lowered during the fish's attempted migration, trapping the fish in the canal for easier harvesting (Fig. 5).

The *loko pu'uone* were coastal brackish water ponds with wide, permeable, irregular walls, or *pu'uone,* built of mud, sand, and coral. The pond was connected to the ocean by a canal that allowed seawater to enter on a rising tide, and had freshwater inputs from streams, aquifers, and groundwater, producing a brackish water environment similar to an estuary and held a variety of fish species.[27] The shoreline fish trap, the *loko 'umeiki*, was very similar to the *loko kuapa*, in the way of being a semicircular stone wall, except that these walls were submerged at high tide and contained many canal-like openings with no *makaha.* Instead, fish were caught by nets held by fishermen at the openings who caught them with the tide.[28] Each fish pond had a

---

[24]Ibid.

[25]Ibid.

[26]Bond and Gmirkin (2003).

[27]See Footnote 21.

[28]Ibid.

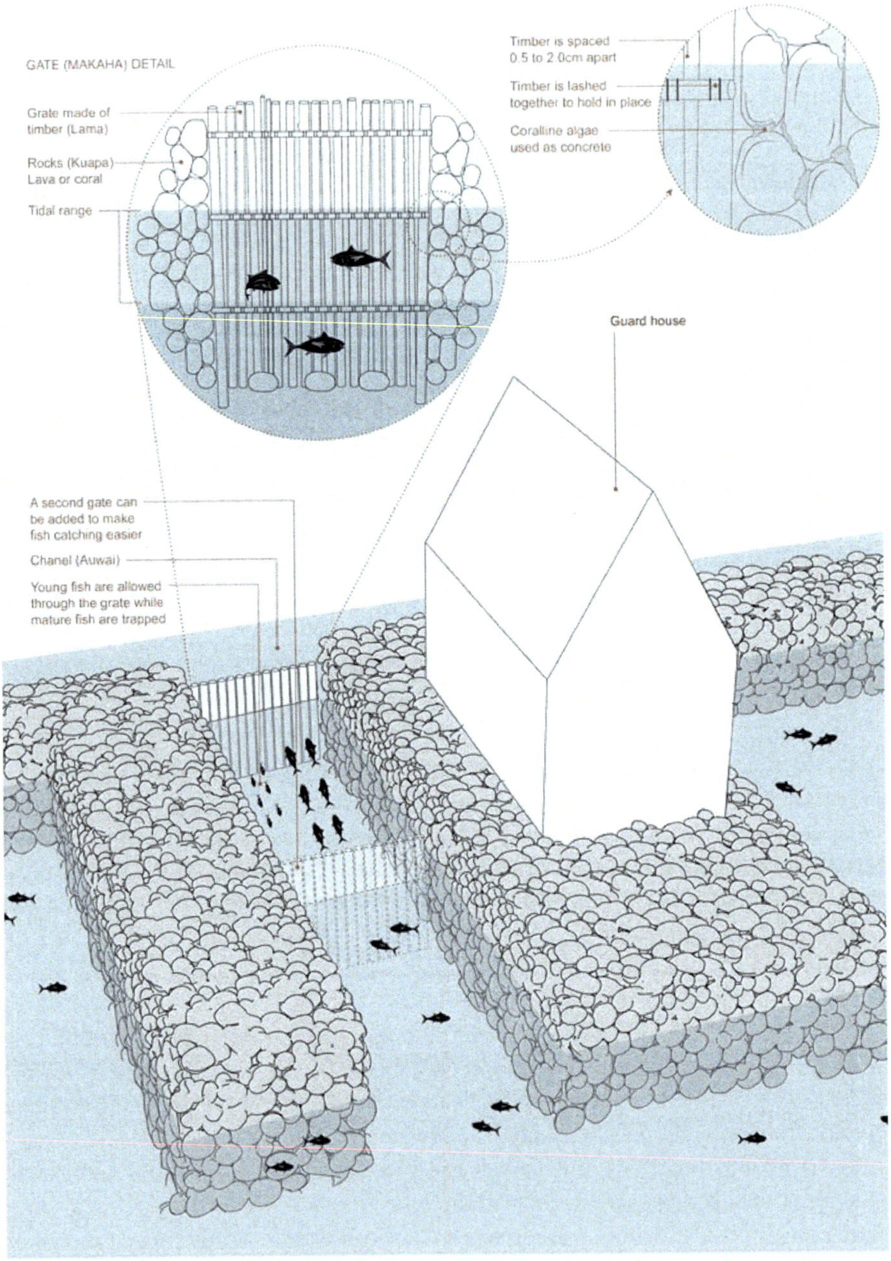

Fig. 5 Gate of the *loko kuapa* fish pond. The rock wall features a canal with a sluice gate made of lashed timber. This gate has narrow openings which allows smaller fish to enter, where they feed and grow, but does not allow them out once they reach maturity

Fig. 6 Huilua Fish pond, in Kahana Bay on the Island of Oahu as seen from the Pu'u Manamana Ridge. This is one of the few fish ponds that remain. *Source* Jesse (2020)

pondkeeper or *kia'i loko* who lived nearby to oversee maintenance. The *loko kuapa* pond additionally had a *hala kia'i* who had a similar but more protective type role to the *kia 'i loko*. In this pond type, simple shacks were placed on the walls adjacent to the *makaha* areas acting as shelters for the *hala kia'i*.[29]

In 1778, three hundred and sixty fish ponds across the Hawaiian Islands supplied the needs of thousands of people. By the early 1900s, as the population of indigenous Hawaiians decreased, fewer than one hundred ponds and traps remained, yet still produced seven hundred thousand pounds of fish yearly. By 2003, only four fish ponds remained (Fig. 6). Many factors contributed to their displacement, including warfare among chiefs who destroyed the fish ponds of enemies, the sandalwood trade which eventually denuded the mountains of trees, causing erosion while simultaneously taking labor from the shoreline, mangrove reforestation that hindered the growth of algae and transformed the ponds into mudflats, natural disaster, disease, land development, and the shift from a subsistence to market economy (Fig. 6).[30]

3.1.2 Applicability to Contemporary Infrastructures

The fish pond typologies show how coastal infrastructures can be built with local ecology and watershed dynamics. Their effective food production and watershed

[29] See Footnote 16.

[30] Joana (2005).

Fig. 7 Kaiadilt fishing traps at Sweers Island, in Northeast Queensland. *Source* Google Earth©2020 Maxar Technologies

management contributed to the success of the indigenous Hawaiian community. The construction of walled ponds also showed how local ecosystems could be used as building materials, such as the colonization of coralline algae to form a jointing concrete and to build productive, aquatic infrastructures. An additional unintended benefit of the coastal fish ponds was a reduction in wave impact and storm surge, which suggests the potential hybridization of wave breaks or dikes with aquaculture systems in contemporary coastal cities for both production and protection.

3.1.3 *Derdernin* and *Ngurruwarr* Fish Traps of the Kaiadilt and Lardil Saltwater People of the Wellesley Islands, Australia

The Wellesley Islands, located south of the Gulf of Carpentaria in Northeast Australia, are populated by three aboriginal groups: the Ganggalida of the mainland, the Lardil and Yangkaal of the North Wellesley Islands, and the Kaiadilt of South Wellesley Islands (Fig. 7). Between winter and summer, these groups migrate from inland camps to coastal areas. In Aboriginal mythology, land and marine systems are shaped and built by their mythic ancestor, *Marnbil*, not by nature.[31] They made a clear distinction between the land and the sea (*mala*). Land, made up of tussock grasslands, eucalyptus woodlands, and scrublands, is known as the "inside country," and the sea and coastal areas, made up of tidal flats, sandy cheniers, boulder beaches, cliffs, and reefs, are

[31] Memmott et al. (2008).

Fig. 8 Australian aboriginal fish traps. (*Source* Reproduced with permission from Reef Catchments (This photograph was taken by Regina Bernard, former Indigenous Project Manager MWINRM Group (Reef Catchments) for the Fishtrap Project 2009, undertaken with the Mackay Whitsunday Isaac Traditional Owner Reference Group))

known as the "outside country."[32] The Lardil and Kaiadilt people who lived on smaller islands both used rock wall fish traps, known as *derdernin* to the Lardil and *ngurruwarr* to the Kaiadilt.[33] Natural rock formations or reefs were also used as traps, with the construction of these being attributed by the Lardil people to the ancestral god *Marnbil*.[34]

Multiple types of rock wall fish traps were constructed, with the *derdernin* and *ngurruwarr* used for harvesting at low tide, the *baljan* for chasing and spearing, and the gated *yilin* trap for chasing and netting (Fig. 8). Enclosed rock wall fish traps were constructed primarily of rocks, but also utilized the surrounding ecology in the construction. Walls were often built from existing rock pieces or elevated mangroves.[35] The layout of rock wall fish traps varied slightly, forming either a U- or a V-like shape (Fig. 9). In the V-shaped trap, which were often linked to form multiple W-like shapes, the point faced the sea and the opening reached the shore. In the V shape, a few rocks at the point could be removed and replaced with a gate,

[32]Memmott and Trigger (1998).

[33]See Footnote 33.

[34]Ibid.

[35]Ibid.

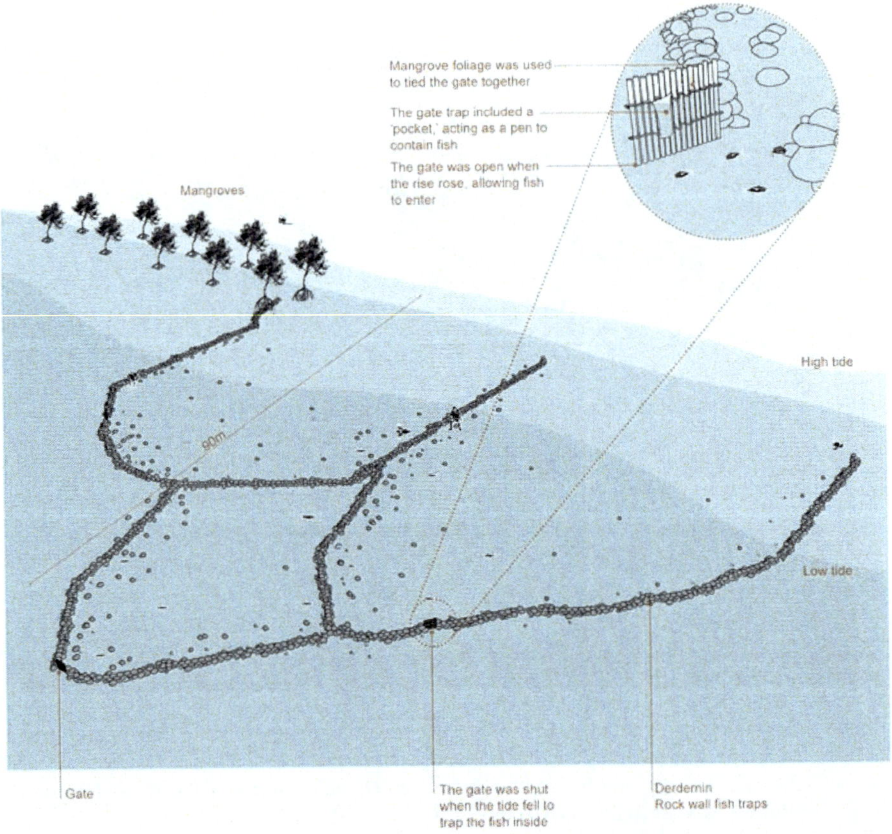

Fig. 9 In the axonometric, the system is shown as a U shape, with the point facing the sea and the opening reaching the shore. The detail shows the gate that opens and closes to allow fish to go in and out

which was opened when the tide rose allowing fish to enter. When the tide fell, the gate, being tied with mangrove foliage, trapped the fish inside.[36] The elaborate gate trap often included a "pocket," which acted as a pen, to contain fish (Fig. 9). The trap walls were low so that during the high tide, which may rise up to three meters, larger fish, dugongs, and other aquatic creatures were allowed to swim over the rock wall, trapping them within the wall when the tide falls (Fig. 10).

The success of a rock wall fish trap was determined by whether they remain stable when hit with the tide. Therefore, when deciding upon a location, the tidal range must be limited and traps positioned on the sheltered side, out of the path of strong, prevailing winds. Oftentimes, the roots of surrounding mangroves grew into the rock wall aiding in its stabilization by reducing some of the tidal force. Fringe reefs, which grow on shallow shores, also functioned like the mangroves, alleviating the tidal force acting on the walls and providing additional stability. Typically, rocks are

[36]Ibid.

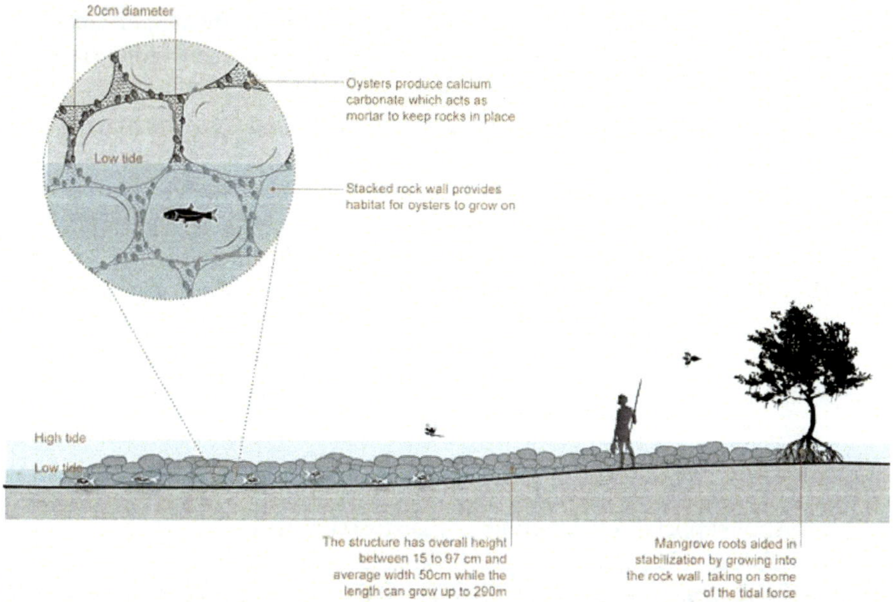

Fig. 10 The section shows the wall structure. The total height would be lower than the high tide to allow water to come in. The mangroves enhance the stability of the system by integrating their roots to the structure

twenty centimeters in diameter, and they are stacked to form the wall. The structure has an overall height between fifteen and ninety-seven centimeters and an average width of fifty centimeters. The length can grow up to two hundred and ninety meters (Fig. 10).[37]

Oysters would grow on the surfaces of the rocks, and mud sediments would pile up against the wall, further strengthening the rock wall itself. There was a symbiotic relationship of stabilization between the rock wall and oysters, as the rock wall provides habitat for oysters to grow on, while the oyster uses calcium carbonate to attach to rocks, which acts as jointing mortar that keeps rocks in place. The oysters form in clusters, creating reefs with older shells. As they die off, the early stage oysters known as spats attach to the older, hollowed shells, growing to contribute to the overall structure (Fig. 10).

Fish were also harvested by spearing prior to the trap becoming emptied. One method involved waiting until the tide had fallen below the top of the wall, then herding fish into groups and directing them toward one or more spearsmen by people hitting the surface of the water. Another method was by spearsmen walking along the walls spearing fish as the tide fell. Dugongs, an animal similar to the more commonly known manatee, were also chased by men on rafts into the shallower water where they could be easily speared.[38] The major difference between the Lardil and Kaiadilt

[37]Kreij et al. (2018).

[38]See Footnote 33.

traps were that the Kaiadilt traps had bigger paddocks and were built on areas where sea weeds grew. These areas were more attractive to turtles and dugong. The Lardil additionally utilized a series of hibiscus rope nets for hunting dugong.

During cool dry seasons, when strong winds occur, inland settlements constructed fish traps in streamlets, which were smaller compared to the coastal rock wall fish traps. To maximize weather conditions in hotter months, settlements were moved to positions along coastal ridges to maximize breeze and access to water making the hunting of dugongs, fish, and turtles easier. Several hundred fish traps are located at over one hundred sites along the island's coast, with the sandy or muddy surface helping to stabilize the rocks. Communities assisted in the building and repairing of the traps in exchange for the fish caught.[39] These fish traps were particularly important for ensuring food security in an environment where tidal surge could destroy other sources of food.

3.1.4 Applicability to Contemporary Infrastructures

The *Derdernin* and *Ngurruwarr* Fish Traps are great examples of aquatic wall construction. Fish traps weathered severe storms through a symbiosis between the preferred oysters' habitat, the wall structure, and the natural production of calcium carbonate by individual oysters upon adhering to a surface. Over time, the walls grew stronger as more calcium carbonate was deposited. Fish traps acted as both an aquaculture infrastructure and a protective wall, acting similar to wave breaker or to a dike. Performing as both a protective and productive infrastructure similar fish pond structures could offer multiple benefits to coastal cities.

3.2 Surrender/Accommodate

The second strategy for sea level rise outlined by the IPCC is "surrender" or "accommodate," which takes the approach to sea level rise of letting water in. This strategy can take place in various scales, methods, and levels of temporality. It can be applied in urban environments through flood zoning restrictions, insurance plans, warning systems, emergency planning, setbacks, and flood barriers. The IPCC's definition of "accommodation" refers to the redesign of physical and political infrastructure to accommodate sea level rise and reduce vulnerability. In addition to urban flood accommodation measures, strategies include updating building codes, raising buildings on stilts, floating houses and gardens, aquaculture, and adapting to salinity intrusion, though changing crop varieties and land use.

In the following section, the **Phumdis or Phumshongs of the Manipuri on Lake Loktak, India**, and the **Ma'a and Mu'ut Taro Gardens and Floating Taro Patches on the islands of Yap and Puluwat, Micronesia** (Fig. 11), are explored

[39]Rowland and Ulm (2011).

Fig. 11 This world map indicates the location of the two indigenous technologies that are analyzed in this section

as adaptive technologies that work similarly to the IPCC strategy of accommodating water, as they are buoyant and employ aquatic vegetation.

3.2.1 Phumdi or Phumshongs of the Manipuri on Lake Loktak, India

Speckled across Lake Loktak, northeast India's largest freshwater lake are thousands of circular, floating meadows known as phumdis or *phumshongs* (Fig. 12). Referred as the "lifeline of Manipur,"[40] the lake's ecosystem is a unique biodiversity hotspot and provides many environmental services such as nutrient recycling, water purification, groundwater recharging, and runoff control.[41] At a width of nine kilometers and length of eleven, almost seventy percent of the lake's surface is covered by several thousand floating meadows, which exist only here (Fig. 13).[42] The *phumdis* are used by the local people for constructing artificial floating islands, known as *athaphum*, topped by huts, *phumsang*.[43] The local Manipuri people create the *athaphum* artificial circular enclosures out of multiple *phumdis* for fish farming. These circles reach one hundred meters in diameter and allow thousands of fishermen to catch an average of one thousand, five hundred tons of fish every year (Fig. 14). The *athaphum* is surrounded with nylon netting that reaches the lake bed, while silt is stirred with

[40]UNESCO (2016).

[41]Rai and Singh (2014).

[42]Singh and Khundrakpam (2011).

[43]Bhardwaj (2017).

Fig. 12 Multiple *phumdis* for fish farming, covering a large portion of the Lake Loktak. *Source* Google Earth©2020 CNES/Airbus

Fig. 13 *Phumdis* at Lake Loktak. *Source* Prasad (2008)

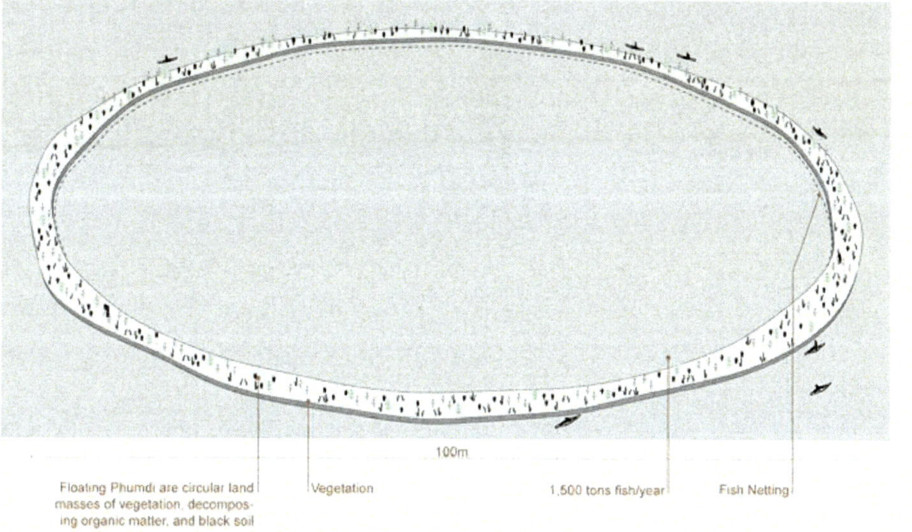

Fig. 14 Axonometric plan of the *phumdi*

bamboo poles to reduce oxygen levels in the water, thereby bringing the fish to the surface to be easily caught.[44]

Phumdi or *phumshongs* are circular land masses of vegetation, decomposing organic matter, and black soil that has thickened into a solid and spongy form, known as peat. A *phumdi* is grown when a dense mass of water hyacinth or other organic matter accumulates silt that is suspended in the water. Grasses and other plants grow on this mass and decompose, building up layers of the peat. The soil is composed of organic carbon, nitrogen, and organic mineral matter, at various stages of decomposition.[45] The first layer, the root zone, is between zero and fifteen centimeters, the second layer, the mat zone, is twenty-five to sixty-five centimeters, and the lowest layer, made of organic matter, is between zero and twenty-five centimeters (Fig. 15).[46] Similar to an iceberg, one-fifth of the structure emerges above water while four-fifths remain submerged below.[47] The *phumdi* can be up to two and a half meters thick, and the high concentration of vegetation matter gives it its buoyancy.[48] The *phumdi* acts as both biofilter and habitat in the aquatic ecosystem. During the dry season, when Loktak Lake's water level drops, the living roots of the islands can reach the lakebed and absorb essential nutrients, resurfacing after spring (Fig. 16).

The largest single mass of *phumdis* lies in the southeastern part of the lake and covers thirty-eight square kilometers. It is home to the *Keibul Lamjao*, the world's only floating national park which was created to preserve the habitat of the endangered

[44]Heggen (2018).

[45]See Footnote 45.

[46]Takhelmayum and Gupta (2011).

[47]Singsit (2003).

[48]See Footnote 45.

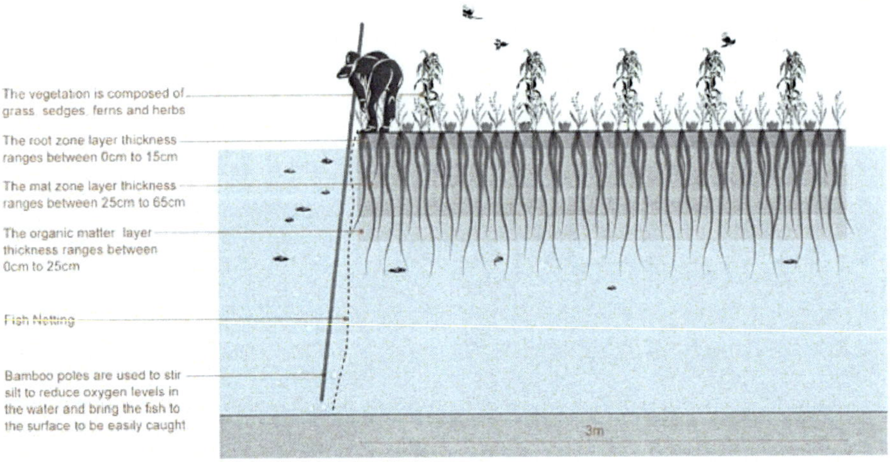

Fig. 15 Detail of the structure of the *phumdi*

Fig. 16 Section of the *phumdi*

brow-antlered deer, the sangai, Manipur's state animal that was once thought to be extinct. The *phumdi* area of the park supports around four hundred species of animals, including the rare Indian python and two hundred species of plants including grass, sedges, ferns, and herbs.[49]

The *phumdi* habitat is threatened by the encroachment of humans and livestock, the annual burning of the *phumdi* by government officials, the inflow of polluted water, and the Ithai Barrage.[50] The Ithai Dam was built in the 1980s to provide power for India's northeast states as well as irrigation and drinking water, yet it has threatened the life of the lake. The lake inflow is fed by almost thirty rivers which contain agricultural chemicals and domestic waste along with high levels of sediment. The only outflow from the lake is through the Ithai Barrage which only releases water.[51] Both the river inflow and the dam outflow conditions have contributed environmental problems such as siltation, proliferation of *phumdi*, eutrophication, prevention of fish

[49]Tuboi et al. (2012).

[50]See Footnote 47.

[51]Ranganathan (2017).

spawning, loss of fish biomass and diversity, and ironically, loss of water available for hydraulic power generation. The dam prevents the natural removal of old *phumdi*, which was typically washed away in monsoon season, while the fertilizer nutrients from the rivers cause an accelerated growth of phumdi, causing a proliferation that is proving harmful to the ecological functions of the lake. The number of *phumdi* slows the flow of water which contributes to siltation; block fishing routes and rotting *phumdi* cause a reduction in overall water quality. The dam also causes water levels to remain high year-round creating a permanent flooding situation and preventing the phumdi from sinking to the lakebed for annual nutrient retrieval.

Nearly one hundred thousand people depend on the lake for their livelihood, and the *phumdis* plays a vital role in this equation. The *phumdis* are important to maintaining the ecology of the lake, and with it the economy of the state. The *phumdi* supports productive fisheries, provides important plant species which are used for fencing and fuel, and is critical to the maintenance of water quality in the lake. The *phumdi* is a good source of high-quality compost, which assists in decreasing the proliferation in the lake.

3.2.2 Applicability to Contemporary Infrastructures

Due to their natural buoyancy, the *phumdi* is able to support populations during water-level changes. Similar floating meadows could be used in urban water bodies to enhance natural biodiversity and natural water purification. This study also exemplifies how lakes could sustainably support large-scale, aquaculture systems. *Phumdis* successfully accommodate different water-level fluctuations, while supporting biodiversity and enhancing food production. However, these natural floating meadows would require salt-tolerant plant species if they were to be used in saltwater environments.

3.2.3 *Ma'a* Floating Taro Islets of the Puluwat (Poluwat) and Floating Taro Patches and Agroforestry System of the Yap, Micronesia

Yap is a cluster of small islands making up one of the states in the Federated States of Micronesia, a country composed of more than six hundred islands in the Pacific Ocean north of Australia.[52] In a tropical climate that is humid year-round, less than one percent of this country is dryland making freshwater limited. This scarcity, coupled with frequent drought, salinity, and poor soil, has inspired unique nature-based innovations centered around the production of taro (Fig. 7).[53] In Palau, another Micronesian island which cultivates Taro, it is believed that "the taro swamp is the mother of life" and women were the main cultivators.[54] Marshland floating taro

[52]Manner (1980).

[53]Ibid.

[54]Ibid.

Fig. 17 Taro islets at Poluwat Atoll, surrounded by the agroforest. *Source* Google Earth©2020 Maxar Technologies

patches in Yap were constructed by older women and young mothers (Fig. 18), using simple tools such as sticks, and without the use of artificial fertilizer. Even so, these fields produced yields higher than commercial farms.[55]

The indigenous people of Yap grow taro free from chemicals, fertilizers, and machinery by creating stone lined canals and pools, hanging baskets, floating islets, floating patches, raised beds, and patches in agroforests. Requiring warm temperatures, taro is a hardy root vegetable with an edible starchy bulb and leaves that grow up to a meter that can withstand the monsoonal rains and tolerate waterlogged soil, and it is slightly tolerant of saltwater.[56] The primary species grown on these islands are giant swamp taro or *Cyrtosperma chamissonis* (also known as *Cyrtosperma merkusii*), and the smaller, less salt-tolerant 'true taro' or *Colocasia esculenta.* The giant swamp taro is suited to agroforestry and marsh integration, being hardy, shade-tolerant, having longevity and the ability to grow both up and down.[57] While the true taro produces large yields quickly and is commonly grown on homesteads, with the ability to only grow upward, *Colocasia* has been interplanted with *Cyrtosperma*, as the *Colocasia's* fast growing provides shade that hinders weeds, and the *Cyrtosperma* grows for longer so it is present after the *Colocasia* is harvested.[58] While many forms of taro agriculture exist on these islands, the wetland cultivation of taro in Micronesia is explored here.

[55]Nakashima et al. (2018).

[56]Moore and Lawrence (2003).

[57]See Footnote 3.

[58]Falanruw, M. C. *Taro Growing on Yap*. Yap Institute of Natural Science.

Fig. 18 A woman repairing a *ma'a* on Puluwat Atoll in 1988. The woman's pride in her taro is seen through the care and attention to details such as the woven coconut frond wrapping the islet. *Source* Reproduced with permission from Manner (2010)

In marshes deep in the mainland island of Yap, taro is grown in baskets, on elevated beds, or floating islets. The baskets, called *kang-kong,* are constructed as either bottomless, built in the patches, or hanging, held in place with sticks. Baskets are made from woven coconut fronds and are filled with mulch or good soil. The taro is planted in the basket, and the roots grow down, to the freshwater below.[59] Elevated beds are also used in coastal areas to raise the taro above saltwater inundated soils, and in areas where the water is too deep. The beds are built from piling organic matter which is then covered in soil. The sides of these beds are typically held in place by woven coconut fronds or husks and supportive sticks.

"Floating" islets, called *ma'a,* and floating patches are present on a few islands. In freshwater marshes deep in mainland Yap, patches float on slow moving water and are built out of the surrounding marsh vegetation.[60] Using only a knife, reeds like *Phragmites* and other marsh vegetation are slashed and piled onto a bed. A ditch is then dug with a stick around the bed and cut from under to disconnect it from surrounding vegetation. Mud and silt are taken from under the mat and layered on top to cover the cut vegetation. The *Colocasia* and *Cyrtosperma* taro are planted on here and grow hydroponically atop the algae-rich mulch bed, absorbing the nutrients

[59]Lopez (2019).

[60]See Footnote 59.

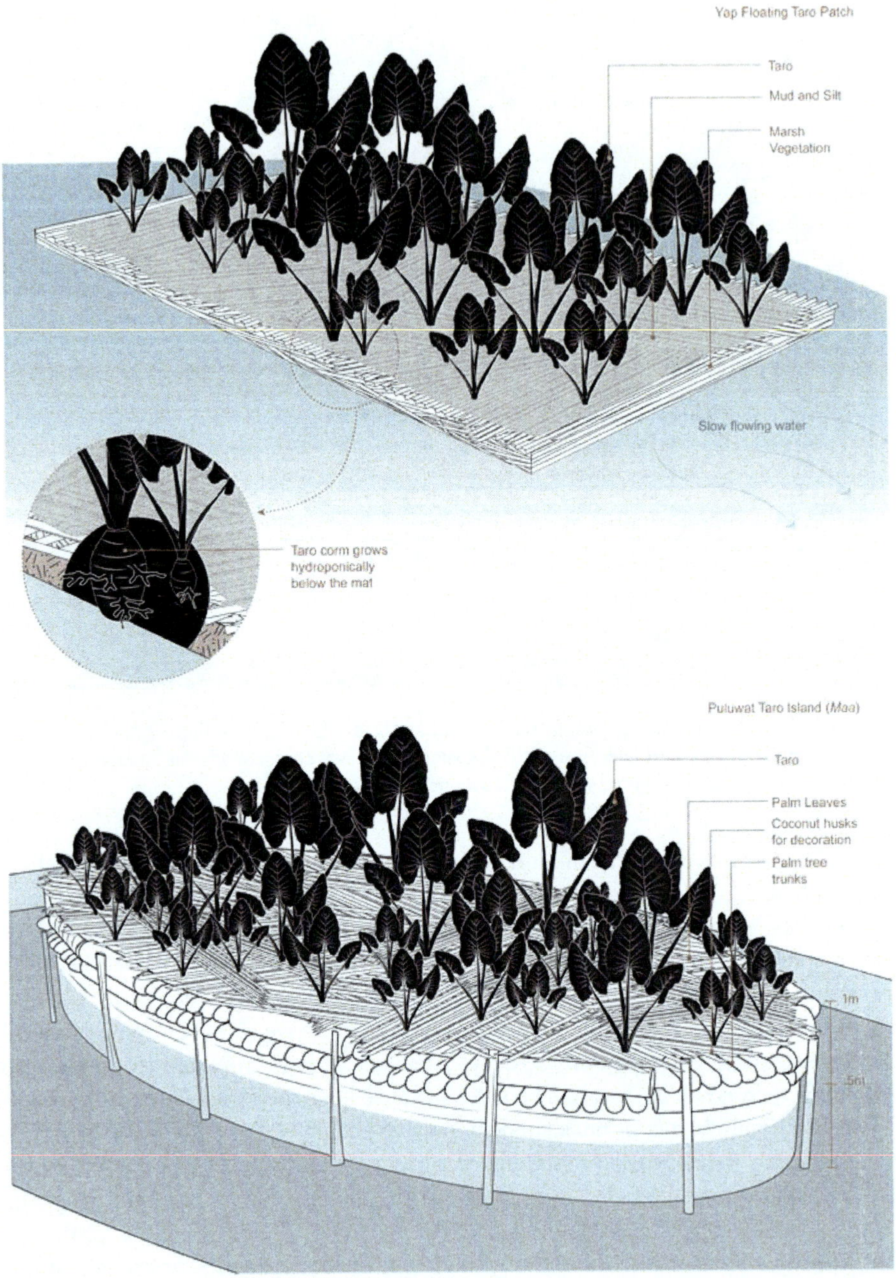

Fig. 19 A diagram of the floating marsh taro patches on Yap and the taro islands on Puluwat

Fig. 20 A *ma'a* seen on Puluwat Atoll in 1988. The *ma'a* are constructed wholly of organic materials (tree trunks and leaves) that are anchored to the bottom of the depression. *Source* Harley I. Manner

from the flowing water. The roots grow across the mat, and the taro corm grows either within or below it (Fig. 19).[61] This system of hydroponic agriculture is comparable to the Aztec floating *chinampas*.

"Floating" taro islets, called *ma'a,* are found on Puluwat (now Poluwat), an island in the Chuuk state of Micronesia, and on Ulithi Atoll. The Puluwat *ma'a* are oval raised beds that sit in freshwater pools in the island's interior swampland and appear to float (Fig. 20). A depression is either found or excavated, and then coconut and pandanus trunks are vertically inserted into the bottom as piles, before being covered with decomposing vegetation. The sides of the bed are surrounded by woven coconut fronds and propped by sticks, acting as a retaining wall. These islands are about one meter high, with soil sitting half a meter above the water table, and on this they grow both *Colocasia* and *Cyrtosperma.*[62] The island sits in the center of the depression, where the water is deeper and less saline, and is surrounded by natural vegetation as protection from salt spray and storm waves (Fig. 19). The islets on Ulithi are constructed in a similar way, except triangular in shape rather than oval.[63]

[61] See Footnote 63.

[62] See Footnote 56.

[63] Morrison et al. (1994).

Outside of the marshes, taro is also a major part of the agroforestry system on Yap. Subsistence agroforestry is very productive, providing food, resources, and ecosystem services such as filtering air and water, preventing erosion, and preserving biodiversity.[64] The agroforest on Yap was part of a three-part system which dominated the landscape: tree gardens, open gardens, and taro patches. In this system, the tree garden's canopy protects the soil from rain-induced erosion, and the decomposing fallen leaves provide nutrients for the soil.[65] Small areas were cleared throughout the agroforest through shifting cultivation, or the practice of burning to clear an area for new crops, which revitalized soil fertility and worked as an insecticide and herbicide. Breaking the canopy and forming "skylights," these open gardens were planted with yams and other crops.[66] These gardens and taro patches were integrated in a series of ditches and raised beds. In the open canopy gardens, ditches were built alongside them to provide drainage and separate the crops from weed's roots.[67] These ditches travelled downstream to low-lying areas where they were widened for a series of connected taro patches with raised land built up next to them where crops could be planted. Earth excavated from the patches was also used for raised land on which houses and paths were built. These paths travel across the island and are paved with shells, coral, and sand which allows them to dry off instantly after a rain event.[68]

The multitude of ways in which taro was grown on these islands are many ways in which agriculture was successful in the presence of saline soils and little space. Traditional taro cultivation understood the interface of freshwater and saltwater. This is also evidenced by the *Mu'ut:* taro patches near the sea that were lined with rocks to prevent saltwater intrusion from below while they received freshwater from rain-fed channels.

3.2.4 Applicability to Contemporary Infrastructures

Innovations like the *mu'ut,* the *ma'a,*, and the other taro cultivation methods could be revived as a system for resilient coastal agriculture based upon the variations in design and plant selection in response to water salinity and nutrient needs. The Yap *mu'ut*, a way in which plants that require freshwater are able to be planted in saline soil, and the use of salt-tolerant species in other areas, can be explored as how to adapt agriculture to salinity intrusion from sea level rise. The use of local vegetation in many forms displays the versatility of using plants as building material. Additionally, the process of excavating soil for the creation of both irrigated planting beds and walls can be replicated. In this process, the lower areas reach the freshwater and the soil that is taken out is used to build up walls and land.

---

64 The Environmental Literacy Council (2015).

65 Falanruw and Ruegorong (2015).

66 Falanruw et al. (1987).

67 Conrad and Newell (1992).

68 See Footnote 3.

Fig. 21 This world map indicates the location of the two indigenous technologies that are analyzed in this section

3.3 Offend/Advance

Advance is a strategy that has a complex history of working against ecosystems. **Advance** is a relatively new sea level rise response strategy to the IPCC, coming to the forefront now due to land scarcity and population growth. This strategy refers to the expansion into the water by creating new land with land reclamation or by the use of dikes.[69]

The filling of wetlands and coastal ecosystems with sand and fill on which to build is an example of this present in many coastal cities. For example, through the historic use of "advance" in China, about fifty percent of coastal ecosystems have been lost, leading to a range of impacts such as loss of biodiversity, decline of bird species and fisheries resources, reduced water purification, and more frequent harmful algal blooms.[70] This strategy has potential to not only destroy protective ecosystems, but actually enhance erosion and grow the coastal floodplain, and so it must be implemented with careful consideration for the ecosystem it inhabits.[71]

In the following section, the **Mulberry Dike and Fish Ponds in Huzhou, China**, and the **Asi artificial islands of the Malaitan on Langalanga Lagoon on Malaita, Solomon Islands**, are explored as infrastructures that combine the creation of new land with the creation of ecosystems (Fig. 21). Rather than creating new land through

[69]See Footnote 2.

[70]Ibid.

[71]Ibid.

Fig. 22 Mulberry dike and fish pond structures at Huzhou, China. *Source* Google Earth©2020 CNES/Airbus

filling ecosystems with foreign materials, these examples expand the materials and services provided by these ecosystems, rather than destroying them.

3.3.1 Mulberry Dike and Fish Ponds in Huzhou, China

The planting of mulberry trees and sericulture, or the rearing of silkworms, began in China over five thousand years ago. The mulberry dike and fish pond system is two and a half thousand years old and integrates this process with fish, livestock, and dikes, called *tanglu* (Fig. 22). Several agricultural production modes working in symbiosis create this complex multi-dimensional eco-agricultural system. The system also functions as part of a larger-scale reservoir for water storage, flood regulation, and drought mitigation.[72] This sustainable closed-loop system incorporates renewable energy, water management, nutrient recycling, and food production through cropping, mulberry tree growing, silkworm rearing, and livestock and fish cultivation (Fig. 23).

Mulberry trees grown on the banks of the fish ponds are the first level in the system. Mulberry tree cultivation provides leaves for silkworms to eat; the silkworm feces and sloughs fall into the adjacent pond providing feed for fish; the fish feces, along with the unconsumed silkworm and mulberry waste, are then decomposed by aquatic microorganisms which produce nitrogen, phosphorus, and potassium, finally returning to the mulberry tree as nutrient-rich manure and beginning the cycle over

[72]The People's Government of Nanxun District (2017).

Fig. 23 Fish ponds with the city of Shenzhen in the background. Fish ponds are also used by farmers in the Pearl River Delta. *Source* Earnest Tse

again.[73] The fish species are chosen by which level they reside in the water, a key detail that contributes to maintaining the biological balance of the pond ecosystem. The silver carp lives in the upper water, eating phytoplankton; the bighead carp lives in the middle of the water, eating zooplankton; the grass carp lives in the middle and lower areas of the water, eating water plants and vegetation; and the black carp lives in the bottom of the water, eating snails and mollusks.[74] Algae and aquatic plants, such as water hyacinth (*Eichhornia crassipes*), grow in the pools and are controlled through consumption by the grass carp. These plants produce oxygen and glucose, adding nutritional benefits to the fish and indirectly the manure and mulberry trees (Fig. 24).[75]

The fish ponds are an element existing within a larger irrigation, drainage, and flood prevention system. In Huzhou, the northern area of the region lies in the Yangtze River Delta in the lowland south of the Taihu Lake, one of the largest freshwater lakes in China, and was vulnerable to flooding.[76] In response to this, the *Zong Pu Heng Tang* system of irrigation and drainage integrated existing and formed rivers—the wide horizontal, *Hengtang*, and narrow vertical, *Zongpu*—into a grid-like formation reaching the lake. Water is stored in the rivers and allows for gradual sedimentation, while overflow is drained to the lake. The banks of dikes within the series of ponds are

[73]Gongfu (1982).

[74]Ibid.

[75]Ibid.

[76]Zhuang (2018).

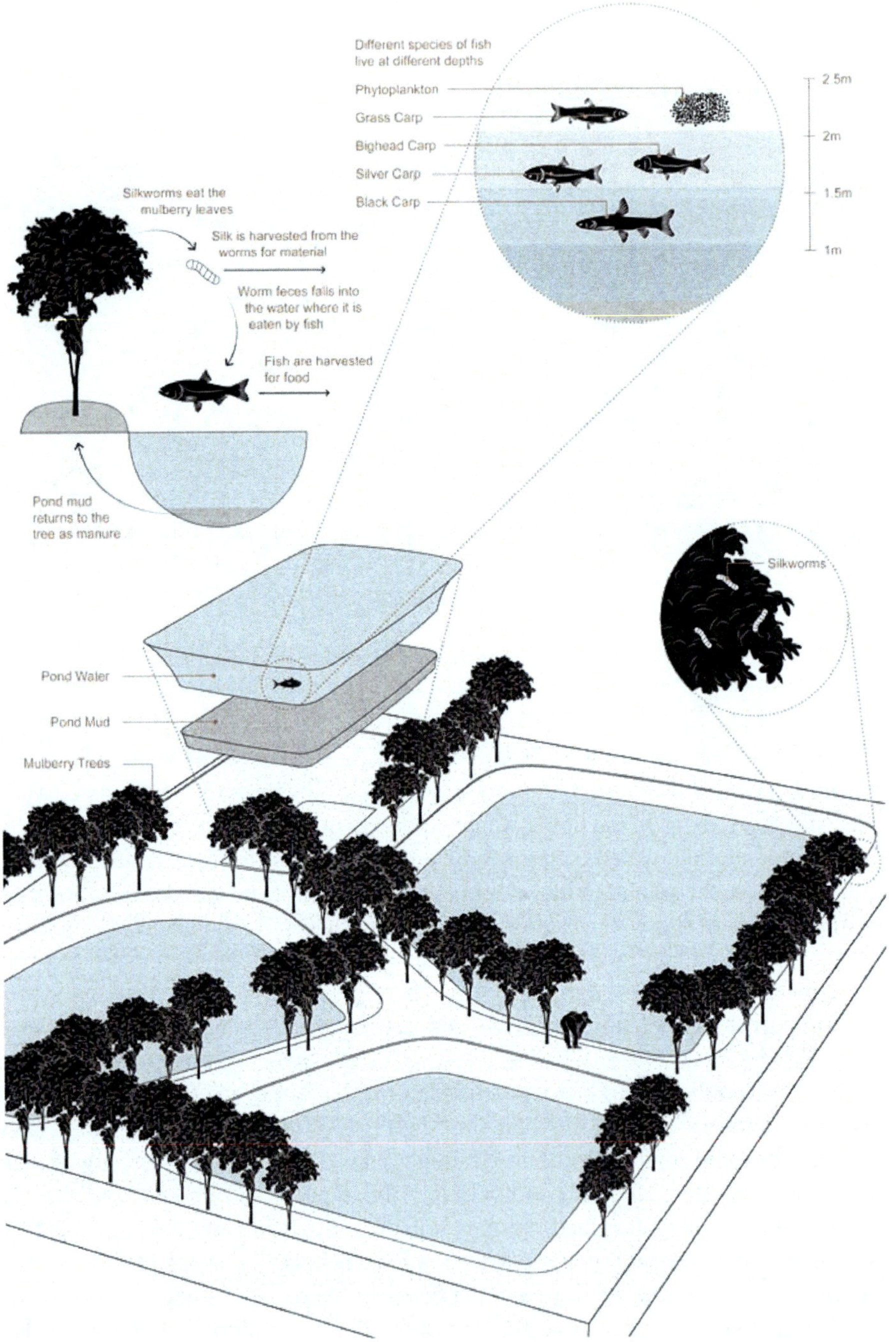

Fig. 24 Mulberry dike and fish pond is a closed-loop system, in which each element provides nutrients for another

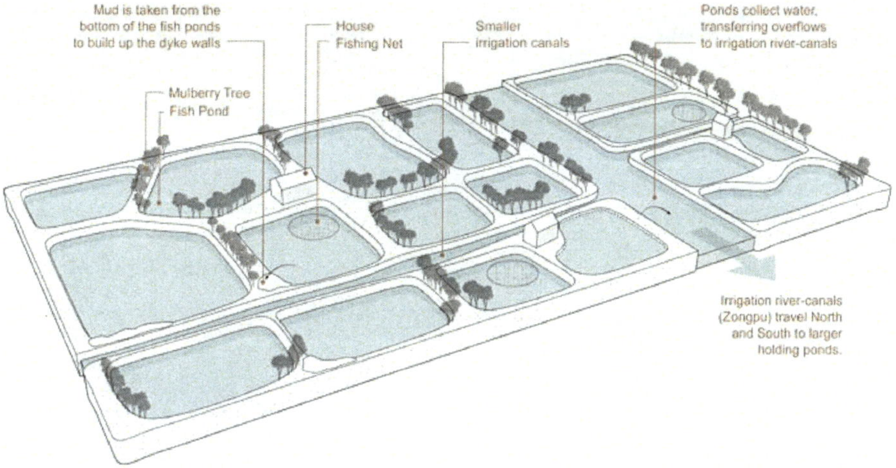

Fig. 25 The fish pond system creates retention ponds for floodwater and dikes for protection decreasing the vulnerability to flooding

created from dredging perennially flooded depressions in the earth, and digging the mud from the fish ponds to the banks on which the mulberry trees are planted. This increases the dike height and water storage capacity, improving flood resilience.[77] This is also the case for the fish ponds at the Pearl River Delta near Shenzhen, as the swamplands were converted into the fish pond landscape (Fig. 25)[78].

3.3.2 Applicability to Contemporary Infrastructures

There are numerous additional benefits from each element of this system, beyond being integral to creating a closed loop. The system maximizes use of local resources and produces commodities (fish and silk), which in turn provide income and employment (Fig. 26). The nutrient-rich mud created in this system is used as both fertilizer and a building material for flood mitigation. It is the result of recycled waste and energy, which replaces the need for chemical fertilizers, pesticides, herbicides, and concrete flood barriers, while cutting costs and creating a zero-emission system. Additionally, the mulberry tree is a great air purifier and can sequester carbon more effectively than most agricultural plants.[79] This integrated system of water and land resources forms a resilient and effective closed-loop, hybrid solution that can inform both water and land management practices, agriculture, sericulture, and aquaculture, in areas that experience frequent flooding. To develop this system within dense urban areas would require the reintroduction of these activities within the urban landscape

[77] Ibid.

[78] Marks (2011).

[79] See Footnote 77.

Fig. 26 The mulberry tree leaves grown on the dikes of the fish pond generate food for silkworms, creating an integrated, sustainable system of fish farming and silk production. *Source* Reproduced with permission from Okic (2020)

and an allowance for floodable zones within the public domain to accommodate aquaculture ponds.

3.3.3 Asi Artificial Islands of the Malaitan on Langalanga Lagoon on Malaita, Solomon Islands

Langalanga stretches north to south along twenty kilometers of Malaita's western coast in the Solomon Islands. The saltwater people who live on the long and narrow Langalanga Lagoon were originally refugees from the mainland. However, researchers argue that the saltwater people started creating the artificial islands fifteen generations ago which is around three hundred years.[80] Today fifteen Asi islands exist, with some of the islands being entirely artificial while others are artificial expansions of natural islands (Fig. 27). These range in size from one thousand square meters to approximately one hundred and twenty thousand square meters, which is the largest and is known as Laulasi Island. It has been estimated that almost twelve thousand Malaita people in the Langalanga Lagoon live on the artificial islands,[81]

[80]Nunn (2009).

[81]Glemarec (2010).

Fig. 27 Multiple artificial islands at Langalanga Lagoon. *Source* Google Earth© 2020 CNES/Airbus

with ten to fifteen families living on some, while on others a lone house stands above the waves (Fig. 28).[82]

To build the islands, stones are carried to a desired location using a log raft. Rafts were placed at the island perimeter to designate the island construction zone. Afterward, they were weighted and sunk using local coral stone. Coral stones were then layered until their surface was above the high tide and then filled with fine, coral rubble and soil to cover the holes and level the new land.[83] The soil was gathered from the river, the beach, or the reef, and it was transferred to the island.[84] At the top, they would plant trees to bind together the coral stones (Fig. 29). A specific tree known as *Alu* exists on every island, and is used mostly for shade, while coconut and lettuce trees on the islands are used for food production. At some islands, Barringtonia trees exist as well.[85] To ensure island durability, the first rock would be placed by the kastom priest to secure the power of the shark, which is believed to be reincarnations of dead ancestors who protected their descendants. Upon completion, a rock from the first clan's island was laid to ensure that the ancestors will protect the Asi.[86]

Two differing theories are recounted to explain the existence of the islands. Firstly, they were built as protection from enemies,[87] and secondly to escape from the

[82]Tuwere (2010).

[83]Guo (2001).

[84]Walter (1930a).

[85]Walter (1930b).

[86]Guo (2015).

[87]Ibid.

Fig. 28 Langalanga artificial island and a traditional house construction. *Source* Reproduced with permission from Scaruffi (2011)

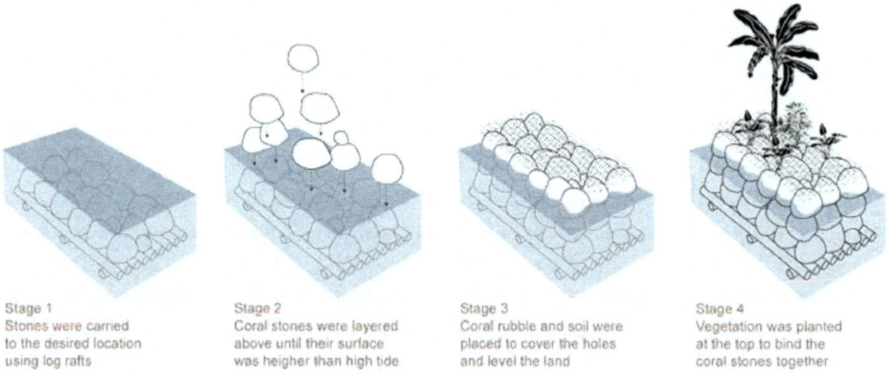

Fig. 29 Construction phases for the artificial islands

malaria epidemic.[88] At the ocean side of the lagoon, reefs and mangroves protect the lagoon from strong waves. Shallow waters, combined with materials from coral reefs, and high biodiversity, created the perfect environment for the construction of artificial islands. Islands were initially built for the size of a family. Later on, they

[88] Bryant-Tokalau (2011).

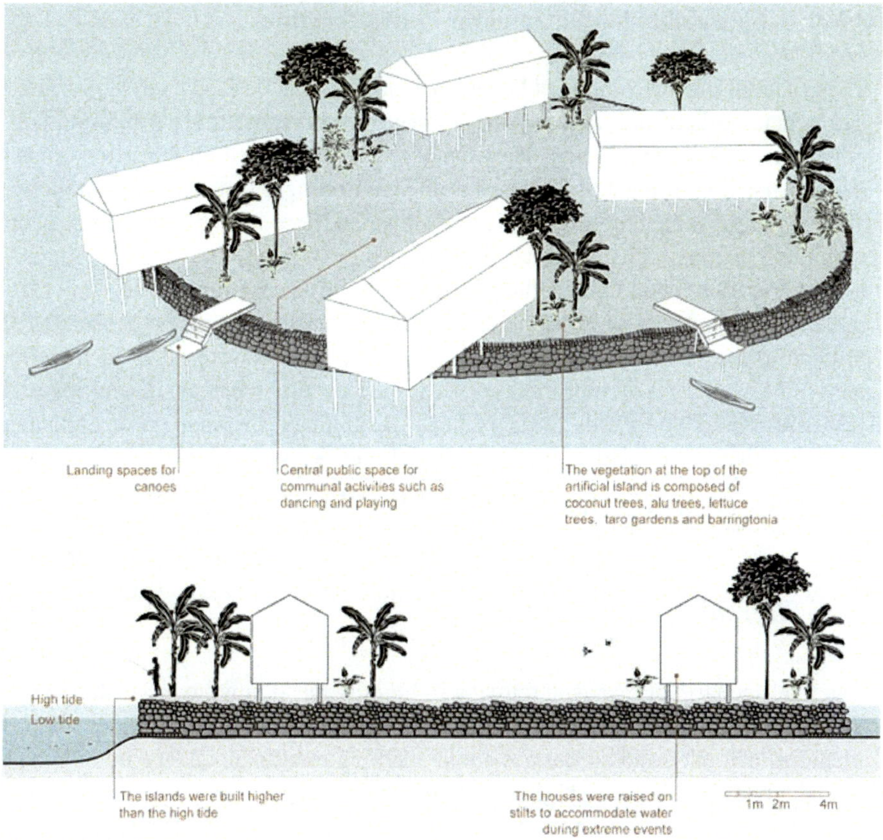

Fig. 30 An axonometric and a section of the artificial islands. The houses are usually found on the perimeter, revealing in this way a vacated space in the middle that is used by the community for gathering. The houses are raised to protect from the water during high tides

would expand to accommodate more families. Multiple dwellings were added on the perimeter, forming a central public space for communal activities such as dancing and playing.[89] The houses were placed on stilts to accommodate water during high tides. For the Langalanga residents, high tides are a natural and welcoming phenomenon. When population and infrastructure density grew, houses were built with two stilts on the island and two in deeper water with additional landing spaces for canoes. One side of the island was often built at a higher elevation for the residents to be able to fish (Fig. 30).

[89] See Footnote 86.

3.3.4 Applicability to Contemporary Infrastructures

These islands are an example of how local resources can be used to create permanent and vegetated land in a lagoon environment. At the same time, it is important to notice the scale of the islands. These small-scale islands required less materials for the construction, protecting in this way the environment from overexploitation and allowing at the same time the natural flow of water. Moreover, the indigenous people, after observations, were able to accommodate both high tides and low tides into their design. As mentioned earlier, the islands are built above high tide levels, creating in this way a layer of protection. The plantation on the side of the ocean was also protecting the islands from storms. In addition, houses were elevated to provide an extra layer of protection in case of extreme events, whereas, during low tides, residents were able to walk to the islands. In summary, the analysis of Langalanga artificial islands provided an important insight for the expansion of land in the water that adapts to the environment and people's lives.

3.4 Retreat

The final scenario of managed retreat will literally mean moving in a different direction. This is a discussion happening most commonly within the circles of scientists, planners, activists, and academics about adapting to climate change by moving out of risk zones and back to higher ground. This strategy does not primarily consist of a physical infrastructure, but rather involves political and social realignments, which are impeded by high upfront costs.[90] Retreat is the strategy employed when all the other adaptation methods have failed to protect a coastal area. The IPCC defines **retreat** as the strategy of protecting people by moving them away from the natural hazard which can be further divided into: migration, the voluntary movement of people; displacement, or resettlement, which is the involuntary and unforeseen movement of people; or relocation, which is managed realignment.[91] In the following section, the **Relocation of the Biloxi-Chitimacha-Choctaw, on Isle de Jean Charles, Southern Louisiana**, is explored as the way in which indigenous people have interacted with this final strategy.

3.4.1 Relocation of the Biloxi-Chitimacha-Choctaw, on Isle De Jean Charles, Southern Louisiana

Since 1930, Louisiana's coastal plain has lost more than five thousand square kilometers, approximately ninety-eight percent of land, to the waters of the Gulf of Mexico

[90] Nonko (2020).

[91] See Footnote 2.

Fig. 31 Remains of the Isle de Jean Charles and the one road that connects the island with the region. *Source* Google Earth©2020 Terrametrics

(Fig. 31). Isle de Jean Charles, the historical homeland of the Biloxi-Chitimacha-Choctaw Indians of Southern Louisiana, is the most profound example of the state's vanishing coast. The narrow island in the bayous of South Terrebonne Parish, Louisiana,[92] has been plagued by a host of environmental problems that have led to increasing flood risk. Unprecedented soil subsidence has led to the once-wooded landscape's slow disappearance into the sea, making the Biloxi-Chitimacha-Choctaw Indians of Southern Louisiana, America's first climate refugees. This has been exacerbated by the lack of soil renewal resulting from the construction of levees that separate the river and the rising sea, and the thousands of oil and gas canals dredged through surrounding marshland that have caused saltwater intrusion, and coastal erosion (Fig. 32). Two years after receiving federal funding to move to higher ground, the tribe remains in a landscape slowly subsiding as they wait for new homes.[93]

When the Biloxi-Chitimacha-Choctaw Indians arrived in the area in the early 1800s, they settled on the highest ridge, located on Isle de Jean Charles. Even though the Isle de Jean Charles was considered uninhabitable swamp land, in 1876 the State of Louisiana began selling the land to private individuals, at a time when it was illegal for a Native American to purchase land.[94] However, the first foreign settler, Jean Marie Naquin, a Frenchman, married Pauline Verdin, a Native American. The 1880 Terrebonne Parish Census listed the first land buyers as residents and included just four families, all descendents of the first settlers who had married

[92]Tribal Government (2019).

[93]Jackson (2018).

[94]See Footnote 100.

Fig. 32 A home on the island that was destroyed by flooding. *Source* Apricot (2007)

Native Americans.[95] By the census of 1910, the area was officially called "Isle á Jean Charles" and had grown to sixteen families, all descendants of the first four families, and totaling seventy-seven people, whose occupations included fishermen, oystermen, or trappers.

For the last nearly ninety years, the Southern Louisiana wetlands have been collapsing due to subsidence. In 1928, the Flood Control Act authorized the building of levees to keep the Mississippi River within its banks, which has prevented sediment-laden spring floods from replenishing the marshes. For millennia, the river has been building land as the gulf retrieves it. Now, without the sediment, it is a battle between the gulf and gravity, with climate change and sea level rise worsening the circumstances.

In 2002, the Morganza to the Gulf Flood Protection System was built by the Army Corps of Engineers to protect communities along the Louisiana coast, but the system was built to pass north of the Isle de Jean Charles determining it was not cost-effective to include the island. In January 2016 after seventeen years of negotiation, the community was awarded a first-of-its-kind forty-eight million dollar relocation grant from the U.S. Department of Housing and Urban Development (HUD), worked out by the Louisiana State Office of Community Development and tribal leaders.[96]

[95] Cottage Films (2014).

[96] Ibid.

For the Biloxi-Chitimacha-Choctaw Indian, land use decisions combined with climate-induced displacement have threatened both unity and sovereignty. Furthermore, as place is the foundation of their way of life, the loss of land has led to depletion of traditional practice and cultural assets. As the relocation process begins, disputes about the original terms of negotiation have arisen as islanders insist the original relocation plan was not presented honestly. They say they were told that if they accepted a new home in the resettlement, they would be allowed to retain ownership and access to their Isle de Jean Charles lands—and the structures built there—that have been part of their tribe's heritage for two centuries, but instead this buyout of their ancestral lands would forfeit their say in protecting them.

4 Futures

Sea level rise and climate change is an unprecedented adversary, one that is adaptive, responsive, and capable of complex interactions along coastlines. Variations in local conditions, communities, and ecosystems coupled with global weather patterns will warrant unique responses, amplifying strengths and counteracting weaknesses. Increasingly in post-disaster scenarios, water is being cast as a menacing and unpredictable force to be mastered by building hard, fortifying infrastructures. Across the globe, homogeneous, single-purpose, high-tech strategies are being deployed irrespective of local communities and conditions, with structures and materials that are incompatible with local ecosystems. This chapter counters this narrative, documenting indigenous systems that have been independently simultaneously evolved in response to sea level rise and climate change. These responses have catalyzed cultural, ecological, economic, agricultural, and climate resilience using local, multi-ecosystem applied nature-based technologies.

The IPCC Special Report on the Ocean and Cryosphere in a Changing Climate from 2019 found that "in practice, hard, sediment-based, and ecosystem-based protection responses are often combined and there is high agreement that such hybrid approaches are a promising way forward."[97] Indigenous innovation has proven, over years of adapting to coastal resiliency, nature-based hybrid approaches are effective. The individual indigenous innovations documented in this chapter exhibit the potential of alternative nature-based technologies of indigenous peoples that have yet to be explored as climate mitigating technologies. In developing this compendium and exhibited in the comparative case studies in this chapter, the phenomenon of simultaneous innovation repeatedly occurs. Simultaneous innovation, which suggests a universal typology in response to a crisis, in the context of indigenous nature-based technologies is described by Watson in Lo-TEK, Design by Radical Indigenism as, "Other unknown cases of simultaneous innovation [that] have emerged in isolated indigenous communities who confront similar environmental constraints." Across the Pacific, indigenous innovations stretching from forest to foreshore, like the

[97] See Footnote 2.

ahupua'a of Hawaii, the vanua of Fiji, the subak of Bali, and the derderin of Australia, simultaneously evolved as cooperative land management systems. These systems cultivate foods, building materials, medicines, and firewood, while regulating coastal protection, biodiversity conservation, and integrated ecosystem management. A unique technology of the vanua and the ahupua'a are the intertidal aquaculture ponds. Although these are each designed independently and distantly, they also resemble the derdernin ponds of Australia.[98] This observation indicates there is ancient knowledge about an optimal response to climatic extremes that could evolve new climate mitigating strategies.

These soft, green, living systems further optimize their symbiotic relationships and reduce their ecological impacts by using biodiversity as a building block composed of local materials and traditional ecological knowledge. Many of these technologies exist in combination as a multi-ecosystem scenario, incorporating multiple technologies that range across a land to sea interface, from underwater ecosystems at low elevation to terrestrial ecosystems at high elevations—whether that be a wall that functions as an ecosystem, wave break, and anti-erosion strategy while providing subsistence fishing needs, such as the Ahupua'a Loko'lai, or a land reclamation technology that utilizes raised, habitable, productive land on a lacustrine environment, such as the Asi Artificial Islands. The resilience of these isolated technologies increases in combination with multiple complementary systems, as has been done for millennia by indigenous communities.

4.1 *Hybridizing Ecosystem-Based Approach is the Best Scenario for Infrastructural Resilience*

Incorporating a mix of adaptive scenarios, including defend, surrender, and offend, these multiple complementary systems are more resilient. Embedding nature-based technologies within ecosystems such as reefs, mangroves, saltmarshes, dunes, rivers, and wetlands amplifies the numerous co-benefits of each technology and offers multiple nested scales of resilience. A significant hybrid example analyzed previously is the Asi artificial island of the Langalanga Lagoon, which utilizes existing reef and mangrove ecosystems with an accommodative infrastructural approach. In the Langalanga Lagoon, a natural reef located at the perimeter of the lagoon composes the first layer of protection, and mangroves then offer a second layer of protection. The finished level of the constructed island above high tide is the third layer, followed by the raised stilt housing which generates the final layer of protection (Fig. 33).

Similar to this, the *Derdernin* and *Ngurruwarr* Fishing Traps of the Lardil and Kaiadilt of the South Wellesley Islands, Australia, utilize both mangroves and reefs. The root systems of mangroves assist in stabilizing the rock wall fish traps by intercepting tidal force. In addition, reefs found at the ocean side of the fish trap were protecting the rock wall from strong waves, while increasing the biodiversity of the

[98] Watson (2019).

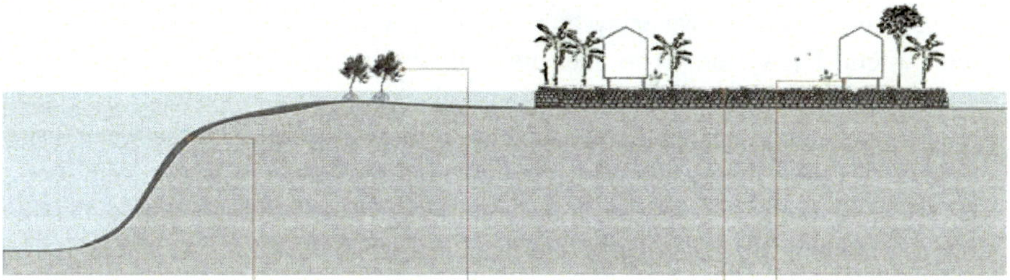

Fig. 33 Ecosystem section of the Langalanga artificial island

area as a reef ecosystem habitat. Moreover, the design of the rock wall took into consideration the strong winds and the tidal forces to determine the best placement for the openings for the fish and the water circulation. Oysters additionally contribute to the protection of the system both by producing calcium carbonate which binds the rocks together and by increasing the biodiversity of the area. At the same time, oysters need the rocks in order to attach and develop, creating in this way a system of symbiosis. The tide was also used as the mechanism for catching fish in the traps, as the rising tide brought the fish in and caught them as it fell. This utilization of the naturally occurring tidal flow required no additional energy input and was predictable and relied upon as a food source especially when storm events disrupted land-based agriculture. In the case of Wellesley Islands, the natural reefs and mangroves combined with the rock infrastructure and the oysters create a hybrid ecosystem approach.

The fish ponds of the Hawaiian Islands were similar to these in the construction of coastal infrastructure for harvesting fish. The ponds were part of the larger *mauka-makai* (mountain-to-sea) *ahupua'a* socioeconomic and land use system, which divided the islands by way of the naturally occurring watershed. The fish ponds, specifically the *loko i'a kalo, loko wai,* and *loko pu'uone*, each used the watershed to feed the ponds and were positioned in different locations on the island to make use of the variations in island topography and land type. The coastal ponds, the *loko pu'uone* and *loko kuapa*, were built from lava rocks and coral and utilized a coralline alga as a cement, making use of local resources and knowledge of ecosystem services. The *ahupua'a* was composed of five biological resource zones: At the highest point was the cloud forest (*wao akua*), a sacred restricted zone; below this came the forest zone (*wao nahele*) which was not cultivated and provided timber, birds, and native plants; then came the agricultural zone (*wao kanaka*) which included upland plantings, agroforestry (*wao la'au*), irrigated taro patches, and a complex system of extensive terraces; and finally the coastal zone (*kahakai*) and the sea zone (*kai*) where the fish ponds were (see Fig. 3)[99]. This watershed system ensured high biodiversity and efficient resource management which resulted in resilient food systems.

[99]Mueller-Dombois (2007).

The use of existing water systems is also seen in the mulberry dike and fish pond system. The system hybridizes the land building strategy of dike building with flood mitigation through implementation with existing water systems. The system dredges pond mud onto the dikes to build up the walls and increase water storage capacity. The overflow water is carried through the rivers to larger water bodies, protecting the crops and villages from flooding. The use of dredged mud also functions as a manure for the mulberry trees which grow on the dikes, the leaves of which contribute to the food chain that makes the mud so nutrient rich. The mulberry-dike system is an excellent example of how an understanding of a hybrid ecosystem can be used to create a low-energy, cost-effective, and self-sustaining system, while at the same time effectively mitigating flooding.

Similar to the *ahupua'a*, the taro growing methods in Micronesia display an understanding of watershed hydrology and vegetation that results in an island-wide productive system. The methods in the system range from forest to coast, each being part of the larger agriculture system on Yap, built within the existing landscape. This system also is a method of land creation, as excavated dirt for taro patches produces raised land on which paths and houses are built. Other methods of taro cultivation float on water, build up to evade water, and build down to prevent salinity intrusion. These cultivation systems are successful without the use of fertilizers and machinery, but rather use nutrient flows, shifting cultivation, and low-tech methods—by means of only a knife and a digging stick.

5 Conclusion—Further Discussion

For indigenous communities, the hybrid multi-ecosystem approach has effectively protected people and the environment for millennia. Indigenous people have learned to live symbiotically with their environments. While these innovations are not all direct examples of coastal resilience, the aquatic infrastructures work within the coastal environment and are multi-functional, symbiotic structures themselves. They were not intended for protection from sea level rise, but they can inform how we can build resiliency by working with the environment, rather than disrupting it further, and simultaneously provide additional services in the form of agriculture and aquaculture. Contemporary coastal cities are overdeveloped in the intertidal zone, destroying protective dunes, foreshore, wetlands, and mangrove ecosystems. Anthropogenic activity has also overexploited natural underwater defense systems, such as coral reef, sea grasses, and oyster beds leaving coastal areas unprotected. Furthermore, incompatible structures and material have polluted the oceans and affected local biodiversity.

Taking inspiration from the multi-ecosystem indigenous infrastructural approaches analyzed above, this chapter argues for a holistic design approach of hybrid adaptation strategies that amplify social, ecological, economic, agricultural, and climate resilience for both the environment and all its inhabitants. Climate change will necessitate an alternative direction, for some communities that will eventually require a managed retreat. Addressing the political and social frameworks needed for this final scenario now is critical. We cannot go backward, fixing all of the hard infrastructures and extractive activities that have exacerbated the need for retreat, but we can go forward, rethinking and rebuilding in a way that supports the resilience of both communities and cities, addressing the inequalities and distance from nature that our current systems support.

References

Apricot K (2007) Isle de Jean Charles. https://www.flickr.com/photos/karenapricot/2834290827. Accessed 06/14/2020

Bhardwaj R (2017) Understanding the seasonal flows of water. Retrieve from https://www.projectsurvivalmedia.org/757-2/. Accessed 19/02/2020

Bond S, Gmirkin R (2003) Restoring a Part of Hawaii's Past: Kaloko Fishpond restoration. Ecol Restor 21(4):284–289. Retrieved from www.jstor.org/stable/43442710

Borger GJ, Ligtendag WA (1998) The role of water in the development of the netherlands: a historical perspective. J Coast Conserv 4(2):109–114

Bradshaw J (2010) Huilua Fishpond walls, Kaneohe, Oahu, a National Historic Landmark on National Register of Historic Places. Wikimedia Commons. Public Domain. https://en.wikipedia.org/wiki/Huilua_Fishpond#/media/File:Oahu-Huiluafishpond-wallcorner.JPG Accessed 06/14/2020

Bryant-Tokalau J (2011) Artificial and recycled islands in the pacific: myths and mythology of "plastic fantastic". J Polyn Soc 120(1): 71–86

Cottage Films (2014) Can't stop the water. https://www.cantstopthewater.com/the-island. Accessed 06/14/2020

Church JA, Clark PU, Cazenave A, Gregory JM, Jevrejeva S, Levermann A, Merrifield MA, Milne GA, Nerem RS, Nunn PD, Payne AJ, Pfeffer WT, Stammer D, Unnikrishnan AS (2013) Sea level change. In: Stocker TF, Qin D, Plattner G-K, Tignor M, Allen SK, Boschung J, Nauels A, Xia Y, Bex V, Midgley PM (eds) Climate change 2013: the physical science basis. Contribution of working group I to the fifth assessment report of the intergovernmental panel on climate change. Cambridge University Press, Cambridge, United Kingdom, New York, NY, USA

Conrad E, Newell L (1992) Proceedings of the session on tropical forestry for people of the Pacific. In: General Technical Report PSW, Ser. 129. Proceedings of the session on tropical forestry for people of the Pacific. U.S. Department of Agriculture, Pacific Southwest Research Station

Costa-Pierce BA (1987) Aquaculture in Ancient Hawaii. Bioscience 37(5):320–331. https://doi.org/10.2307/1310688

Ellen RF, Parkes P, Bicker A (2000) Indigenous environmental knowledge and its transformations: critical anthropological perspectives. Harwood Academic Publishers, UK

Falanruw MC et al (1987) Vegetation survey of Yap, Federated States of Micronesia. In: Resource bulletin PSW, vol 21. U.S. Department of Agriculture, Forest Service, Pacific Southwest Forest and Range Experiment Station. https://doi.org/10.2737/psw-rb-21

Falanruw MVC, Ruegorong F (2015) Dynamics of an island agroecosystem: Where to now? In: Cairns M (ed) Shifting cultivation and environmental change: indigenous people, agriculture and forest conservation. Routledge, New York, pp 367–386 (Chapter 19)

Garroutte E, Defining M (2020) Radical Indigenism and Creating an American Indian Scholarship. Retrieved date 23 Jan 2020. Retrieved from https://www.semanticscholar.org/paper/Eva-Marie-Garroutte-Defining-%E2%80%9C-Radical-Indigenism-%E2%80%9D-Garroutte/265622ddca0cf464d527f99f66d886a243f9507f

Glemarec Y (2010) Project/programme proposal—adaptation fund. Retrieve from https://www.adaptation-fund.org/wp-content/uploads/2015/01/AF_Proposal_Solomon_Islands_-%20revision%2011Jan2011.pdf. Accessed 02/04/2020
Gongfu Z (1982) The Mulberry Dike-Fish Pond complex: a Chinese ecosystem of land-water interaction on the Pearl River delta. Human Ecol 10(2):191–202. https://doi.org/10.1007/BF01531240
Guo P (2001) Landscape, history and migration among the Langalanga, Solomon Islands. ProQuest Dissertations Publishing
Guo P (2015) Island builders: landscape and historicity among the Langalanga, Solomon Islands. Pluto Press, p 189
Haasnoot M, Kwakkel JH, Walker WE, ter Maat J (2013) Dynamic adaptive policy pathways: a method for crafting robust decisions for a deeply uncertain world. Glob Environ Change 23(2):485–498. https://doi.org/10.1016/j.gloenvcha.2012.12.006
Heggen R (2018) Floating Islands, 2nd edn. In: An activity book. Retrieve from https://books.google.com/books/about/Floating_Islands_2nd_Ed.html?id=MbtTDwAAQBAJ. Accessed 11/02/2020, p 481
Hoad FT (Ed) (2003) The concise oxford dictionary of english etymology, 1 edn. Oxford University Press (Online)
Jackson T (2018) On the Louisiana coast a native community sinks slowly into the Sea Retrieve from https://e360.yale.edu/features/on-louisiana-coast-a-native-community-sinks-slowly-into-the-sea-isle-de-jean-charles. Accessed 02/21/2020
Jesse (2020) Hiding in the foliage of the Hawaiian rain forest. Adobe stock. https://stock.adobe.com/images/hiding-in-the-foliage-of-the-hawaiian-rain-forest/192426743 Accessed 06/14/2020
Joana V (2005) Restoring The Loko I'a. Retrieved from https://mauifishpond.com/no-ka-oi-magazine-article/ Accessed 01/03/2020
Kikuchi WK (1976) Prehistoric Hawaiian Fishponds. Science 193(4250):295–299. https://doi.org/10.1126/science.193.4250.295
Kreij A, Scriffignano J, Rosendahl D, Nagel T, Ulm S (2018) Aboriginal stone-walled intertidal fishtrap morphology, function and chronology investigated with high-resolution close-range Unmanned Aerial Vehicle photogrammetry. J Archaeol Sci 96:148–161. https://doi.org/10.1016/j.jas.2018.05.012
Levy JK, Chernisky J (2005) Water resources IMPACT. Sustain Water Resour Manage Hawaii 7(2):20–22
Lopez DR (2019) Societal dimensions of environmental science: global case studies of collaboration and transformation, 1 edn. CRC Press Inc., Milton
Manner H (1980) Taro (Colocasia Esculenta (L.) Schott) In the Atolls and Low Islands of Micronesia. Retrieve from https://www.doc-developpement-durable.org/file/Culture-plantes-alimentaires/FICHES_PLANTES/taro/taro%20in%20micronesia_College%20of%20Tropical.pdf. Accessed 10/02/2020
Manner H (2010) Farm and forestry production and marketing profile for giant swamp taro (*Cyrtosperma chamissonis*). In: Specialty crops for Pacific Island Agroforestry. https://pdfs.semanticscholar.org/7e80/7f7153d7adaea0dd1e30b5a99487a9df16d3.pdf. Accessed 06/14/2020
Marks R (2011) China: its environment and history. Rowman and Littlefield
Memmott P, Trigger DS (1998) Marine tenure in the Wellesley Islands region, Gulf of Carpentaria. In: Rigsby B, Peterson N (eds) Customary marine tenure in Australia. Oceania Publications, Sydney, NSW, Australia, pp 109–124
Memmott P, Robins R, Stock E (2008) What exactly is a fish trap? Methodological issues for the study of Aboriginal intertidal rock-wall fish traps, Wellesley Islands region, Gulf of Carpentaria, Australia. In: Connolly J, Campbell M (eds) Comparative island archaeologies. Archaeo Press, Oxford, UK, pp 47–67
Moore LM, Lawrence JH (2003) TARO Colocasia esculenta. Retrieve from https://plants.usda.gov/plantguide/pdf/cs_coes.pdf. Accessed 12/02/2020

Morrison J, Geraghty P, Crowl L (1994) Science of Pacific Island people, vol 2. Land use and agriculture. Institute of Pacific Studies

Mueller-Dombois D (2007) The Hawaiian ahupua ‘a land use system: its biological resource zones and the challenge for silvicultural restoration. Bishop Mus Bull Cult Environ Stud 3:23–33

Nakashima D, Krupnik I, Rubis J (eds) (2018) Indigenous knowledge for climate change assessment and adaptation. Cambridge University Press, Cambridge. https://doi.org/10.1017/9781316481066

Nonko E (2020) NYC's coastline could be underwater by 2100. Why are we still building there? Retrieve from https://ny.curbed.com/2020/1/2/21046581/new-york-city-climate-change-managed-retreat-development. Accessed 03/01/2020

Nunn PD (2009) Responding to the challenges of climate change in the Pacific Islands: management and technological imperatives. Climate Res 40(2/3):211–231. https://doi.org/10.3354/cr00806

Nunn PD (2007) Climate, environment and society in the Pacific during the last millennium, vol 6. Elsevier, London, Amsterdam, pp 53–58

Nunn PD, Runman J, Falanruw M, Kumar R (2017) Culturally grounded responses to coastal change on islands in the Federated States of Micronesia, northwest Pacific Ocean. Reg Environ Change 17(4):959–971. https://doi.org/10.1007/s10113-016-0950-2

Okic E (2020) Mulberry leaves, grown around the fish pond. Shezhong Village, Linghu County, Zhejiang Province, China. Retrieve from https://turtle.photoshelter.com/image/I00002sBMhI7Oavc. Accessed 01/04/2020

Oppenheimer M, Glavovic BC, Hinkel J, van de Wal R, Magnan AK, Abd-Elgawad A, Cai R, Cifuentes-Jara M, DeConto RM, Ghosh T, Hay J, Isla F, Marzeion B, Meyssignac B, Sebesvari Z (2019) Sea level rise and implications for low-lying Islands, coasts and communities. In: IPCC special report on the ocean and cryosphere in a changing climate

Oreskes N, Conway EM (2013) The collapse of western civilization: a view from the future. Daedalus 142(1):40–58. https://doi.org/10.1162/DAED_a_00184

Prasad S (2008) CSP_4615. Retrieve from https://www.flickr.com/photos/11124467@N02/3397975930. Accessed 11/02/2020

Rai PK, Singh MM (2014) Wetland Resources of Loktak Lake in Bishenpur District of Manipur, India: a review. Sci Technol J

Ranganathan P (2017) One last dance: the fate of Manipur's Dancing Deer. Retrieve from https://www.currentconservation.org/issues/one-last-dance-the-fate-of-manipurs-dancing-deer/. Accessed 10/02/2020

Rowland MJ, Ulm S (2011) Indigenous fish traps and weirs of Queensland. Queensland Archaeol Res 14:1–58. https://doi.org/10.25120/qar.14.2011.219

Scaruffi P (2011) Pictures of Solomon Islands. Retrieve from https://www.scaruffi.com/monument/oceania/solomon.html. Accessed 03/04/2020

Schuttenhelm R (2019) In face of rising sea levels the Netherlands ‘must consider controlled withdrawal’. Retrieved date 12 Jan 2020. Retrieved from https://www.vn.nl/rising-sea-levels-netherlands/

Singh AL, Khundrakpam ML (2011) Phumdi proliferation: a case study of Loktak lake. Manipur Water Environ J 25(1):99–105. https://doi.org/10.1111/j.1747-6593.2009.00197.x

Singsit S (2003) The dancing deer of Manipur. Retrieve from https://web.archive.org/web/20120219231812/; http://oldwww.wii.gov.in/publications/newsletter/autumn2003/fromthewild.htm. Accessed 19/02/2020

Takhelmayum K, Gupta S (2011) Distribution of aquatic insects in phumdis (floating island) of Loktak Lake, Manipur, northeastern India. J Threat Taxa 3(6):1856–1861. https://doi.org/10.11609/JoTT.o2526.1856-61

The Environmental Literacy Council (2015) Forest ecosystem services. Retrieve from https://enviroliteracy.org/land-use/forests/forest-ecosystem-services/. Accessed 22/01/2020

The People's Government of Nanxun District (2017) Zhejiang Huzhou Mulberry dyke and Fish-pond System. Retrieve from https://www.fao.org/3/CA3193EN/ca3193en.pdf. Accessed 09/01/2020

Tribal Government (2019) Bienvenue, Halito, Welcome to Isle de Jean Charles. Retrieve from https://www.isledejeancharles.com/island. Accessed 02/20/2020
Tuboi C, Babu MM, Hussain SA (2012) Plant species composition of the floating meadows of Keibul Lamjao National Park, Manipur. NeBIO 3(4)
Tuwere JW (2010) When the king tide comes. Retrieve from https://www.redcross.int/EN/mag/magazine2010_2/18-19.html. Accessed 03/03/2020
UNESCO (2016) Keibul Lamjao conservation area, Retrieve from https://whc.unesco.org/en/tentativelists/6086/. Accessed 09/02/2020
Walter GI (1930a) The island builders of the Pacific. Seeley Service and Company, London, p 60
Walter GI (1930b) The island builders of the Pacific. Seeley Service and Company, London, p 52
Watson J (2019) Lo—TEK design by radical Indigenism Taschen, Germany
Zhuang Y (2018) Rice fields, water management and agricultural development in the prehistoric Lake Taihu region and the Ningshao Plain. UCL Press p 87

Julia Watson Designer, activist, and academic, Julia Watson is a leading expert on indigenous technologies and climate resilience. Her new book with Taschen, Lo-TEK: Design by Radical Indigenism is a bestseller and has been featured in The New York Times, The Guardian, Dwell, Architectural Digest, Topos, and more. Julia teaches at Harvard and Columbia University and her landscape and urban design studio focuses on rewilding. She has written for Topos, Landscape Architecture Frontier, ioARCH, Kerb, Scroope, Nakhara Journal, Water Urbanisms East and A Spiritual Guide to Bali's UNESCO World Heritage. She's a fellow of TED, Summit REALITY, Pop!tech, Christensen Fund, Charles Eliot, Olmsted award. Born in Australia, she regularly travels to indigenous cultures and sacred sites across the globe.

Despina Linaraki Architect engineer, Despina Linaraki is a Ph.D. researcher at SeaCities, Cities Research Institute, in Queensland. She has completed a Master of Science, 2015 from Columbia University in New York, at Advanced Architectural Design and the Development of Global Cities. Furthermore, she holds a Master of Architectural Engineering, 2013 from Technical University of Crete, in Greece. Since 2014 she is a registered Architect in Greece. Having lived experience in different coastal cities around the world, she is researching nature-based solutions for the adaptation of low-lying islands and coastal cities to sea level changes and floods through the interdisciplinary of architecture with the fields of biology and geology.

Avery Robertson is a designer, researcher, writer, and editor passionate about the intersection of the built environment and the ecological environment. She completed a Bachelor's of Science in Architecture Studies with a focus on sustainable design before going on to obtain her Master's in 2018 in Sustainable Urban Environments, both from Northeastern University. Her current work focuses on bringing sustainability practices and climate science to the public through design and communications.

Exploiting Sediment- and Morpho-Dynamics in Coastal Adaptation Strategies to Sea-Level Rise: A Case Study of the Vietnamese Mekong Delta

Thang Viet Nguyen, Kelly Shannon, and Bruno De Meulder

Abstract This chapter explores potential strategies to engage with sediment dynamics and coastal geomorphology in response to sea-level rise through a case study of the Vietnamese Mekong Delta. The delta has been formed by deposition of sediments during the last several thousand years. With a low elevation and significant land subsidence, it is among the most vulnerable deltas in the world under the impacts of climate change, sea-level rise and particularly flooding. Its entire coastline is characterized by a dynamic process of accretion and erosion. The chapter firstly provides an overview of geomorphology—concentrating on sedimentation and erosion—along the coast of the Mekong Delta, where most of the massive sediment discharge of the Mekong River is trapped on the subaqueous delta area and two-thirds of the sediment is transported southeastwards to the southernmost Ca Mau Cape. Then, the chapter draws lessons from an implemented local scale project on Vietnam's East Sea (also referred to as the South China Sea), where bamboo fences are used to reduce coastal erosion and stimulate sedimentation. Permeable T-fences trap sediment and create calm water conditions for further deposition along the muddy coast. In turn, restoration of the eroded coastal floodplain creates preconditions for rehabilitation of the mangrove forest. Finally, the last section examines a coastal strategy, proposed in the Mekong Delta regional plan, to enhance controlled and increased sedimentation along the East Sea coast in order to drastically up-scale the natural land-gaining process, which can be further consolidated by the systematic planting of mangroves. By engaging with the complex circulation system of sediments along the coast, this strategy can initiate a new protective/productive landscape amidst a constructed ecology that will not only strengthen the ecological structure of the Mekong Delta, but also increase its resilience in the context of sea-level rise.

T. V. Nguyen (✉)
SeaCities, Griffith University, Griffith, Australia
e-mail: thang.nguyen4@griffithuni.edu.au

K. Shannon · B. De Meulder
KU Leuven, Leuven, Belgium
e-mail: kelly.shannon@kuleuven.be

B. De Meulder
e-mail: bruno.demeulder@kuleuven.be

J. Baumeister et al. (eds.), *SeaCities*, Cities Research Series,
https://doi.org/10.1007/978-981-15-8748-1_5

However, it must be noted that sedimentation in the Mekong Delta is increasingly threatened by sediment supply deficit.

1 Introduction

In the context of the SeaCities project, the Mekong River Delta is an incredibly interesting case study. Since its formation, the relation of water and land has been continually shifting and given the predicted consequences of climate change, today it is an extremely vulnerable territory. The Mekong River Delta has an area of 49,500 km^2 (40,604.7 km^2 of which is in southern Vietnam and the remainder in Cambodia) and is the third largest delta in the world. The Mekong river, running ~4700 km from the Himalayas to the East Sea (also known as the South China Sea), is one of the largest river systems in the world, in terms of the discharge of water, as well as in terms of discharge of sediment (Hein et al. 2013). The processes of sedimentation and erosion over millennia has formed an enormous delta plain which hosts incredible fertility and biodiversity. Gradients of wetness and dryness determine occupation. Ultimately, the land is saturated with water and the water is full of land (De Meulder and Shannon 2019) (Fig. 1).

With a low elevation and significant land subsidence, the Mekong Delta plain is among the most vulnerable deltas in the world under the impacts of climate change, sea-level rise and particularly flooding. A study by Minderhoud et al. (2019) concludes that the Mekong Delta plain has a mean elevation of 0.82 m, meaning that it has one of the lowest elevations of all mega deltas in the world. The tidal and coastal areas of the Mekong Delta have elevations of only 0.3–0.7 m (Balica et al. 2014). Moderate estimates of absolute sea-level rise [~40 cm by 2100 (Church et al. 2013)] may result in a quarter of the delta falling below mean sea level by the end of the century (Minderhoud et al. 2019). Minderhoud et al. (2017) also found the delta subsiding at increasing rates, with present delta-average rates exceeding 1.1 cm per year due to massive groundwater extraction. Therefore, large parts of the delta may face submersion already during the coming decades. According to Zoccarato et al. (2018), the lower Mekong Delta is threatened by permanent inundation when the rates of natural compaction of the delta plain (up to ~2 cm per year at the coastline) is no longer counterbalanced due to reduced sedimentation rates in scenarios of sediment supply depletion and sea-level rise. An important effect of the possible submersion due to subsidence is that flows of water throughout the delta, which are already slow due to very limited differences in topography, will come to a standstill in many places and consequently aggravate sanitation and water pollution issues.

The Mekong Delta is naturally endowed with valuable ecological assets. Over the past three centuries, hydraulic interventions have resulted in the delta becoming a key productive region for food, sea products, fruits, flowers and tree nurseries. The hydraulic interventions turned what in essence was a quagmire into a mosaic of paddy fields, regulated by a sophisticated and immense network of canals and drains. The water system is the essence of the Mekong Delta and literally the register upon

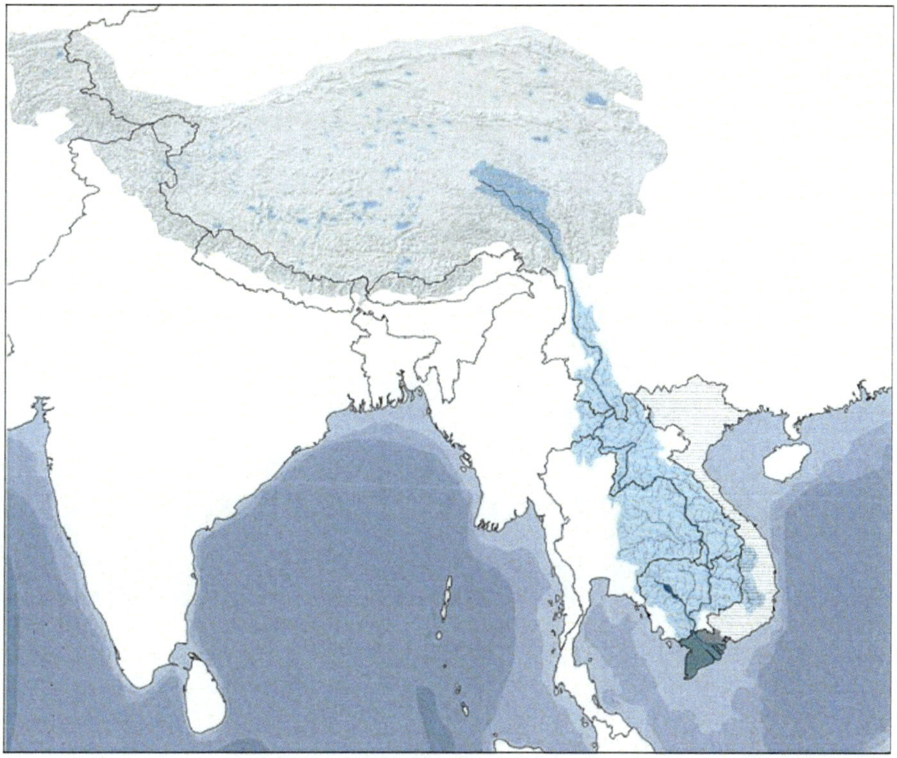

Fig. 1 The location of the Mekong Delta and the basin of Mekong River from Tibetan plateau. *Source* RUA (2016)

which everything is grafted. During the past four decades, Vietnam moved from a country which could not sufficiently feed its population, to a major exporter of aqua-agriculture products. The Vietnamese portion of the delta is home to approximately 17.3 million Vietnamese inhabitants (GSO 2019) and has undergone accelerating urbanization and development. Nonetheless, it remains a predominantly agricultural region. The overall pattern of settlements in the Mekong Delta is characterized by Terry McGee's definition of 'desakota,' from Bahasa Indonesian *desa* for village and *kota* for town or city (as referenced in De Meulder and Shannon 2019) (Fig. 2).

The urban population in the coastal zone accounts for approximately 37% of total urban population of the Vietnamese Mekong Delta (SISP 2017). In 2020, it is estimated that approximately 2 million people live in urban settlements (as classified by Vietnamese regulations: An urban settlement has at least 4000 residents) in the coastal zone, with 18 cities/towns (i.e., urban population ranging from 50,000 to 250,000 residents) located within 40 km from the coastline (SISP—RUA 2017). These settlements are embedded within 720 km of coastline and are driven by expanding economic activities including aquaculture, energy, tourism, fishing and agriculture. It is also an area strongly influenced by the natural dynamics of the coastal environment: tide, saline intrusion, flooding, shoreline erosion and sedimentation.

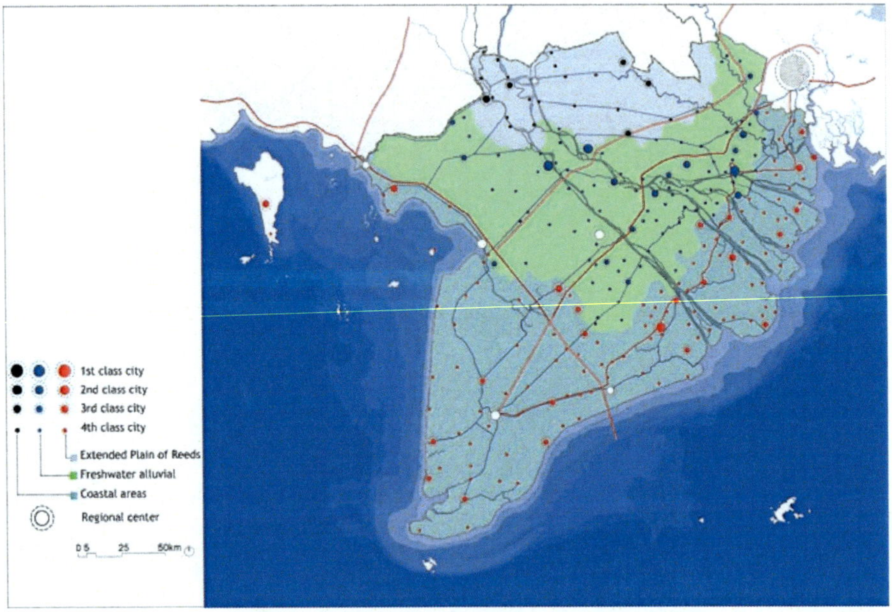

Fig. 2 Urban network in the Vietnamese Mekong Delta. *Source* RUA (2016)

On the other hand, coastal development in the Mekong Delta has increasingly created negative footprints on valuable ecosystems. Mangrove forests are rapidly disappearing, mostly due to land reclamation, with conversion of these important ecosystems to both shrimp ponds and agricultural fields (GIZ 2018). Sea dykes are being increasingly built along parts of the muddy East Sea and Gulf of Thailand coasts for protection from marine flooding and for shrimp farms, generating a 'mangrove squeeze' (Phan et al. 2015). With the regional change of sediment supply, the diminishing of mangrove belt in the coastal zone contributes to the delta's increased vulnerability in the face of sea-level rise.

Although for the last several thousand years the Mekong Delta has continually extended its shoreline seaward (~30 m per year) by sedimentation (Liu et al. 2017a), nowadays, about 377 km (i.e., more than half) of the 720 km Mekong Delta's shoreline is eroding—with 70 km of these having extreme rates of more than 20 m of erosion annually (GIZ 2018). According to Marchesiello et al. (2019), the recent observed patterns of erosion along the Mekong Delta's coast are due to a combination of natural and anthropogenic processes, including: natural redistribution of subaqueous sediments (due to hydro-morphodynamic processes), fluvial sediment supply deficit, land subsidence and mangrove squeeze. Karlsrud et al. (2017) and Marchesiello et al. (2019) considered that subsidence rates are particularly significant (2 cm/year) along the eastern peninsula between Bac Lieu and Soc Trang, explaining the observed shoreline retreat despite favorable sediment transport, and along the western Ca Mau peninsula (1 cm/year), explaining a regime change from slightly accreting to eroding.

A number of factors beyond Vietnam's boundaries, particularly upstream dams and reservoirs, are completely altering the water regimes. Volumes of both water and sedimentation loads are decreasing rather drastically. In the pursuit of fresh water, deeper and deeper wells are dug throughout the entire delta for agriculture and drinking water—resulting in the collapse of aquifers and subsidence. As a result, even larger parts of the lower Mekong Delta now lie below sea level and are increasingly prone to both flooding and drought, the impacts of global climate change and sea-level rise (GIZ 2018). Given these challenges, it is considered that 'building with nature' principles are beneficial in formulating coastal protection and adaptation strategies in the Mekong Delta (RUA 2016; GIZ 2018).

The future of the Mekong Delta, specifically in relation to the contemporary challenges of climate change, must be grounded in an understanding of both its geology and its context-embedded cultural practices of agriculture, settlement and floodwater management (De Meulder and Shannon 2019). This chapter explores potential strategies to engage with sediment dynamics and coastal geomorphology in response to sea-level rise—concentrating on sedimentation and erosion—along the coast of the Mekong Delta. It seeks to draw lessons from an implemented project of bamboo fences to reduce coastal erosion and stimulate sedimentation and a territorial-scale coastal strategy, as proposed in the Ministry of Construction's Revised Mekong Delta Regional Plan to 2030, Vision to 2050. The aim is to innovatively manage the decreased levels of sedimentation along the East Sea coast in order to create new protective and productive landscapes. The Mekong Delta's shallow continental shelf, which depends on, and has an effect on, seabed sediment composition and accumulation patterns, provides yet untapped opportunities for transportation, aquaculture, mining, fishery and tourism (RUA 2016).

2 Overview of Geomorphology Along the Coast of the Mekong Delta

2.1 Sediment Accumulation and Formation of the Mekong Delta

Most of the Mekong River's sediment originates from the Himalayan highlands. Volumes of sediment are transported over ~4700 km in river systems and most of the sediment reaches the ocean as fine silt and clay (i.e., muddy sediment finer than 64 μm) with some sand trapped in shallow coastal settings (Nittrouer et al. 2017). Milliman and Syvitski (1992) estimated the annual sediment load was ~160 million tons in the pre-dam era. According to Liu et al. (2017a: 73), 'sediment accumulation in the past 7500 years has built the Mekong Delta ~220 km seaward and formed ~50,000 km^2 of subaerial surface and 11,000 km^2 of modern subaqueous delta. The long-term average shoreline growth rate has been ~30 m per year, and the net land gain rate has been 7 km^2 per year. Mekong-derived sediment has also accumulated along the shelf to the southwest, forming a > 200 km long mud-dominated deltaic coast. Most subaqueous delta deposition has occurred within the last ~1000 years.' The present-day Mekong Delta has been shaped by sedimentation over the time scale of millennia (as shown in Fig. 3), resulting in a broad, asymmetric subaerial delta plain bounded by a long, narrow subaqueous delta clinoform—the region of greatest sediment accumulation in the deltaic system (Eidam et al. 2017). An asymmetric progradation process implies the southwestward sediment transport and reflects an increased wave influence (Li et al. 2017). The offshore portion of the Mekong Delta rests on a very shallow and gently sloping (1:15,000) continental shelf (Liu et al. 2017a).

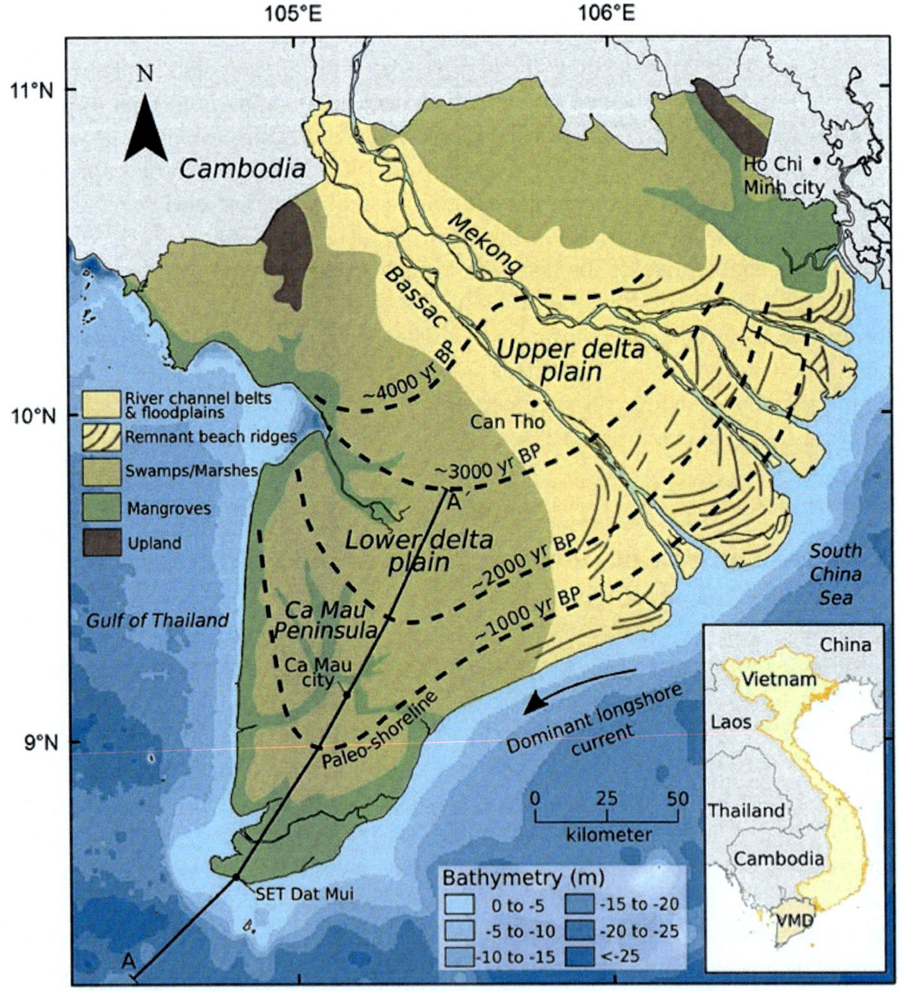

Fig. 3 Map of the Vietnamese Mekong Delta showing the main depositional paleoenvironments. *Source* Zoccarato et al. (2018), https://creativecommons.org/licenses/by/4.0/

2.2 Subaqueous Delta and Longshore Sediment Transport

The Mekong River discharges into the East Sea and forms a subaqueous deposit known as a clinoform (Nittrouer et al. 2017). It is estimated that 80% of the sediment discharge of the Mekong River into the East Sea is trapped on the subaqueous delta area (Xue et al. 2012). This subaqueous delta clinoform, which extends >300 km along the inner shelf adjacent to the Mekong Delta, has a shallow topset region (<5 m deep) extending ~8–10 km seaward, an inclined foreset region (~5–20 m deep) extending ~5–6 km, and a bottomset region (>20 m deep) (Nittrouer et al. 2017, Liu et al. 2017a). Most of the sediment remaining in the ocean accumulates on the inclined foreset region at local rates of >10 cm per year (Nittrouer et al. 2017). According to Eidam et al. (2017), the Mekong clinoform is distinguished among large river deltas in that the transition (or subaqueous 'rollover point') between the broad, flat topset and steeper foreset lies at a depth of only 4–6 m, while for many large river deltas, the subaqueous rollover depth is commonly located at 25–40 m depth.

While approximately one-third of the Mekong's sediment accumulates near the mouths of the Mekong's distributary channels, two-thirds of the Mekong's sediments accumulate in distal areas extending to the Ca Mau Peninsula and beyond (Nittrouer et al. 2017), as a result of sediment transport along the coastline to the southwest and to the southernmost spit of Cape Ca Mau (Xue et al. 2010). The Mekong Delta, classified as a mixed-energy system (wave- and tide-dominated), is affected by two different circulation systems: to the east is the southern East Sea, characterized by a regular semi-diurnal tide with 3.5 m tidal range; to the west is Gulf of Thailand, characterized by an irregular diurnal micro-tide with 0.8–1.0 m tidal range. The shelf current is controlled by the monsoon regime. For the East Sea, currents shift direction twice annually during May (from southwestward to northeastward) and October (from northeastward to southwestward). Current velocity along the Gulf of Thailand side is much smaller than along the East Sea, in both northeast monsoon and southwest monsoon seasons (Xue et al. 2014). There is wave-driven sediment transport for the Mekong-derived sediment: during the high river discharge season (May–October), a considerable part of riverine sediment is delivered to the Mekong River mouth and temporally deposited there; during the low river discharge season (November–April), previously deposited sediment is resuspended by strong mixings associated with the strong northeast monsoon and subsequently transported southwest along-shelf by coastal circulation (Xue et al. 2014).

According to a model of suspended sediment transport in the Mekong Delta by Manh et al. (2014), the sediment load entering the coast from the Mekong River was 42–50 million tons per year in the period 2009–2011. Modelling suspended sediment dynamics on the subaqueous delta by Thanh et al. (2017) demonstrates that in the high river flow season, 90% of the annual sediment volume is discharged and mostly deposited in front of the river mouths. During the low river flow season, much less sediment is exported from the river; however, the alongshore sediment transport rate is nine times higher than the sum of riverine sediment outputs, due

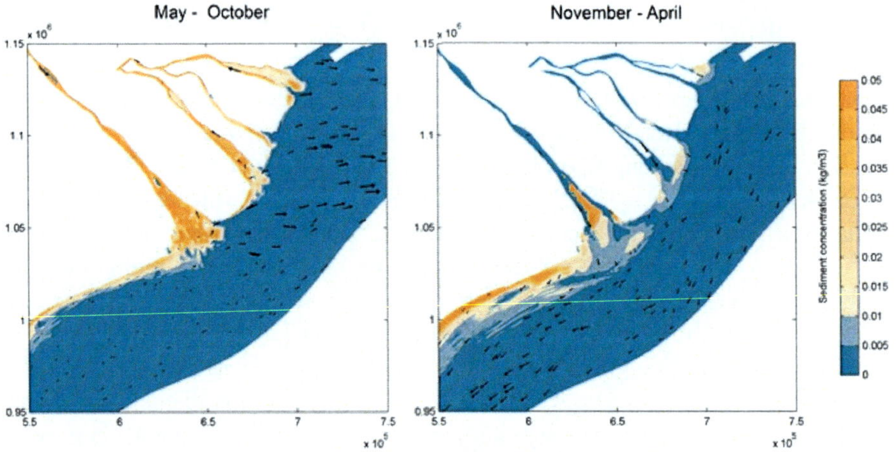

Fig. 4 Simulated sediment concentration in the surface layer of spring tide in the high flow season (left) and in the low flow season (right). *Source* Thanh et al. (2017)

to wave and seasonal current-driven process of resuspension of the sediment layer previously deposited in the high discharge season (see Fig. 4). It is estimated that between 40 and 66% of Mekong-derived sediment is transported southwestwards by the strong seasonal monsoon-derived currents and tidal currents (Unverricht 2014; Liu et al. 2017a, b).

2.3 Shoreline Erosion and Sediment Supply Reduction

The shorelines of the Mekong Delta alternately prograde and erode, both in space and time (Ogston et al. 2017). However, erosion trends along the Mekong Delta coast are accelerating, with erosion occurring over about half of the entire coastline (Anthony et al. 2015). Li et al. (2017) divided the Mekong shoreline into four categories: increasing accretion (17% of total shoreline length), decreasing accretion (36%), increasing erosion (28%), and decreasing erosion (19%). Through a study of the past 43 years of Landsat images, Liu et al. (2017a) indicated that the mode of sedimentation in the delta shifted, starting in 2005. In their study, from 1973 to 2005, the Mekong Delta's seaward shoreline growth decreased gradually from a mean of 7.8 to 2.8 m per year; from 2005, it became negative, with a retreat rate of −1.4 m per year. Thus, suggested by Liu et al. (2017a), in about 2005, the subaerial Mekong Delta transitioned from a constructive mode to an erosional (or destructive) mode. This notable shift was coincident with the onset of river damming, and since 2006, the Mekong Delta's subsidence rate has increased (Minderhoud et al. 2017). Liu et al. (2017a) inferred that dam construction and land subsidence might be major contributors to delta erosion. Anthony et al. (2015) argue that a decreasing river sediment supply to the coast is the prime cause of erosion, and most likely due to both existing dam retention of sediment and to massive channel-bed sand mining in the delta, an activity on the increase over the last decade. Subsidence accelerated by groundwater extraction is highest along parts of the muddy East Sea coast which is most severely affected by erosion (Anthony et al. 2015). Anthony et al. (2015) suggest that seasonal to persistent depletion of mud along the muddy East Sea and Gulf of Thailand sectors of the delta's coast results in less wave energy dissipation, and, consequently, in shoreline erosion. It is evident that with the construction of more dams, uncontrolled sand mining, delta subsidence, increasing storms and sea-level rise, the Mekong Delta will likely face more destructive changes, with regard to erosion of both coastlines and underwater deposits (Liu et al. 2017a).

The formation and progradation of the Mekong Delta have resulted from sediment accumulation, and sediment supply is vital for the survival of the delta in the conditions of land subsidence and sea-level rise. However, many studies have shown a decreasing discharge of sediment from the Mekong River in the past three decades (Tamura et al. 2020). Milliman and Farnsworth (2011) reported that the sediment discharge decreased to 110 million tons per year as a result of dam construction in the basin, compared to the pre-dam estimate of 160 million tons per year (Milliman and Syvitski 1992). According to Liu et al. (2017a), more recent estimates for post-dam sediment discharge vary from less than 67 million tons per year to 145 million tons per year. Based on hydrodynamic surveys in the lower distributary channel of the Mekong, Nowacki et al. (2015) suggested that sediment discharge could be as low as ~40 million tons per year from the entire Mekong River to the East Sea. (Table 1) This reflects the impacts of sharply increased human activities in the river basin, such as upstream dam construction and sand mining (Anthony et al. 2015). The rates of sediment accumulation on the inner shelf will be affected by sediment supply reduction (Allison et al. 2017).

In summary, the overview of geomorphology along the coast of the Mekong Delta reveals that long-term sedimentation has formed its progradation seaward. Its mud-dominated subaqueous delta has a broad shallow flat topset region (<5 m deep, ~8–10 km wide, also mentioned as the subaqueous delta platform in this chapter), an inclined foreset region (~5–20 m deep, ~5–6 km wide) along the East Sea coast, where sediment transport southwestward occurs seasonally (November–April) and results in the accumulation of two-thirds of Mekong sediment in distal areas of subaqueous delta extending to the southernmost Cape Ca Mau and beyond. The shorelines of the Mekong Delta alternately progrades and erodes, both in space and time. However, in 2005, sedimentation in the delta shifted from a constructive mode to an erosive mode,

Table 1 Estimation of sediment discharge from the Mekong River

Researchers and the year of publication	Milliman and Syvitski (1992)	Milliman and Farnsworth (2011)	Nowacki et al. (2015)
Sediment discharge by the Mekong River to the East Sea (in million of tons)	160	110	40

due to decreased sediment supply. The understanding of geomorphology, sediment dynamics and their influential factors in the Mekong Delta is necessary in the pursuit of nature-based coastal protection and adaptation strategies to address the challenges of sea-level rise, coastal erosion and flooding, as well as new opportunities in the shallow subaqueous delta.

3 Lessons Learned from Erosion Reduction and Sedimentation Stimulation by Using Bamboo Breakwater Along the East Coast of the Mekong Delta

3.1 Introduction

In the German Corporation for International Cooperation (*GIZ*) Coastal Zone Management Project for the Mekong Delta, a new approach to coastal protection was piloted along the coast of Soc Trang and Bac Lieu provinces (see Fig. 5). A coastal

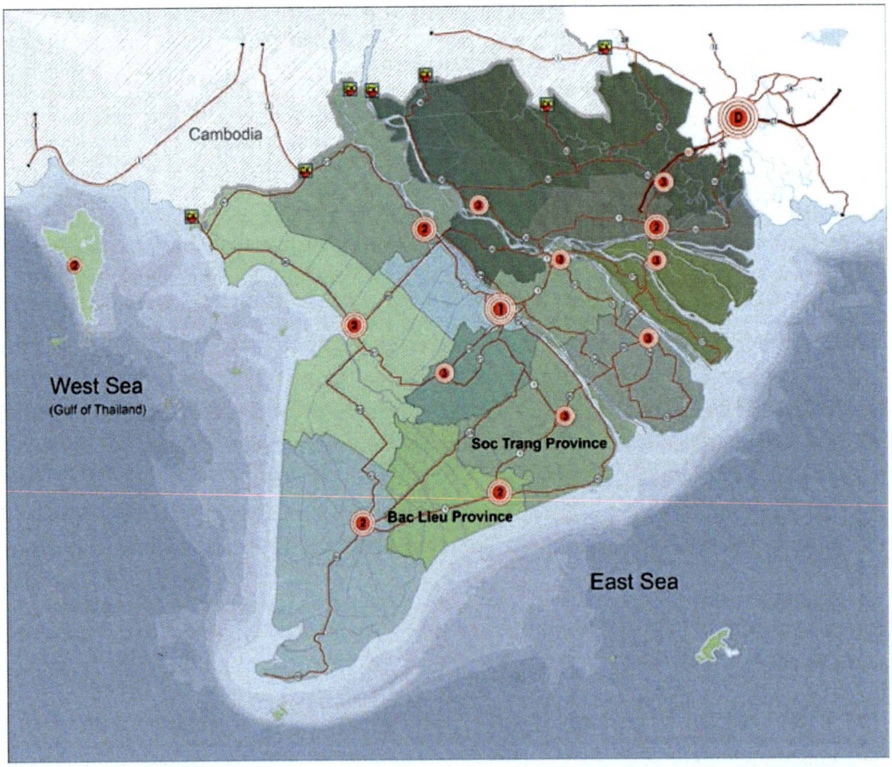

Fig. 5 Location of Bac Lieu and Soc Trang provinces in the Vietnamese Mekong Delta. *Source* SISP (2017)

protection strategy was developed which uses floodplain management as a sustainable and effective way of mitigating coastal erosion and enhancing flood protection (Albers et al. 2013) in places where mangrove forests and coastal floodplain (tidal flats) were adversely affected by erosion. The tested 'area coastal protection' strategy combines various coastal protection elements, including reconstructed coastal floodplains (with sedimentation accelerated by bamboo breakwaters), mangrove forests and, where necessary, an appropriate earth dyke (Schmitt and Albers 2014). Schmitt and Albers (2014) consider the strategy one of 'ecosystem-based adaptation' since biodiversity and ecosystem services are part of the overall adaptation strategy to help people adjust to the adverse effects of climate change. The design and construction of structural protection measures was based on hydrodynamics and shoreline development modeling, as well as physical modeling to ensure effectiveness and avoid negative effects such as downdrift erosion (Albers et al. 2013).

3.2 Application of T-shaped Bamboo Fences and Mangrove Rehabilitation

The T-shaped bamboo fences were constructed in 2012 in several selected sites along the muddy coast of Soc Trang and Bac Lieu provinces (Fig. 6). Sedimentation and mangrove rehabilitation were accelerated by T-fences and mangrove planting (Fig. 7). Clearly observed reductions in wave energy (erosion) and increased sedimentation indicated that the strategy could be successful for the mangrove-mud coasts of the East Sea of the lower Mekong Delta (Schmitt and Albers 2014). In 2016, entire areas of the T-fences were covered with mangroves. This re-establishes a natural protection system (GIZ 2016). The implementation of the pilot project demonstrated the effectiveness of ecosystem-based approach for coastal protection in the Mekong Delta. Moreover, the GIZ document *'A Decision Support Tool for Coastal Protection for the Mekong Delta' (CPMD)* (2018) developed guidelines for T-shaped bamboo fences and mangrove forest rehabilitation through a toolbox of coastal waterworks and recommendations for other appropriate coastal protection segments of the Mekong Delta.

Fig. 6 T-shaped bamboo fences on the coast of Bac Lieu province, the Mekong Delta. *Photo* Cong Ly, GE Wind. *Source* Albers et al. (2013)

Fig. 7 Changes at the site on the coast of Bac Lieu province after installation of the T-shaped bamboo fences. May 2012 (top left), September 2012 (top right), December 2012 (bottom left) and September 2013 (bottom right). *Photo* by Steurer. *Source* Schmitt and Albers (2014)

The principle of the T-fences for the analyzed project is based on the knowledge transfer of the success of well-known structures from the Wadden Sea where floodplain management has become an important active coastal protection measure (GIZ 2018). In the Wadden Sea, land reclamation created new, fertile areas for agricultural cultivation over the course of several centuries. However, during the last six decades, land reclamation by floodplain management has been used in Germany and the Netherlands as an active measure aimed at coastal protection that includes the construction of sedimentation fields surrounded by brushwood fences at tidal flats (Schmitt and Albers 2014).

When applied along the muddy coast of the Mekong Delta, permeable T-shaped bamboo fences, working as breakwaters, create calm water conditions for sediment deposition that restore the eroded coastal floodplains and become pre-conditions for rehabilitation of the mangrove forest (GIZ 2018). The T-shaped fences' wave transmission effect is sufficient to reduce wave heights significantly and stimulate sedimentation on the landward side (GIZ 2018), therefore accelerating the process of sedimentation and regeneration of mudflats. It was reported by Schmitt and Albers (2014) that mangrove forests were able to colonize newly accreted land along the coastline of Soc Trang and Bac Lieu. After initially high sedimentation rates in year one, the process of natural regeneration, which was supported with low density (4000 trees/ha) and mixed species (Rhizophora apliculata and Avicennia sp.) plantings, was observed to slowly reclaim the newly created mudflats within the T-fences (GIZ 2016). Since mangroves trap additional sediments with their roots, the subsequent stabilization of the soil results in soil surface elevation and the expansion of land toward the sea (Schmitt and Albers 2014).

The general layout of T-shaped bamboo fences is explained in CPMD document by GIZ (2018: 115):

> The general arrangement of the permeable bamboo fences consists of a long-shore part, which dampens the incoming wave energy and a cross-shore part, which decreases the long-shore currents (Figs. 8 and 9). The long-shore elements close the eroded gap in the mangrove forest by connecting the remaining headlands. The reduction in wave height and thus in orbital velocities under waves leads to accelerated sedimentation rates. The results from benchmarks in Bac Lieu province showed a deposition of approximately 17 cm of sediments within 7 months. The reduction of wave action on the landward side of the fences also accelerates the consolidation of the mud and thus increases the stability of the sediments against erosion.
>
> The actual placement of the T-fences is done in such a way that they do not interfere with prevailing currents and thus will not cause downdrift erosion. The definition of the general layout needs comprehensive knowledge about the overall hydraulic and morphodynamic boundary conditions.

Bamboo was chosen due to its strength, availability and low cost (Albers et al. 2013). It is light, easy and cheap to transport, easy to manipulate and apply, low-tech, locally available and organic: Once it has done its job, it can rot and disintegrate, it is ecological and does not pollute the environment in any way (while being productive and protective). The bamboo used for wave-breaking barriers in Khok Kha (Samut Sakhon Province, Thailand) remained effective for 5–7 years, which is considered more than sufficient for the re-creation of eroded floodplains and rehabilitation of mangroves in the Vietnamese context, based on the first results of sedimentation monitoring and the time it takes for mangroves to naturally regenerate (Schmitt and Albers 2014). However, the application of T-shaped bamboo fences has clear limits: if the location exceeds a certain degree of exposure to waves and a certain duration of submergence, the effort for maintenance significantly increases and the

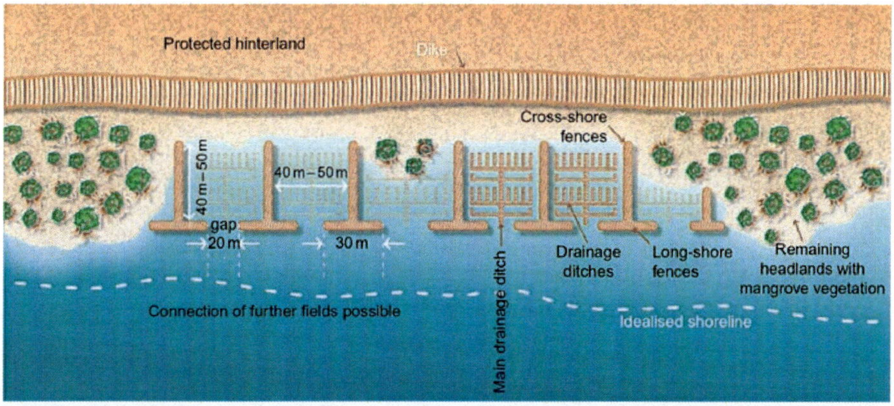

Fig. 8 Land reclamation using cross-shore and longshore bamboo fences. *Source* Albers et al. (2013)

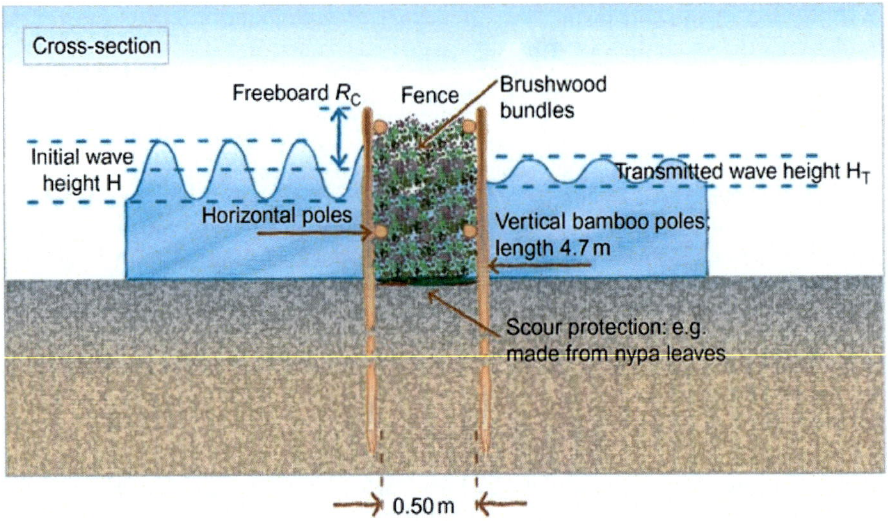

Fig. 9 Construction of a fence in a sedimentation field (cross section). *Source* Albers et al. (2013)

application may become impractical (GIZ 2018). In more exposed locations with a steeper beach profile, a combination of the T-fences with concrete breakwaters could offer a solution (GIZ 2018).

Schmitt and Albers (2014) suggested that restoring eroded floodplains for mangrove rehabilitation in the 'area coastal protection' strategy may also contribute to counteracting sea-level rise to some extent, by assisting the sediments for mangroves to migrate seaward. Presently, mangroves are unable to migrate inland due to existing dikes along the coast. However, since mangroves are important elements of coastal protection their seaward migration is an alternative way to increase their density. The ability of mangroves to respond to sea-level rise was found in several studies; however, their effectiveness remains dependent on sediment supply. According to Krauss et al. (2014), mangroves act as 'ecosystem engineers,' promoting vertical accretion, thus allowing the forest to keep pace with sea-level rise. A report by McIvor et al. (2013) concluded that mangrove surface elevations have kept pace with sea-level rise over thousands of years in some locations, with rates of surface elevation rise ranging between 1 and 10 mm/year in different locations and settings, depending largely on external sediment inputs and the growth of subsurface roots.

In the Mekong Delta, the most robust coastal protection system consists of wide floodplains vegetated with mangroves, not the construction of a technically sound dyke (as stand-alone measure) which is very time-consuming, technically demanding (due to soft foundation soil) and very expensive (GIZ 2016). Dykes require the massive transport of materials, which often creates environmental issues at their material extraction sites. Hard structures often fail in stopping erosion since the subsoil of mangrove-mud coasts and foreshores are too soft and too prone to consolidation: The structures simply sink into the mud or collapse otherwise (GIZ 2016).

When using mangroves as part of an ecosystem-based approach to coastal protection, an earthen dyke is often sufficient, because a healthy mangrove forest will protect the dyke from erosion (Schmitt and Albers 2014). The mangroves reduce wave heights and dissipate the wave energy that leads to reduced required dyke heights and toe protection or revetments (GIZ 2016). The dynamic forces to which the dykes are exposed are reduced, and hence, they can be constructed by more simple means and with earth. Experience from all over the world has shown that ecosystem-based flood defense is more sustainable and cost-effective than conventional coastal engineering (Temmerman et al. 2013). It was concluded by Albers and Schmitt (2015) that in the mud coasts and the low-lying areas of the Mekong Delta, a protection strategy which includes floodplain (foreshore) management, a dynamic mud-coast mangrove system and, where necessary, an appropriate earth dyke can ensure that coastal protection measures are effective and can respond timely to spatial and temporal changes in sediment- and morpho-dynamics. This ecosystem-based coastal protection strategy is proposed as an appropriate alternative to the Vietnamese government plan (SIWRP 2012) to upgrade and newly build a total 617 km of sea dykes with +4.7 m elevation along the East Sea and +2.8 m elevation along the Gulf of Thailand to respond to sea-level rise.

4 A Potential Strategy of Accelerated Sedimentation for New Protective/Productive Landscape on the Subaqueous Delta Platform of the Mekong Delta

4.1 Introduction

The sedimentation stimulation and erosion reduction which was successfully piloted in two provinces along the muddy shorelines of the Mekong Delta's East Sea coast embodies a potential for up-scaling. Large-scale accelerated sedimentation for a large expanse of the delta was proposed by the Mekong Delta Regional Plan. The Revised Mekong Delta Regional Plan 2030 and Vision 2050 was commissioned by the Vietnamese Ministry of Construction and jointly made by Research Urbanism Architecture-RUA (international consultant for planning concepts led by Professors Bruno De Meulder and Kelly Shannon) and the Southern Institute for Spatial Planning-SISP (local consultant) during 2014–2017. The following explanation of selected concepts of the plan is adapted from the planning concept report by RUA (2016).

The revision of the regional plan for the Mekong Delta worked from the premise that accelerated climate change, and the way it reconfigures the geography of the territory, is an opportunity to realign the plan and consequently development. The Mekong Delta Regional Plan is a first step-stone to consciously re-engage with the dynamics that nature, particularly water, imposes on the territory. In doing this, the Mekong Delta Regional Plan develops three major urban and rural morphologies which strongly correspond to agro-ecological zones, which are defined and derived from soil types, sedimentation, erosion and subsidence, as well with the 2050 predictions of inundation and saline intrusion.

Among these morphologies, a 'coastal' type would be developed for the East Sea area and Ca Mau peninsula to form part of a more dynamic and robust coastline (where the coastline is understood as a very broad zone, given the extremely low topography of the Mekong Delta). It would include massive mangrove reforestation to allow for a better mix of saline and freshwater systems, work as a 'soft' infrastructure providing protection against the increasing threat of storm surges and as well allow for a more sustainable aquaculture economy (Fig. 10). Mangrove reforestation can speed up the gaining and consolidation of land by natural sedimentation processes at the coast of the East Sea and the Gulf of Thailand. Settlements would be developed to work with micro-topography differences and sectionally respond to the East Sea's parallel and alternating rows of sand ridges and back swamps, and Ca Mau Peninsula's successive peat layers and subsiding lands. A new settlement strategy in the coastal area would primarily comprise of clearings within the mangrove reforestation areas, where floating buildings form collective spaces for gathering and production. Various forms of renewable energy (wind, solar and tidal) would be integrated into the coastal ecologies.

Fig. 10 Coastal necklace of massive mangrove afforestation proposed in the Mekong Delta Regional Plan. *Source* RUA (2016)

4.1.1 *Exploring the Concept of Accelerated Sedimentation for New Protective + Productive Landscape on the Subaqueous Delta Platform*

The overview of geomorphology discussed earlier in this chapter has shown how the Mekong Delta has prograded seaward by sedimentation and seasonal sediment transport southwestward along the shallow subaqueous delta. In Fig. 4, a simulation of sediment concentration by Thanh et al. (2017) highlights the flows of suspended sediment in the surface layer, within ~10 km from the current shoreline, on the shallow topset region (<5 m deep) of the subaqueous delta. Given the huge amount of sediment discharged annually from the Mekong River, the Mekong Delta Regional Plan proposed to enhance a controlled and increased sedimentation along the East Sea in order to drastically speed- and scale-up the natural land-gaining process. It capitalized on the flows of discharge to build land by 'catching' sediment before it settles on the subaqueous delta or moves further out to sea. To achieve this, the systematic implantation of obstacles (breakwaters) on the shallow subaqueous delta platform (topset region) would break the existing streams' deposits. As efficient coastal protection devices, breakwaters slow down the water current and halt it locally. It is anticipated that this would strongly enhance sedimentation (as it can traditionally be observed at breakwaters) and consequently induce an accelerated formation of sand banks that ultimately result in a land-gaining process. This would then be further consolidated by the systematic planting of mangroves. The artificially induced 'natural reclamation' would not only strengthen the ecological structure of the Mekong Delta, but also increase its resilience in the context of sea-level rise since it would naturally build an enormous mangrove forest upon the gained foreshore land. It replaces the notion of a coastline (and the 'hold-the-line' concept of coastal protection) into a very wide transition area with dynamic and varying constellations of islands and peninsulas that grow with consecutive waves of sediment and vegetation until both processes intertwine and unify (Fig. 11).

The proposal of accelerated sedimentation on the subaqueous delta platform can potentially become a new type of land reclamation that simultaneously increases coastal protection and creates productive landscapes. The land reclamation program explicitly designs, besides space for ecological development (coastal protective mangrove reforestation), various gradients of water and land that can accommodate new fish and shrimp farms, oyster banks and other seafood nurseries that can be combined with new sustainable coastal settlements, renewable energy farms and new state-of-the-art fishery ports. In other words, the gradual and long process of land creation will, besides reconstructing ecologies, generate a multitude of sustainable and renewable resources (fish and seafood farming, wind energy, etc.). Sustainable settlements can develop to exploit these resources in new experimental ways. The future of such coastal settlements is envisioned to be predominantly water based. In this way, the engagement with sediment- and morpho-dynamics can initiate new protective and productive landscapes amidst a constructed ecology on the shallow subaqueous delta platform (Table 2). It is suggested by the plan that the long process of land creation can start with local interventions in order to gain experience and test

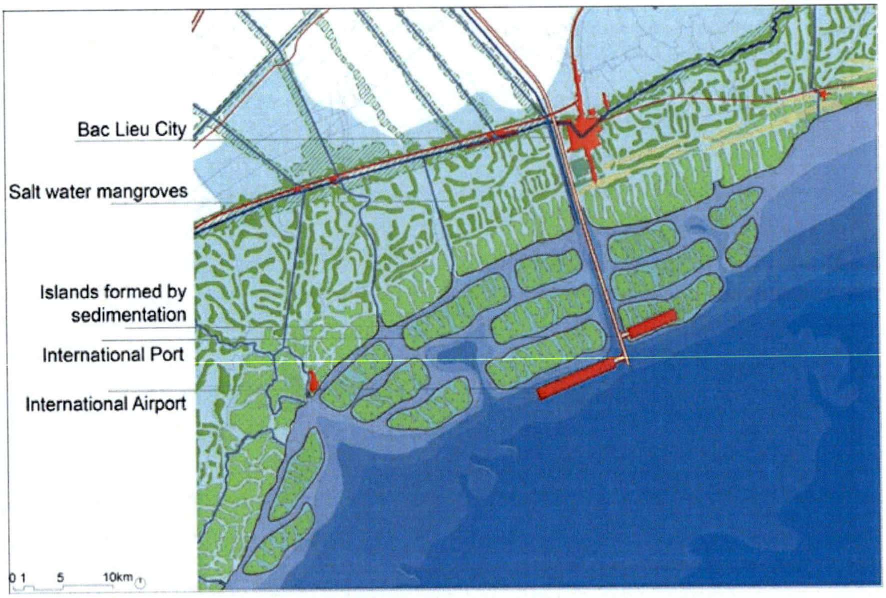

Fig. 11 Conceptual image of islands formed by accelerated sedimentation as proposed in the Mekong Delta Regional Plan. *Source* RUA (2016)

Table 2 Different types of benefits of land reclamation program

Benefits of land reclamation	Elements of the program
Coastal protection/ecological development	Coastal protective mangrove reforestation
	breakwaters
Productive landscapes amidst a constructed ecology Generation of sustainable and self-renewable resources	Fish and seafood farms
	Renewable energy farms
	Water-based settlements
	Infrastructure: fishery ports, sea ports

variables. As suggested by the analysis of the test programs with bamboo fences, mixed techniques of bamboo fences and concrete blocks might be indicated where currents are stronger. Bamboo fences and permeable textile nets, as used in fisheries, might be applicable in other conditions.

On first sight, the concept for the new form of land reclamation might appear farfetched, but it merely capitalizes on the natural processes that exist throughout the region. The Mekong Delta is one gigantic, unstable mud plain that oscillates between being ground and water, land and sea. There is no real or precise shoreline, but rather an entire territory that could be considered as one extended shoreline. The sea dyke that was previously planned is proposed to be replaced by a larger territory with a substantial, evergrowing depth. The heroic 'hold-the-line' paradigm of dykes is exchanged for a new landscape that, by almost replicating the Mekong Delta itself, works as an infrastructure. It provides protection from the sea and acts as a filter between salt, brackish and fresh water while supporting a wide range of coastal activities. A mosaic of appropriations anchor themselves in the natural conditions, ranging from new ecological shrimp farms and tidal energy installations to fisherman villages and other sustainable coastal settlements.

The proposed land reclamation program not only engages with the sediment dynamics along the East Sea coast, but also increases synergies by incorporating the large infrastructure of a new seaport (and possibly airport in further future) and existing infrastructure projects. The location, southeast of Bac Lieu is an area dominated by wide mud flats under the direct influence of the sediment load from the Mekong River. The project assembles a number of the Vietnamese government's major projects, including tidal and wind energy parks, an international deep seaport and the start of the TransAsian highway (De Meulder and Shannon 2016).

4.2 Technical Considerations

The proposed process can be compared with the well-known techniques of dune formation and reconstruction through the placing of windbreaks and land consolidation through carefully choreographed vegetation. The main technical issue of the land-gaining program is related to the design of the 'obstacles' (breakwaters); specifically, the distribution pattern that would need to be tailored in relation to dominant water current directions and their variations and the indicated openings and distances between the obstacles.

It is important that the distribution pattern, shapes and dimensions of breakwaters would be designed not only to accelerate sedimentation for creating land, but also to reduce erosion in the shorelines of Ca Mau Peninsula (prone to downdrift erosion). In this way, the offshore breakwaters (emerged or submerged depending on design solutions) can be both protective for the shoreline, by reducing erosive energy of waves (simultaneously stimulating sedimentation along shoreline like the case of GIZ project) and 'productive' for island creation, by trapping sediment on the shallow subaqueous platform. The engineering of these breakwaters will need to work with the hydrodynamics, sediment dynamics and morpho-dynamics on the muddy deposit bed (~15–20 m thick on topset, according to Liu et al. 2017a) of the subaqueous delta.

The optimization of the parameters of breakwaters would steer a design that can be adapted to the dynamics of the sea and desired land formation while simultaneously targeting desirable alterations between water and land, deep and shallow water, navigable areas and inaccessible natural domains, etc. These variations would create optimal conditions for a variety of habitats (e.g., shrimps, oysters and other seafood, fish, birds, plants, wood, etc.). Differentiations in the shape and depth of the breakwaters, distribution pattern and orientation and distances in between the breakwaters will generate a variety of water qualities and landscapes, including new waterways of various characters.

Inevitably, hard engineering and its associated costs are necessary for such a large-scale and long-term intervention. Its implementation would have to be calibrated to subsidence and sedimentation rates and the time necessary for land creation. Testing of models on site and experiments could set in motion an iterative operation and lead to perfection of the procedures of land creation. Further studies would need to be made with regard to the impacts on marine ecology and to the shoreline development further southwest and beyond the southernmost cape of the Mekong Delta. It would be necessary to identify trigger points to strategically position the 'obstacles' (breakwaters) to begin the process of enhancing sedimentation. Monitoring the progress of engaging with sediment dynamics in pilot areas would help to define the tactics for the next steps and respond to any changes in sediment- and morpho-dynamics, in light of the changing sediment budget in the Mekong Delta (as reported by Kondolf et al. 2018).

4.3 The Threat of Sediment Supply Deficit

The effectiveness of coastal protection and adaptation strategies engaging with sediment dynamics depends not only on innovative design and engineering solutions, but also demands a reliable available sediment budget. However, as previously mentioned, the Mekong Delta is increasingly threatened by declining sediment supply from the Mekong River, particularly due to the rapid pace of upstream dam construction. By April 2016, 35 dams had been commissioned for hydropower, irrigation reservoirs, drinking water supply, and further 226 dams are under construction with even more planned (WLE-Lower Mekong 2016; Allison et al. 2017). The construction of all the planned dams would greatly reduce sediment delivery to the delta, with estimated reductions ranging from 60 to 96%, once temporary sediment storage in the channel is exhausted (Kondolf et al. 2018). Zoccarato et al. (2018) predicted that a sediment reduction of 50–100% over the twenty-first century would result in a loss of elevation amounting to ~1.2–2 m (by the end of the century) of the delta's present coastline due to ongoing natural compaction of the Holocene sediment sequence; this high rate up to ~2 cm/year exceeds predicted values of absolute sea-level rise for the Mekong Delta of ~3.5 mm/year (Zoccarato et al. 2018). Therefore, sediment supply to the delta plain is a key factor for the survival of the Vietnamese Mekong Delta (Allison et al. 2017; Zoccarato et al. 2018).

Nittrouer et al. (2017) suggest that in the future an extreme impact of dam construction and water diversion is likely—with a consequence of near complete loss of water and sediment discharge in the Mekong Delta. In such a scenario, the delta and distal shorelines would become undernourished, and most likely mangrove forests would be eroded and destroyed if coastal processes (e.g., waves and currents) remained similar or became more energetic (Nittrouer et al. 2017). As such, serious deficit of sediment supply would negatively impact not only the sedimentation process of the delta (both subaerial and subaqueous parts), but also mangrove forests, which are expected to play important roles in coastal protection and adaptation to sea-level rise. According to Besset et al. (2019), mangroves cannot accomplish their land-building and coastal protection roles under conditions of a failing sediment supply and prevailing erosion. Comparing the width of the mangrove belt and shoreline changes in the Mekong Delta, Besset et al. (2019) find that the width of the energy-dissipating mangrove fringe alone does not play a determining role, either in the context of erosive oceanic forcing, or in the context of decreasing sediment supply to the delta. This reflects a tipping-point effect wherein once sediment supply to the coast is in chronic deficit (a deficit aggravated by delta-plain trapping to compensate for accelerated subsidence), the vertical growth of shorefront mudflats is no longer assured (Besset et al. 2019). This finding should be taken into careful consideration for the ecosystem-based protection strategies using mudflats and mangrove fringes.

4.4 Potential Application

In response to sea-level rise, the potential strategy of creating land using the dynamics of nature in the shallow sea can be applied in locations where sediment supply and sediment transport are abundantly available. Similar ideas have also been explored by some other researchers. In 2019, MIT's Self-Assembly Lab and Invena, a Maldivian organization, completed experiments in the Maldives for their Growing Islands Project (https://www.dezeen.com/2019/05/13/the-growing-islands-mit-self-assembly-lab). According to the Self-Assembly Lab, they are working 'to create a system of underwater structures that use wave energy to create sand accumulation in strategic locations; over time, the accumulation of sand will grow into new islands or help rebuild existing beaches, creating an adaptable solution to protect coastal communities from rising sea levels.' (Project website: https://selfassemblylab.mit.edu/growingislands). Scape, the landscape architecture office of Kate Orff, developed, in the slipstream of Hurricane Sandy and the Rebuild by Design program (New York), the concept and project of Living Breakwaters, where accumulation of mineral material generated over time by artificial oyster banks naturally grows breakwaters that, while enhancing sedimentation around them, simultaneously enhance beneficial ecological conditions (Orff 2016). Evidently, similar ideas and principles can be adopted to site specific designs that respond to various types of sediment, seabed topography, geomorphology, waves and currents, etc.

As previously mentioned, the proposed strategy would create both protective and productive landscapes amidst a constructed ecology in the interface between land and sea. Clearly, this can become a relevant approach in SeaCities and coastal developments. It could create synergies between protective and productive functions of ecosystem-based elements (e.g., mangroves, floodplains, seagrasses, dunes, sand bars, islands, marine habitat) and human-made elements (e.g., shoreline nourishment, breakwaters, artificial reefs, renewable energy farms, aquaculture farms, ports, multi-purpose floating platforms). Together, the natural and human ecologies can adapt to sea-level rise while creating opportunities with sustainable sea resources and their dynamics.

5 Conclusion

Since the majority of the world's population lives in deltas and along coasts, new conceptualization and frontiers of Sea Cities is necessary. Humankind must balance the compulsive need to control coastlines (as simplistic as unsuitable hold-the-line systems) and develop novel ways to engage in coastal dynamics, where coastlines are understood as extensive areas of overlap between land and sea. A genuine bracing of the dynamics, where the geography is constantly in the making allows for novel interplays of ecology, economy and occupation. New forms of landscape as infrastructure and settlement types must be developed within self-renewing ecologies that support a multitude of activities.

The case study of the Vietnamese Mekong Delta in this chapter presents an ecosystem-based approach in coastal protection and adaptation strategies which exploits sediment- and morpho-dynamics in response to sea-level rise. In a way, the approach advocates the enhancement of what nature does by itself, nothing more. Such novel response is to extend seaward the transition zone between land and sea using the dynamics of nature to grow land from sedimentation and mangrove processes, rather than the conventional 'hold-the-line' concept of coastal protection. This extended transition zone on the shallow subaqueous delta provides not only protective, but also productive opportunities for constructing ecology (mangroves), generating renewable resources (energy, aquaculture) and developing infrastructure (ports) and water-based settlements. The hybridization of programs and sectional richness is representative of 'polytechnicity' advocated by Lewis Mumford (Mumford 1934), where there is a merging of the technical, civic, economic and ecological performances. The new pilot project and revisioning of the Mekong Delta can be considered as relevant examples of the SeaCities 'expand' strategy.

References

Albers T, Dinh CS, Schmitt K (2013) Shoreline management guidelines: coastal protection in the lower Mekong Delta. Deutsche Gesellschaft für International Zusammenarbeit (GIZ) GmbH, Soc Trang, Vietnam. 124 pp, https://coastal-protection-mekongdelta.com/download/library/40.Shoreline-Management-Guidelines_EN.pdf. Accessed on 02.02.2020

Albers T, Schmitt K (2015) Dyke design, floodplain restoration and mangrove co-management as parts of an area coastal protection strategy for the mud coasts of the Mekong Delta. Vietnam Wetlands Ecol Manage 23:991–1004. https://doi.org/10.1007/s11273-015-9441-3

Allison MA, Nittrouer CA, Ogston AS, Mullarney JC, Nguyen TT (2017) Sedimentation and survival of the Mekong Delta: a case study of decreased sediment supply and accelerating rates of relative sea level rise. Oceanography 30(3):98–109. https://doi.org/10.5670/oceanog.2017.318

Anthony EJ, Brunier G, Besset M, Goichot M, Dussouillez P, Nguyen VL (2015) Linking rapid erosion of the Mekong River delta to human activities. Sci Rep 5:1–12. https://doi.org/10.1038/srep14745

Balica S, Dinh Q, Popescu I, Vo QT, Pham QD (2014) Flood impact in the Mekong Delta Vietnam. J Maps 10(2):257–268. https://doi.org/10.1080/17445647.2013.859636

Besset M, Gratiot N, Anthony EJ, Bouchette F, Goichot M, Marchesiello P (2019) Mangroves and shoreline erosion in the Mekong River delta, Viet Nam. Estuar Coast Shelf Sci 226:106263. https://doi.org/10.1016/j.ecss.2019.106263

Church JA, Clark PU, Cazenave A, Gregory JM, Jevrejeva S, Levermann A, Merrifield MA, Milne GA, Nerem RS, Nunn PD, Payne AJ, Pfeffer WT, Stammer D, Unnikrishnan AS (2013) Sea level change. In: Stocker TF, Qin D, Plattner G-K, Tignor M, Allen SK, Boschung J, Nauels A, Xia Y, Bex V, Midgley PM (eds) Climate change 2013: the physical science basis. Contribution of working group I to the fifth assessment report of the Intergovernmental Panel on Climate Change. Cambridge University Press, Cambridge and New York, pp 1137–1216. https://www.ipcc.ch/site/assets/uploads/2018/02/WG1AR5_Chapter13_FINAL.pdf. Accessed on 02.02.2020

De Meulder B, Shannon K (2016) Polysynthetic reclamation. The system, vol 47, pp 86–91. https://archis.org/publications/volume-47-the-system/

De Meulder B, Shannon K (2019) The Mekong Delta: a coastal Quagmire. In: Mossop E (ed) Sustainable coastal design and planning. CRC Press, Boca Raton, pp 293–314

Eidam EF, Nittrouer CA, Ogston AS, DeMaster DJ, Liu JP, Nguyen TT, Nguyen TN (2017) Dynamical controls on shallow clinoform geometry: Mekong Delta, Vietnam. Continental Shelf Res 147:165–181. https://doi.org/10.1016/j.csr.2017.06.001

GIZ (2016) Integrated Coastal Protection and Mangrove Belt Rehabilitation: Pre-feasibility study for investments in coastal protection along 480 kilometers in the Mekong Delta. Deutsche Gesellschaft für Internationale Zusammenarbeit (GIZ) GmbH. https://www.unique-landuse.de/images/publications/vereinheitlicht/Pre-feasibility-study-for-investment_coastal_EN.pdf. Accessed on 02.02.2020

GIZ (2018) A decision support tool for coastal protection for the Mekong Delta (CPMD). Deutsche Gesellschaft für Internationale Zusammenarbeit (GIZ) GmbH. https://coastal-protection-mekongdelta.com/download/library/Coastal%20Protection%20for%20the%20Mekong%20Delta%20(CPMD)_EN.pdf. Accessed on 02.02.2020

GSO (General Statistics Office) (2019) Statistical yearbook of Vietnam 2019. Hanoi. https://www.gso.gov.vn/default_en.aspx?tabid=515&idmid=5&ItemID=19691

Hein H, Hein B, Pohlmann T (2013) Recent sediment dynamics in the region of Mekong water influence. Glob Plan Change 110:183–194. https://doi.org/10.1016/j.gloplacha.2013.09.008

Kondolf GM, Schmitt RJP, Carling P, Darby S, Arias M, Bizzi S, Castellettic A, Cochrane TA, Gibson S, Kummu M, Oeurng C, Rubin Z, Wild T (2018) Changing sediment budget of the Mekong: cumulative threats and management strategies for a large river basin. Sci Total Environ 625:114–134. https://doi.org/10.1016/j.scitotenv.2017.11.361

Karlsrud K, Vangelsten BV, Frauenfelder R (2017) Subsidence and Shoreline Retreat in the Ca Mau Province—Vietnam. Causes, Conseq Mitig Options Geotech Eng J SEAGS AGSSEA 48(1): 26–32

Krauss KW, McKee KL, Lovelock CE, Cahoon DR, Saintilan N, Reef R, Chen L (2014) How mangrove forests adjust to rising sea level. New Phytol 202:19–34. https://doi.org/10.1111/nph.12605

Li X, Liu JP, Saito Y, Nguyen VL (2017) Recent evolution of the Mekong Delta and the impact of dams. Earth Sci Rev 175:1–17. https://doi.org/10.1016/j.earscirev.2017.10.008

Liu JP, DeMaster DJ, Nguyen TT, Yoshiki S, Nguyen VL, Ta TKO, Li X (2017a) Stratigraphic formation of the Mekong River Delta and its recent shoreline changes. Oceanography 30(3):72–83. https://doi.org/10.5670/oceanog.2017.316

Liu JP, DeMaster DJ, Nittrouer CA, Eidam EF, Nguyen TT (2017b) A seismic study of the Mekong subaqueous delta: Proximal versus distal accumulation. Cont Shelf Res 147:197–212. https://doi.org/10.1016/j.csr.2017.07.009

Manh NV, Dung NV, Hung NN, Merz B, Apel H (2014) Large-scale suspended sediment transport and sediment deposition in the Mekong Delta. Hydrol Earth Syst Sci 18:3033–3053. https://doi.org/10.5194/hess-18-3033-2014

Marchesiello P, Nguyen NM, Gratiot N, Loisel H, Anthony EJ, Dinh CS, Nguyen T, Almar R, Kestenare E (2019) Erosion of the coastal Mekong Delta: assessing natural against man induced processes. Cont Shelf Res 181:72–89. https://doi.org/10.1016/j.csr.2019.05.004

McIvor AL, Spencer T, Möller I, Spalding M (2013) The response of mangrove soil surface elevation to sea level rise. Nat Coast Prot Ser Rep 3. Cambridge Coastal Research Unit Working Paper 41. 59 pp. https://www.conservationgateway.org/ConservationPractices/Marine/crr/library/Documents/mangrove-surface-elevation-and-sea-level-rise.pdf. Accessed 02.02.2020

Milliman JD, Farnsworth KL (2011) River discharge to the coastal ocean: a global synthesis. Cambridge University Press, Cambridge, New York, 394 pp. https://doi.org/10.1017/CBO9780511781247

Milliman JD, Syvitski JP (1992) Geomorphic/ tectonic control of sediment discharge to the ocean: the importance of small mountainous rivers. J Geol 100:525–544. https://doi.org/10.1086/629606

Minderhoud PSJ, Erkens G, Pham VH, Bui VT, Erban L, Kooi H, Stouthamer E (2017) Impacts of 25 years of groundwater extraction on subsidence in the Mekong Delta, Vietnam. Environ Res Lett 12:064006. https://doi.org/10.1088/1748-9326/aa7146

Minderhoud PSJ, Coumou L, Erkens G, Middelkoop H, Stouthamer E (2019) Mekong Delta much lower than previously assumed in sea-level rise impact assessments. Nat Commun 10:3847. https://doi.org/10.1038/s41467-019-11602-1

Mumford L (1934) Technics and Civilization. Harcourt, Brace and Company, Inc., New York

Nittrouer CA, DeMaster DJ, Eidam EF, Nguyen TT, Liu JP, Ogston AS, Phung PV (2017) The Mekong continental shelf: the primary sink for deltaic sediment particles and their passengers. Oceanography 30(3):60–70. https://doi.org/10.5670/oceanog.2017.314

Nowacki D, Ogston AS, Nittrouer CA, Fricke AT, Pham DTV (2015) Sediment dynamics in the lower Mekong River: transition from tidal river to estuary. J Geophys Res 120:6363–6383. https://doi.org/10.1002/2015JC010754

Orff K (2016) Toward an Urban Ecology. The Monacelli Press, New York

Ogston AS, Allison MA, Mullarney JC, Nittrouer CA (2017) Sediment- and hydro-dynamics of the Mekong Delta: From tidal river to continental shelf. Cont Shelf Res 147:1–6. https://doi.org/10.1016/j.csr.2017.08.022

Phan LK, van Thiel de Vries JS, Stive MJ (2015) Coastal mangrove squeeze in the Mekong Delta. J Coast Res31(2): 233–243.https://doi.org/10.2112/JCOASTRES-D-14-00049.1

RUA (Research Urbanism Architecture) (2016) Planning concepts for revised Mekong Delta regional plan 2030 and vision 2050

Schmitt K, Albers T (2014) Area coastal protection and the use of bamboo breakwaters in the Mekong Delta. In: Nguyen DT, Takagi H, Esteban M (eds) Coastal disasters and climate change in Vietnam: engineering and planning perspectives. Elsevier Insights, Amsterdam, pp 107–132. https://doi.org/10.1016/B978-0-12-800007-6.00005-8

SISP (Southern Institute for Spatial Planning under Vietnamese Ministry of Construction) and RUA (Research Urbanism Architecture) (2017) Comprehensive report of revised Mekong Delta regional plan 2030 and vision 2050

SIWRP (Southern Institute of Water and Resource Planning under Vietnamese Ministry of Agriculture and Rural Development) (2012) Master plan of water resources for impacts of climate change—sea level rise in the Mekong Delta in the period 2012–2020 and orientation to 2050

Tamura T, Nguyen VL, Ta TKO, Bateman MD, Gugliotta M, Anthony EJ, Nakashima R, Saito Y (2020) Long-term sediment decline causes ongoing shrinkage of the Mekong megadelta, Vietnam. Sci Rep 10:8085. https://doi.org/10.1038/s41598-020-64630-z

Temmerman S, Meire P, Bouma TJ, Herman PM, Ysebaert T, De Vriend HJ (2013) Ecosystem-based coastal defence in the face of global change. Nature 504: 79–83. https://doi.org/10.1038/nature12859

Thanh VQ, Reyns J, Wackerman C, Eidam EF, Roelvink D (2017) Modelling suspended sediment dynamics on the subaqueous delta of the Mekong River. Cont Shelf Res. https://doi.org/10.1016/j.csr.2017.07.013

Unverricht D, Nguyen TC, Heinrich C, Szczuciński W, Lahajnar N, Stattegger K (2014) Suspended sediment dynamics during the inter-monsoon season in the subaqueous Mekong Delta and adjacent shelf, southern Vietnam. J Asian Earth Sci 79:509–519. https://doi.org/10.1016/j.jseaes.2012.10.008

WLE–Greater Mekong (Water, Land, and Ecosystems–Greater Mekong) (2016) Dams in the Mekong River Basin: commissioned, under construction and planned dams in April 2016. Vientiane, CGIAR Research Program on Water, Land and Ecosystems–Greater Mekong. 1-page map sheet. https://wle-mekong.cgiar.org/wp-content/uploads/Mekong_A0_2016_Final.pdf. Accessed on 02.02.2020

Xue Z, Liu JP, DeMaster D, Nguyen LV, Ta TKO (2010) Late Holocene evolution of the Mekong Subaqueous Delta, Southern Vietnam. Mar Geol 269:46–60. https://doi.org/10.1016/j.margeo.2009.12.005

Xue Z, He R, Liu JP, Warner JC (2012) Modeling transport and deposition of the Mekong River sediment. Cont Shelf Res 37:66–78. https://doi.org/10.1016/j.csr.2012.02.010

Xue Z, Liu JP, DeMaster D, Leithold EL, Wan S, Ge Q, Nguyen VL, Ta TKO (2014) Sedimentary processes on the Mekong subaqueous delta: clay mineral and geochemical analysis. J Asian Earth Sci 79:520–528. https://doi.org/10.1016/j.jseaes.2012.07.012

Zoccarato C, Minderhoud PSJ, Teatini P (2018) The role of sedimentation and natural compaction in a prograding delta: insights from the mega Mekong Delta, Vietnam. Sci Rep 8:11437. https://doi.org/10.1038/s41598-018-29734-7

Online News

Aouff RS (2019) MIT's Self-Assembly Lab proposes new way of growing islands and coastlines. Dezeen https://www.dezeen.com/2019/05/13/the-growing-islands-mit-self-assembly-lab/. Accessed on 22.02.2020

Self-Assembly Lab: Growing Islands. https://selfassemblylab.mit.edu/growingislands. Accessed on 22.02.2020

Thang Viet Nguyen is a Ph.D. researcher at SeaCities, Cities Research Institute, Griffith University in Queensland, Australia. He previously has extensive experience in the field of regional and urban planning in Vietnam, ranging from regional planning for the Mekong Delta region, Ho Chi Minh City region, city planning of different scales to master planning for urban development projects. In these plans, he contributed key roles as team leader, principal planner, technical manager. He had strong collaboration with international consultants in many planning and research projects in Vietnam. The Mekong Delta Regional Plan 2030 and Vision 2050, in which he was a local team leader in cooperation with RUA, was awarded the Vietnamese National Planning Prize in 2018. Urbanization policies, sustainability, climate change adaptation, water urbanism and community participation are among his professional interests.

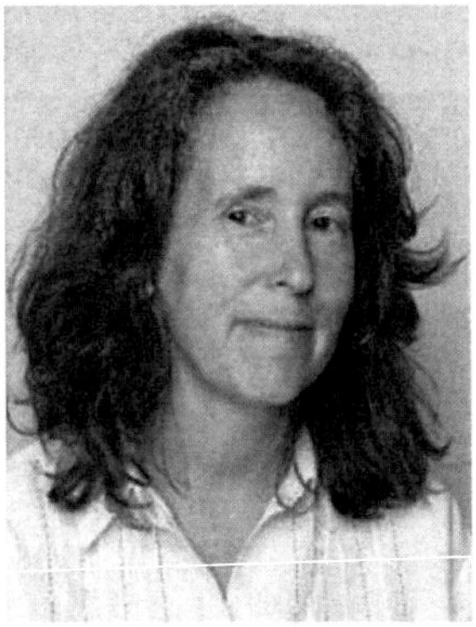

Kelly Shannon is a Professor of Urbanism at KU Leuven. Her research is at the intersection of analysis, mapping and new cartographies and design. Most of her work focuses on the evolving relation of landscape, infrastructure and urbanization in Asia. A particular interested focuses development of landscape urbanism strategies that work with water and topography. Together with Bruno De Meulder, she has worked extensively in the Vietnamese context, developing various masterplans for cities and regions all which accentuate locational assets and landscapes to frame urbanization. She is presently the director of the Master of Human Settlements and Master of Urbanism, Landscape and Planning programs.

Bruno De Meulder is a Professor of Urbanism at KU Leuven. His teaching and research travels between various practices of urban design and theory and history of urbanism. His work continuously combines design with research in his own post-industrial European context and within a number of post-colonial situations that are characterized by fast and dynamic development. Together with Kelly Shannon, he has worked extensively in the Vietnamese context, developing various masterplans for cities and regions all which accentuate locational assets and landscapes to frame urbanization. He is presently the vice head of the Department of Architecture.

Water as Leverage: Design Studies for Khulna, Chennai and Semarang

Anne Loes Nillesen, Mona zum Felde, Eva Pfannes, Han Meyer, and Olv Klijn

Abstract Deltas are attractive locations for settlements due to their favourable conditions for food production and trade, which historically led to many deltas being urbanised. The urbanisation, polderisation and industrialisation of many deltas increases their vulnerability to climate change. The 'Water as Leverage' program aims to formulate innovative, bankable and implementable design proposals for urban water related challenges. This paper presents three design studies for the cities of Khulna, Chennai and Semarang, which are part of the 'Water as Leverage for Resilient Cities Asia' program, and seeks to address the complex water management and urban challenges with system-based design proposals. The three projects show how several water management measures on different scales can be combined in integral design proposals that, in addition to addressing climate change, also improve the urban quality and liveability of the cities.

1 Introduction

Challenges related to water management become more urgent as the number of people living in vulnerable areas is rising. This will only increase in the future with rising sea levels, a changing climate and continuous urbanisation processes. Experts from different fields around the world are addressing different challenges related

A. L. Nillesen · M. zum Felde (✉)
Defacto Urbanism, Rotterdam, The Netherlands
e-mail: mona@d.efac.to; mzfelde@gmail.com

A. L. Nillesen
e-mail: anneloes@d.efac.to

E. Pfannes
OOZE Architects & Urbanists, Rotterdam, The Netherlands

H. Meyer
Delft University of Technology, Delft, The Netherlands

O. Klijn
FABRICations, Amsterdam, The Netherlands

J. Baumeister et al. (eds.), *SeaCities*, Cities Research Series,
https://doi.org/10.1007/978-981-15-8748-1_6

to urbanisation, climate change and water management in an attempt to improve safety, liveability and resilience in vulnerable urbanised areas. While addressing these complex issues it becomes more and more apparent, that sectorial approaches are not sufficient to respond to the complexity of these environments. As a response, integral frameworks and solutions were developed with the aim to deal with this complexity.

This paper presents three case studies from Asia, which use an integral system-based design approach to understand and address various water management challenges in the cities of Khulna, Chennai and Semarang. A comparison of these three design proposals, which is presented in Sect. 4, lays an emphasis on a more balanced water cycle by addressing multiple water management challenges simultaneously. To achieve this, nature-based solutions (NBS) are proposed as (part of) the interventions; they do not only address water management but also provide additional benefits to the cities and their inhabitants.

The following Sect. 2 introduces water management challenges in coastal areas around the world and introduces the 'Water as Leverage for Resilient Cities Asia' program, which is the umbrella program that the three case studies are part of. In Section 3 the approaches, main issues and the proposed concepts and interventions are described separately for each of the three case studies. The last Sect. 4 concludes with main findings from the three case studies.

2 Water Management Challenges in Urbanised Areas

2.1 *Coastal Towns and the Delta Paradigm*

Floods occur in different frequencies in many parts of the world (Fig. 1). This becomes an issue when floods occur in areas where land use functions are vulnerable to flood exposure and insufficient land is available to accommodate the water without causing damage. Especially vulnerable are urbanised areas which are frequently subject to flooding, because of the great damage they can cause and the lives that can be at risk (Revi et al. 2014). Areas in which a high population density and a high frequency of flood events coincide the most, are often delta areas.

Deltas have always been attractive places for settlements because of the favourable conditions they have due to their location at the edge between the sea and inland fresh water sources. The fertile floodplains and high nutrition loads in the sea that have diversity in food resources and strategic locations for trade (Meyer and Nijhuis 2014) have been ideal to provide for communities and over time resulted in high density delta areas. According to the UN-Habitat State of the World's Cities of 2006, around 50 percent of the world's urbanised areas are located in deltas (UN-Habitat 2006). However, due to their specific characteristics such as proximity to the sea and rivers, and the rapid urbanisation and cultivation of the often flat tidal plains, delta areas also face very characteristic challenges (Meyer and Nijhuis 2014). Among

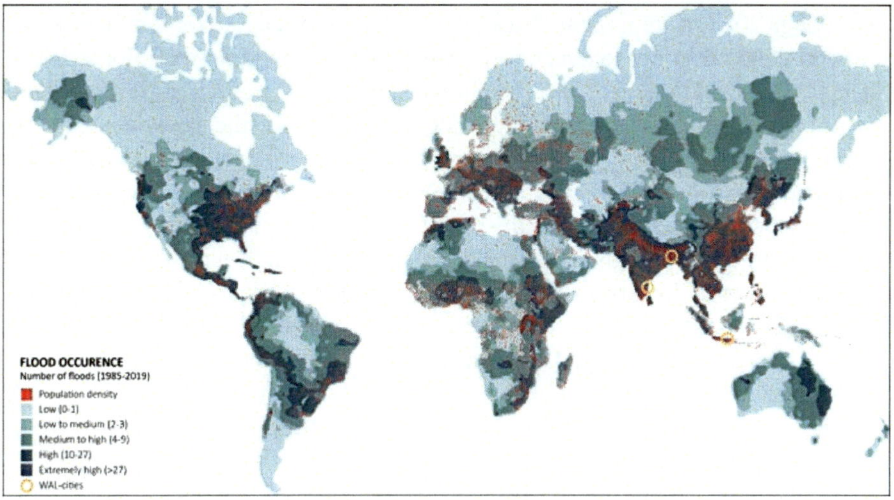

Fig. 1 Worldwide flood occurrence and population density, showing that many densely populated areas are positioned in places with a high flood occurrence (image: Nillesen 2019, data source flood occurrence: World Resources Institute (WRI) Aqueduct, data source population density: Population explorer 2017)

some typical challenges are for instance: unsustainable urban growth, flood risk, subsidence, salinisation, freshwater shortage, cyclones, lack of sedimentation and erosion (Kothuis et al. 2015; Meyer and Nijhuis 2014; Water as Leverage 2018). Many of these challenges will be further increased by climate change (Seneviratne et al. 2012).

These specific areas, in which many diverse challenges related to flood risk management and (rapid) urbanisation come together, call for integral approaches in order to protect and improve people's lives and livelihoods. As risks are increasing and urban challenges are rising, it is necessary to find sustainable solutions to ensure resilient urban environments in the future.

2.2 *The 'Water as Leverage for Resilient Cities Asia' Project*

Over the years, The Netherlands has built a strong reputation with integral design approaches for flood risk management challenges. These approaches were used both nationally and internationally in order to both respond to (reactive) and prepare for (proactive) flood risk challenges. Examples of reactive programs are for instance the Dutch Dialogues in New Orleans after Hurricane Katrina (Meyer et al. 2009), Houston after the flooding caused by 2008 Hurricane Ike and 2017 Hurricane Harvey, and the New York rebuild by design challenge after Hurricane Sandy (Ovink and

Boeijenga 2018; Kothuis et al. 2015). The Netherlands Delta Programme was established in 2007 in order to develop strategies that address long-term flood risk challenges (Delta Committee 2008) and contains important research by design examples of pro-active strategy development, as does the 2017 Resilience by design competition for the San Francisco Bay area. The 'Water as Leverage' (WAL) initiative continues this tradition of proactive integral design strategies (Water as Leverage 2018) for water management with special emphasis on including bankability as an integral part of the design (Netherlands Enterprise Agency, n.d.).

The 'Water as Leverage for Resilient Cities Asia' (WAL) is an initiative of the Special Envoy for International Water Affairs of the Kingdom of The Netherlands, Henk Ovink, and the Netherlands Enterprise Agency, that aims to address "urban water-related challenges in an innovative and inclusive way" (Netherlands Enterprise Agency 2020). As part of this program, multidisciplinary teams were selected and asked to formulate innovative, bankable and implementable design proposals for the cities of Khulna (Bangladesh), Semarang (Indonesia) and Chennai (India).

In no other part of the world do flood risk and urban challenges come together as drastically as they do in Asia, especially in the coastal regions. For the selection of the three cities as case studies, different aspects have been taken into consideration. Based on an analysis on water, climate and urban related issues, three cities have been identified as hotspots for transformative opportunities. Besides their individual challenges and opportunities, the cities also have the potential to function as a pilot, so that solutions may be reproduced in other cities throughout Asia which are facing similar challenges. Before finalising the selection of Khulna, Semarang and Chennai as case studies, coalitions have been formed with local and international financial, governmental and non-governmental institutions and organisations. This, to ensure a continuous dialogue, innovative financing solutions and a sense of ownership for local partners that would guide the projects from the beginning to the end (Water as Leverage 2019).

For each city, two multi-disciplinary teams with different compositions were selected. All teams followed a workshop-based process, organised and facilitated by the WAL core team, with support of Partners for Resilience (PfR) and 100 Resilient Cities. The workshop process included a series of three local workshops (organised in the towns of Khulna, Semarang and Chennai) that brought together the multidisciplinary experts and local stakeholders, as well as two regional workshops where the results of the different design teams were shared. The progress was monitored by an international review board and the design teams and stakeholders were further introduced to potential financiers (Netherlands Enterprise Agency 2020).

2.3 *Water and Climate Related Challenges in the Three WAL-Cities*

As an introduction to the work of the design teams, a ‘Call for Action’ (Water as Leverage 2018) document was prepared, summarising the different water and climate related challenges of the three cities based on an initial analysis. The teams would later deepen these systems analyses, connecting the water management and flood risk challenges to urban and economic development and growth phenomena. The scheme below is extended based on results of the team’s analyses and shows the range of water and climate related extreme events and long-term effects present in the three cities (Fig. 2).

The scheme shows the variety of extreme events the three cities are exposed to. These, together with the long-term effects of which most have started to become visible already, make the cities extremely vulnerable to social and economic disruption, damages and fatalities (Water as Leverage 2018). Different urban challenges make it difficult to recover from extreme events, and climate change is likely to exacerbate intensity and occurrence of extreme events and the extent of long-term effects. Without concepts that address water-related challenges as a whole and relate it to the urban context, even daily life will be dominated by these issues.

As indicated in the scheme, not all of the identified challenges are addressed in the design proposals. Based on the team’s analyses of the water systems, the different short- and long-term challenges were prioritised based on their probability and exposure, and therefore their potential to reduce the vulnerability and disruption of the residents’ daily lives. In Semarang for example, the drought issue is one of the prioritised challenges, because water scarcity increases extraction of groundwater, which causes the city to sink and directly relates back to an increase in flood risk.

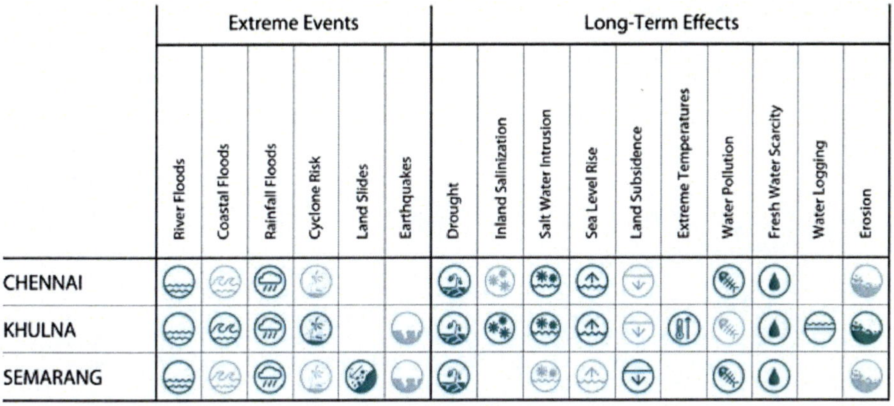

Fig. 2 Diagram showing the different water management challenges per city, distinguishing the ones addressed by the design teams (dark blue) and the ones not addressed (light blue). (Image: based on a diagram by Architecture Workroom Brussels (Water as Leverage 2018), extended by Defacto Urbanism)

3 Water as Leverage (WAL) Design Projects

Despite the differences in size, location and therefore also in external factors of the three cities, they all face a similar range of water management challenges. However, their different settings (Fig. 3) also reflect a range of local challenges, which were identified by the different teams.

Based on the systems analysis, the teams developed their own integral design strategy. This section describes the projects (the position of the city in the delta, the team composition, the main challenges addressed and the resulting design strategy and proposed interventions) developed for Khulna, Chennai and Semarang.

3.1 *Khulna as a Water Inclusive Enclave*

The city of Khulna (Fig. 4) is located in the tidal floodplain of the Ganges-Brahmaputra Delta (Nishat 2019) in the south-west of Bangladesh. With its location on the edge of the coastal zone in a flat, low-lying terrain, Khulna is one of the most vulnerable cities to climate change and sea level rise (Roy et al. 2018). As for many other delta cities the question is, how can Khulna adapt to these changing conditions while accommodating future urban growth?

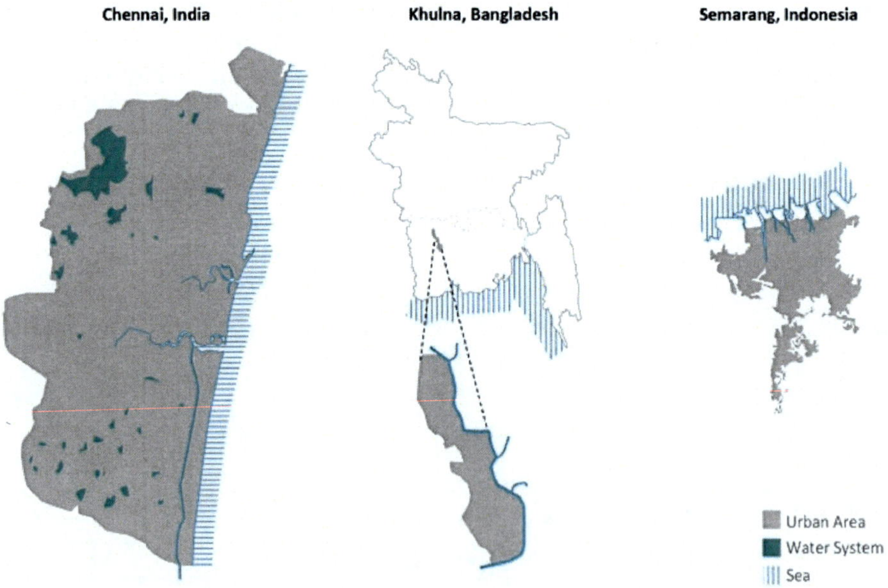

Fig. 3 A comparison of the three WAL cities (image: Architecture Workroom Brussels (Water as Leverage 2018: 85), edited by Defacto Urbanism)

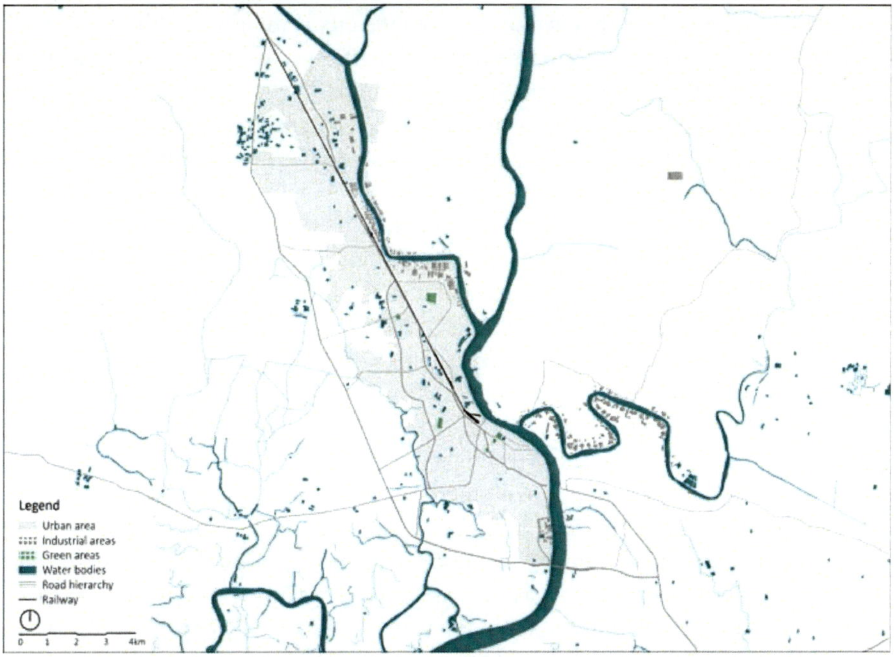

Fig. 4 The map of Khulna shows main water bodies, infrastructure, and the urban area of the city (image: Defacto Urbanism)

3.1.1 Multidisciplinary Team

The multidisciplinary team consists of different parties with different roles within the project. Defacto Urbanism was the design lead of the project coordinating the research and design; CDR-International, Royal HaskoningDHV and Nelen & Schuurmans were responsible for the hydraulic analysis and technical aspects of the proposal, for which Nelen & Schuurmans built a (storm)water-model for the city of Khulna. Dev Consultants and Khulna University provided local knowledge and connections and together with Royal HaskoningDHV performed the stakeholder analysis. Wageningen University & Research provided knowledge with regard to agriculture practices and transformations.

3.1.2 Challenges Addressed with the Project

Khulna is facing many challenges related to the local, regional and (inter-)national scale. This project explores how the water management system of Khulna and the connection between the water and the city's inhabitants can be restored and extended to face current and future flood risk and urbanisation challenges.

In the past, rainwater was stored in ponds within the city of Khulna. However, over time the urbanisation (densification), polderisation and industrialisation transformed the water into a pollutant that is drained from the city (Prakash and Singh 2013). We see a water management development in which the surplus of water is drained, and freshwater is supplied through pipelines and pumps.

The Ganges and Gorai river basin used to be a balanced and dynamic system, in which the river naturally moved eastward (Nishat 2019). Over time, human interventions have put this natural system out of balance, resulting in different challenges in the three zones of the tidal floodplain along the Gorai (see Fig. 5). The upstream freshwater zone is suffering from drought. The transition zone faces issues regarding polderisation, subsidence, pollution, salinisation and (urban) drainage issues. The downstream coastal zone is subject to salinisation, erosion, pollution and deforestation (Winterwerp and Giardino 2012).

The city of Khulna is located in the transition zone and while it is currently still relatively safe from storm surges and floods, the city already experiences shortages in freshwater availability and as a result an increase in salinity of surface water. Fresh water for drinking and agriculture is supplied by pipeline from the Modhumoti river and pumped from deep aquifers. With rising sea levels, more extreme rainfall (ADB 2011) and a continuous increase in salinity, Khulna will gradually become more exposed. At the same time, the city is experiencing extreme urban growth (Ahmed et al. 2014), putting extra pressure on the water system.

Different developments like land subsidence (ADB 2011), urban flooding, salinisation and pollution of surface water, exhaustion of ground water (ADB 2011; Prakash and Singh 2013) and unsustainable urbanisation have already created a downward spiral with negative interplay between these developments. Without a new approach to these challenges, the already unbalanced system will be stressed further, aggravating the trend of migration, urban flooding, poverty, and a loss of crops, lives and ecology.

During this research-by-design project, a system-based approach is used. This means that the analysis focusses on the understanding of functional relations of the water system at different scale levels and between the water system and other thematic (such as urbanisation, agriculture, and ecology). The goal is to address integral water related challenges with system interventions that restore or improve the water system, instead of only addressing symptoms resulting from the dysfunctional water system. To gain understanding of the water-related challenges and complex systems of the city and region, an analysis is performed including desk studies, field trips, workshops with experts and stakeholders, design explorations and hydraulic modelling. The process of analysis, interpretation and propositions is repeated in a circular design and analysis process while zooming in from the regional scale to the city of Khulna.

3.1.3 Developed Vision/Strategy

The system-based approach for Khulna to become a water inclusive city in the delta is based on the restoration of the natural system and the prevention of a lock-in situation in the future, which will be the consequence of the continuation of the applied high

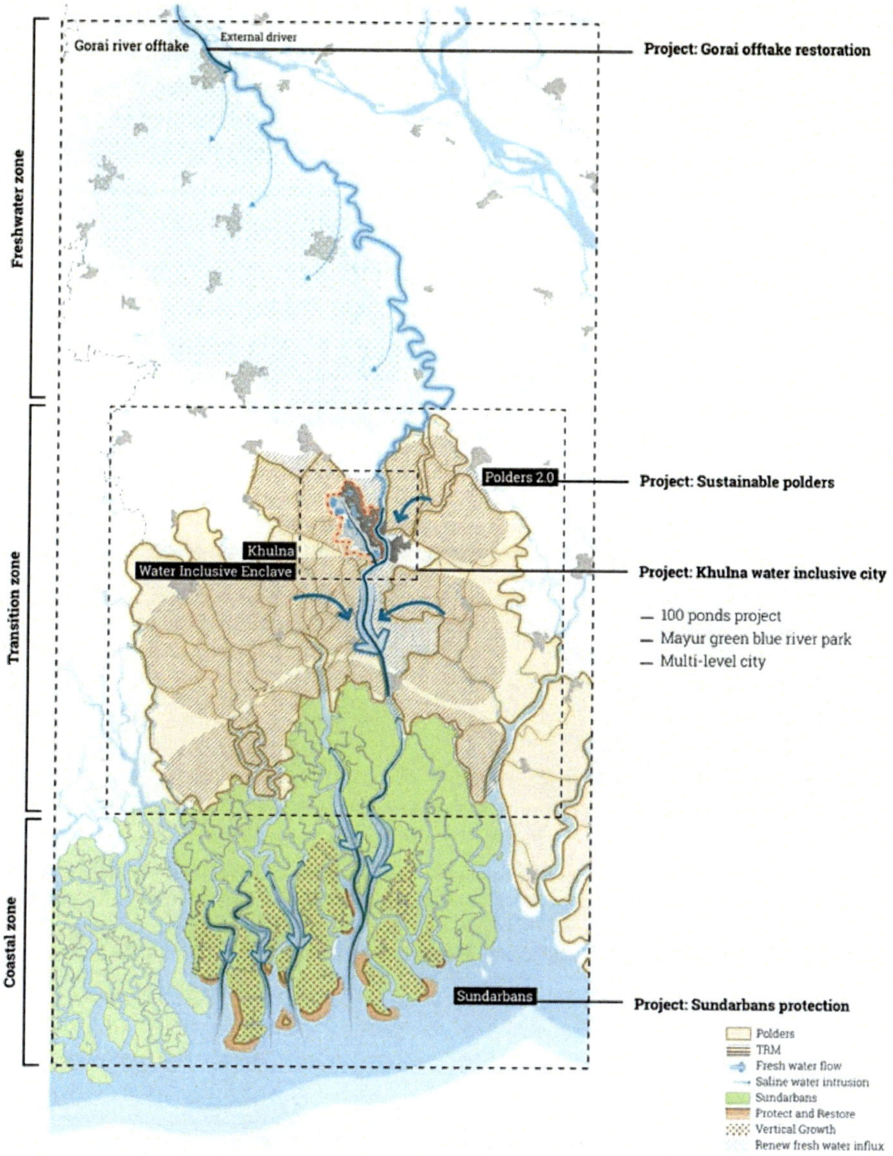

Fig. 5 Different projects are interlinked on the regional scale and combined contribute to the main strategy (image: Defacto Urbanism)

maintenance water management structures (such as polders). This approach has been translated into two main strategies:

(1) The regional strategy focuses on restoring the natural system between the Gorai river offtake and the Sundarbans with different strategies tailored to the needs of the three zones (see Fig. 5).

(2) The strategy for the city of Khulna is based on the 'store, buffer and drain' concept (van Dijk et al. 2002) and consists of different interventions in different parts of the city, that together form a cohesive strategy for urban water management (Fig. 6). The interventions have been selected based on their performance towards water management (as evaluated with a hydraulic model developed based on rainfall predictions and elevation data) as well as their feasibility within the Bangladesh context and their qualitative contribution to the city.

Both strategies make use of NBS. Although these take time to be implemented, they will restore the balance of the tidal floodplain and increase the natural capacities, resulting in a more sustainable future for the area. The proposed regional scale strategy and interventions are mainly based on combining best practices and existing research outputs. They provide context regarding the type of regional interventions (and supporting policy directions) needed to support the proposed integral Khulna strategy. The strategy for Khulna derives from the research by design

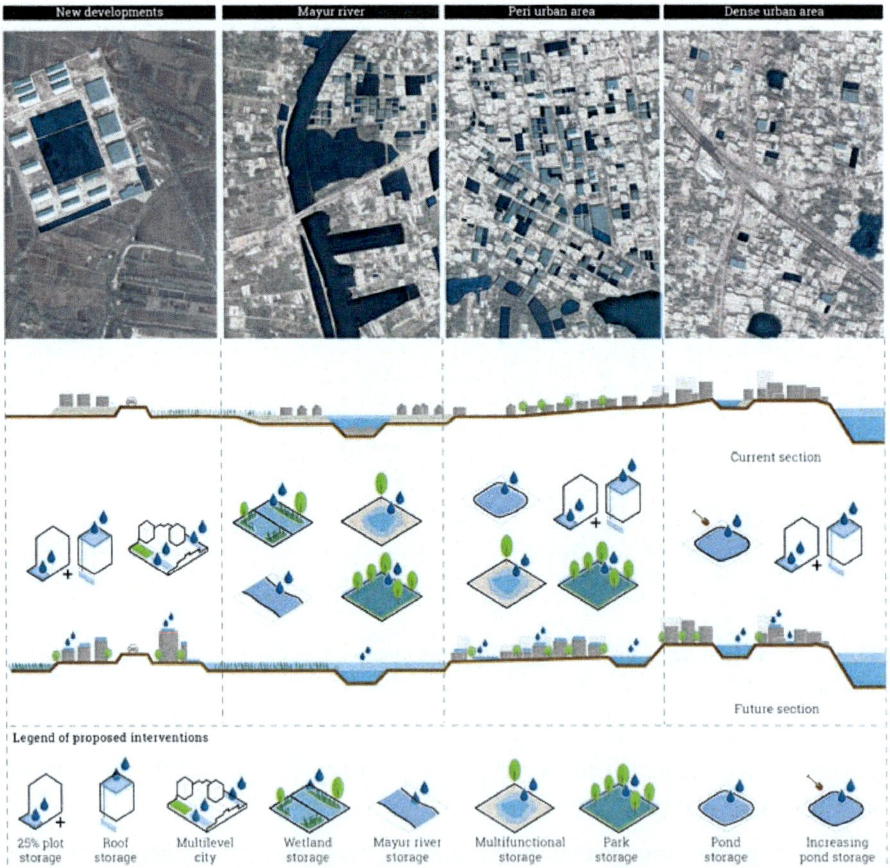

Fig. 6 Diagram showing water storage principles proposed for different parts of the city (image: Defacto Urbanism)

approach performed as part of this study, and it is a first design proposition that will have to be elaborated further in a subsequent pre-feasibility study.

3.1.4 Interventions

Gorai river offtake restoration project (upstream of Khulna)

The interventions start with the upstream Gorai river offtake restoration, that will reconnect the Gorai to the Padma river (Hore et al. 2013). Interventions upstream of Khulna aim at ensuring adequate water flow in the Rupsha River during the dry season. This will not only help to reduce the tidal influence, but also prevent further increase of salinity intrusion in soil and groundwater and reverse the riverbed siltation and elevation process. Freshwater flowing from the Gorai River is currently reduced, due to continuous siltation at the offtake area between the Ganges and the Gorai River. In addition to this regional scale intervention, international cooperation between Bangladesh and India is required to regulate Ganges water flows on a national level (GED Bangladesh 2018).

Sustainable polders (transition zone)

In the transition zone, the effects of climate change, waterlogging and salinisation are causing a decrease in potential crop yields, possible crop types, ecological functions, liveability, and protection against floods. There is an urgent need for measures that address all these issues in a sustainable way to safeguard food security. Tidal River Management (TRM) is a building with nature strategy that can be used effectively to elevate the polders (Seijger et al. 2019), using the sediments transported by the river, indirectly benefitting agriculture, ecology, and climate resilience in the Khulna District. TRM can have social, ecological, and economic potential if executed deliberately with attention for customisation to the right place and time, with the right stakeholders involved, and managed and monitored appropriately. Since the applicability of TRM is explored in a parallel study (Strategic Delta Planning research project (2014–2020) (IHE Delft 2020)), the specific application has not been elaborated in this study.

Sundarbans (coastal zone)

The Sundarbans are not only an internationally acknowledged and important Ramsar site (Nishat 2019), but are also important for local communities and bring economic benefits to the District of Khulna. There are many developments which pose threats to the sustainability of the Sundarbans. The main drivers are the increasing salinity levels in the northern part of the Sundarbans, as a result of lack of freshwater flow and rising sea levels, and extensive deforestation. Although the tidal flats may keep pace with sea level rise, the lateral erosion threatens the outer tidal flats and islands in the Sundarbans. This loss of tidal flats has a detrimental effect on the ecology of the system (Mukhopadhyay et al. 2018).

The intervention proposed for the Sundarbans aims to enhance natural siltation by influencing the local hydraulic conditions, using poles and brushwood to dampen the waves and capture sediment between the barriers and the land (Wetlands International, n.d.). Once the near shore bed level rises enough, mangroves can regenerate naturally, developing a natural defence that will protect the hinterland from further erosion and strengthen its function of diminishing cyclone impacts. The project should include a surveying and monitoring program, as well as awareness training, to ensure that construction of natural barriers will promote mud-flat sedimentation. Simultaneously, a buffer zone (with coinciding awareness programmes) between the agriculture polders and mangrove forest protects the forest from the north. The aim of the training and awareness program is to reduce fertiliser use in the buffer zone and reduce illegal forest cutting.

Water inclusive Khulna city

For the city of Khulna, the urban water management strategy aims to both restore the historical capacity and sponge function of the city and the relation of the residents with water, as well as to extend the water management capacity, using the 'store, buffer and drain' concept as the guiding principle. The proposed three main interventions for implementation, corresponding to different spatial typologies within the city (see Fig. 6), together form an integral framework for urban resilience.

The interventions include the re-use of rainwater (rooftop water can be stored and filtered for household use, pond storage for watering parks or agriculture use), the creation of temporary and permanent water storage areas and new regulations for future urban expansion to ensure a water inclusive urban development. The projects are urgent; if no action is taken soon, the city will continue its unsustainable growth and expensive countermeasures, that are difficult to implement, will be necessary to achieve the same impact.

Ponds Project—100 × 1 Urban Acupuncture

For the dense old town of the city, urban acupuncture, using severalmany small scale interventions throughout the city to transform the city (Lerner 2014), is proposed to increase the water storage capacity. Due to the lack of space, interventions on large areas are not possible, however the multiplicity of small interventions can achieve the same result (Fig. 7). Water tanks as well as the restoration, protection and excavation of urban ponds are the main part of this project, which aims to bring back a characteristic feature of Khulna. This will need to happen through building regulations, such as land use restrictions and reverting the current low land prices for ponds, and the transformation and demarcation of ponds into public spaces (to support maintenance). To kick-start this process, the '100 ponds' project is proposed, aiming to restore and excavate a series of 100 ponds. Depending on the position of the pond, the bottom can be prepared for maximal infiltration or water storage. For the ponds that store water, water quality and the prevention of mosquito breeding are important attention points. Water that cannot be stored within the old town will be drained towards the fringes of the city.

Fig. 7 The image illustrates how urban acupuncture increases the water storing capacity in the existing dense urban fabric, by deepening existing ponds, adding new ponds in public or private open areas and incorporating roof storage into new buildings (image: Defacto Urbanism)

Mayur river restoration

Through the suburbs on the fringes of the city flows the Mayur river. The river is used for drainage for the city, however it is diminishing rapidly due to encroachments formed by land reclamations for urbanisation. It will transform into a narrow culvert (Khulna University 2015; Prakash and Singh 2013) without protection from these developments. Therefore, this project focuses on the protection and extension of the Mayur rivers drainage capacity.

With strategic investments, land use planning (restrictions indicated by the Khulna Development Authority) and the new public function as a linear park (to support implementation), a strategic area can be protected from urbanisation processes to function as a buffer zone for stormwater (Fig. 8). In the future, the Mayur will be able to store rainwater falling in the suburbs and stormwater which is drained from the old town up to a 1:50-year event. In more extreme events, the Mayur can drain naturally into the Gorai river, as the Gorai offtake restoration has lowered the riverbed and restored the drainage capacity of the Mayur towards the Gorai.

Besides the function for urban drainage, the Mayur is used for bathing and fishing by low income residents. During drought periods, the stored water can be used for irrigation of agricultural areas. Since the river is also running centrally through the predicted city outline of 2040, which is expected to extend westward rapidly (Khulna City Corporation projection map; Ahmed et al. 2014), it can function as a green-blue spine in the future, providing a place for recreation and heat mitigation.

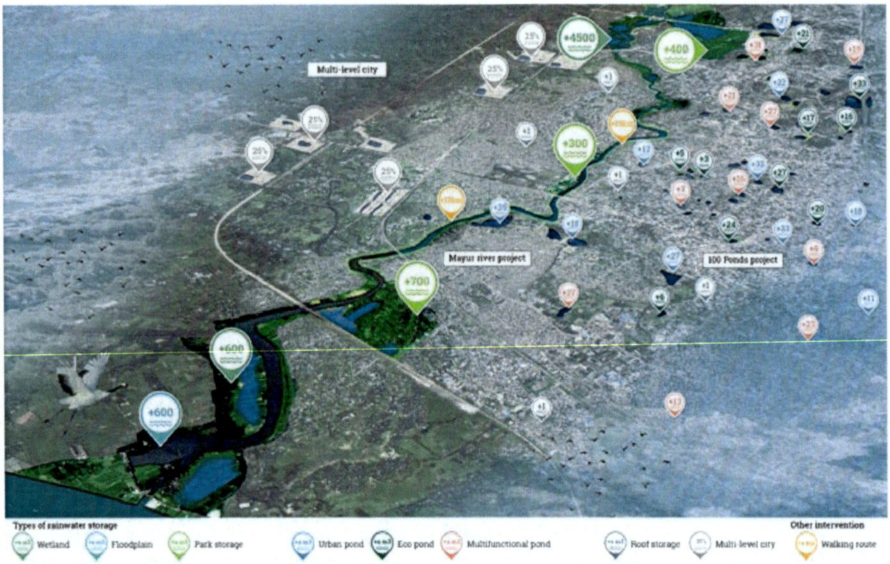

Fig. 8 The bird eye shows the different interventions that are proposed in different parts of the city to increase the water retention capacity. In the existing urban fabric east of the Mayur, urban acupuncture can be applied to create many small-scale water storages that together have a substantial retention capacity. The Mayur river restoration can increase the water storage capacity to also be able to store the runoff from the existing city. The Multi-level city can be applied west of the Mayur where the urban expansion will take place, and can be implemented at the same pace as urban developments. (Image: Defacto Urbanism)

Multi-level city—step by step resilience

The west side of the city is currently dominated by agricultural land, but the city is continuously expanding into this area. For these new developments, new building regulations are being proposed, which combine compact urban settlements with water storing landscapes in the multi-level city concept (Fig. 9).

This project aims to store, buffer and reuse rainwater even during extreme events of up to 1:100-year events, without draining water into the already pressured urban water management system. Due to the plot-based concept, the water capacity can grow at the same pace as the urbanisation of Khulna. For individual developments, a water-plot ratio should apply, prescribing the water ratio that needs to be stored per individual plot. Projects with more than one building require an integral plan to ensure that all built up areas can temporarily store their own rainwater. The water storage capacity will have to be circa 25% of the storage area for frequent events (stored in gardens and public spaces) and 40% for extreme events (stored in parking garages, agriculture lands and gardens). To achieve these storage capacities while maintaining critical functions, the latter are assessed on their vulnerability and vitality and are elevated according to a maximum flood frequency.

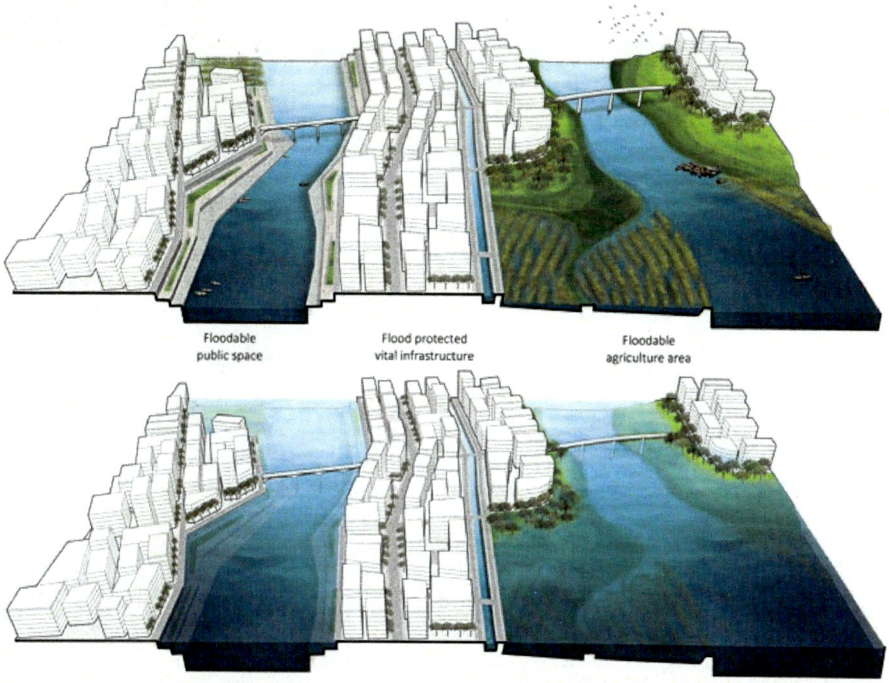

Fig. 9 Impression of the multi-level city with normal and extreme water levels (image: Defacto Urbanism)

4 City of 1000 Tanks, Chennai

With 53% growth in population over the last two decades, Chennai (Fig. 10) is expected to host 14 million people by 2035 from ~10 million today. While the population is increasing, water availability is decreasing. Maximum temperatures in Chennai could increase by 2 °C by 2040, while rainfall (currently 1389 mm per year) is projected to decrease by 14% (CCC&AR and TNSCCC 2015). However, it is also anticipated that the city will experience more intense cloud bursts which can lead to flood disasters.

4.1 Multidisciplinary Team

The City of 1000 Tanks team includes OOZE Architects & Urbanists (team-lead, overall integrated water & engagement strategy & designs), Delft University of Technology (TU Delft) (regional planning), Indian Institute of Technology (IIT) Madras (engineering), IHE Delft Institute for Water Education and Dr. Verena Balz (governance), Care Earth Trust (local governance and ecology), Madras Terrace (local project management), Paperman Foundation (solid waste management and

Fig. 10 The map of Chennai shows main water bodies, infrastructure, and the urban area of the city (image: Defacto Urbanism)

PR), Goethe Institute (cultural engagement), Rain Centre (rain water consultant), Biomatrix Water (wastewater & NBS engineering), HKV (flood risk assessment) and Information and Resource Centre for the Deprived Urban Communities (IRCDUC) (engagement vulnerable populations) among others.

4.2 Challenges and Opportunities

As Chennai is located along the coast, groundwater extraction due to drought has led to saline intrusion. This will be further exacerbated as sea level rise facilitates saltwater moving further inland and filling a dry aquifer. Chennai risks running out of water in the next decade, given projected increases in population and the depletion of the water table. With decreasing rainfall yields and increasing water demand, Chennai has begun to invest in desalination plants and distribution channels. Desalination exponentially increases the cost of water production and is carbon intensive. Not only does desalination involve large capital investments but it also negatively impacts sensitive coastal ecosystems by increasing salinity due to brine disposal. It further impacts marine life and in turn the livelihoods dependant on it. Rather than solving the water problem, desalination only exacerbates the problem by putting an extra load of wastewater on the city that requires treatment.

As part of this project, an analysis of produced sewage per capita against the city's available sewage treatment infrastructure revealed that 50% of the city's sewage is untreated and expelled to natural water bodies and storm drains, eventually polluting the shallow aquifers and making water unsafe for consumption (Fig. 11). In order to be treated by the existing centralised sewage treatment plants, all sewage must first be pumped through nearly 3000 km of sewer mains, often uphill with many unreliable pumps that fail. Chennai's relatively flat topography makes the system inefficient and prone to leaks and other mishaps along the way. Existing stormwater management systems with drainage pipes expel most of the rainwater to the sea, rather than taking advantage of it for water sourcing. Overwhelmed by the amount of rainfall during monsoons and clogged by silt and sewage, these systems have failed to protect citizens during floods.

Historically, Chennai has relied on its temple tanks (and traditional water bodies called 'erys') for measuring the groundwater table while regulating its supply between seasons (Ganesan 2008). With rapid urbanisation leading to the building over essential water bodies, the dynamic and infrastructural role of temple tanks has been lost and have become ornamental rather than functional.

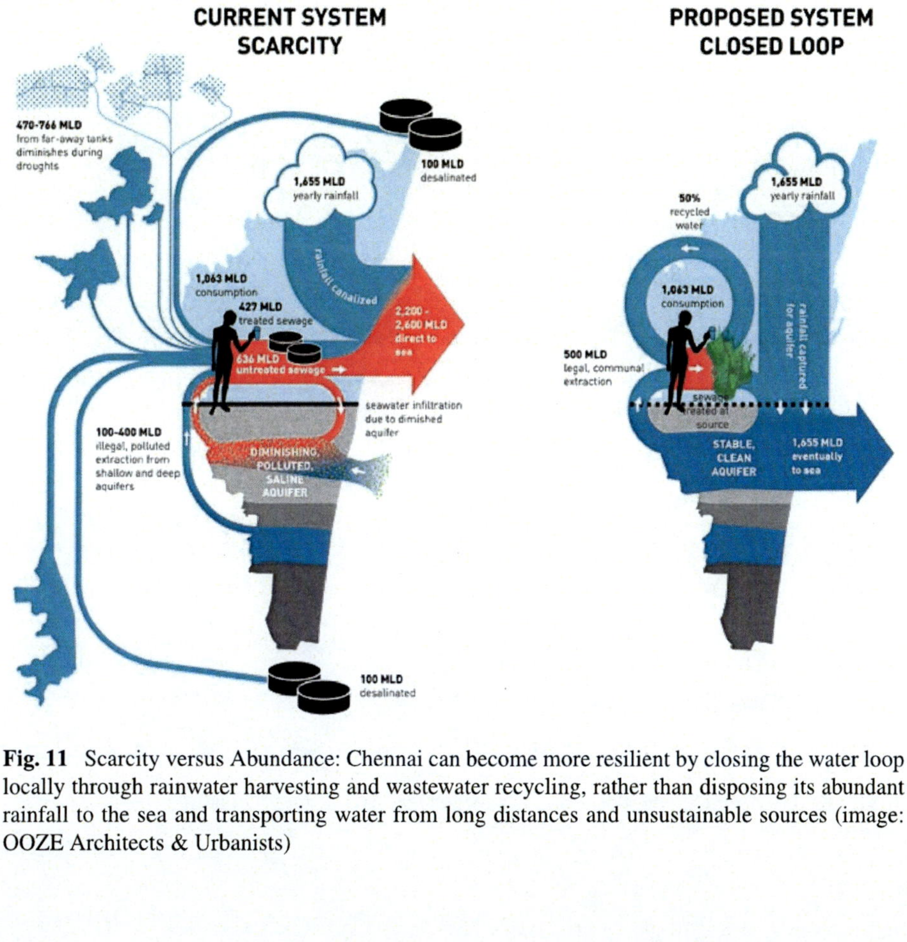

Fig. 11 Scarcity versus Abundance: Chennai can become more resilient by closing the water loop locally through rainwater harvesting and wastewater recycling, rather than disposing its abundant rainfall to the sea and transporting water from long distances and unsustainable sources (image: OOZE Architects & Urbanists)

This intensive research was carried out through the means of desk studies, historical analyses, analysis of policy documents and handwritten records, field visits and documentation of other water-related projects in Tamil Nadu. Open source data, data from research institutes and studies, Google terrain and GIS has been used for mapping. The most important sources of information were interactions with key actors and experts through working sessions, workshops and interviews.

4.3 Developed Vision/Strategy

City of 1000 Tanks is a comprehensive strategy to achieve a city-wide, sustainable water management model in Chennai (Fig. 12). Future proofing and managing water as a resource means going back to and rewiring natural dynamics inside the city. At the heart of this strategy sits the newly developed 'Water Balance Model', integrating and linking all different flows of Chennai's hydrological system. The entire water cycle—from rainfall and consumption to wastewater treatment and groundwater recharge—is part of an interconnected parametric equation, that informs how to make the most efficient and economical use of the available water resources locally. This can make the city more resilient to climate change induced droughts by increasing

Fig. 12 Chennai: City of 1000 Tanks (image: OOZE Architects & Urbanists)

groundwater reserves and prevent saline intrusion from sea-level rise. Simultaneously, it can mitigate risks associated with high-frequency floods as well as sewage pollution.

Designing with Nature-Based Solutions

Water security is intended to be achieved by collecting rainwater, treating wastewater and runoff pollution with decentralised NBS and recharging both to the underground aquifer for later extraction (Fig. 13). Putting nature to work, plant-based, low-cost technologies such as constructed wetlands, bioswales and detention parks are integrated into the urban context. NBS have often light, even ephemeral designs, that minimise the use of "heavy" construction approaches, potentially involving steel and concrete. The light construction results in low embodied energy and carbon. It uses ubiquitous materials that can be sourced locally, which delivers low material-kilometres. It has shorter construction timelines and has opportunities for prefabrication. Some systems can be dismantled and redeployed readily to a new location. NBS have high aesthetic values, as green structures are doing the work currently typically done by grey structures. Public acceptance is high in urban environments, as NBS are using natural materials and growing ecologies in a benign approach. Longer retention times than conventional technology enables the treatment of "exotic" pollutants, such

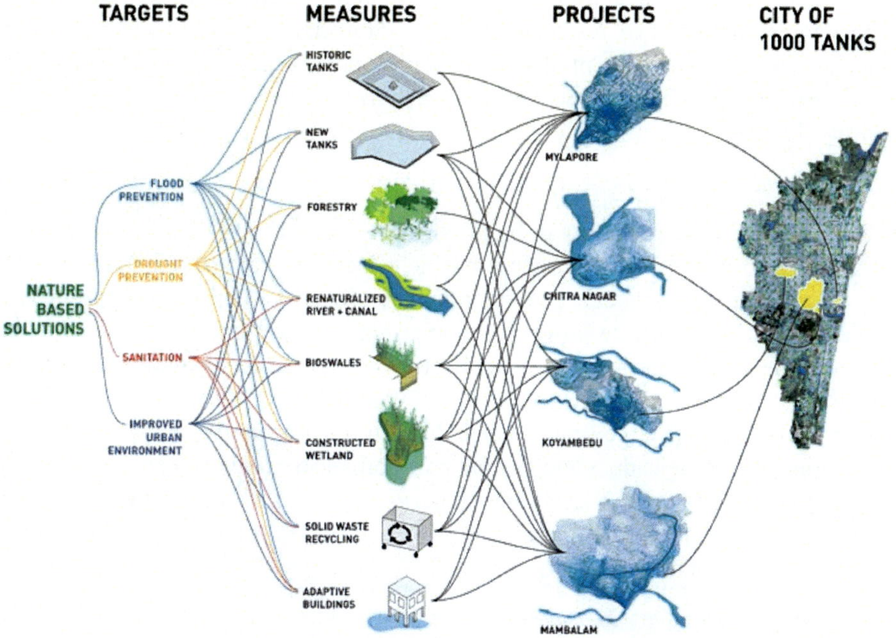

Fig. 13 Programmatic strategy for the integration of NBS measures in the city's different watersheds (image: OOZE Architects & Urbanists)

as phenols, antibiotics, endocrine disruptors and some heavy metals (phytoremediation). They are ensuring no drop of water goes to waste, thus improving productivity in water sourcing, reducing consumer water costs, economic losses caused by droughts and floods, and the urban heat island effect; whilst increasing biodiversity and sequestering and reducing carbon emissions, NBS provide an alternative water source to high energy consuming desalination plants.

Spatial integration of NBS in the urban context of Chennai

The warm climate in India with constant temperatures above 20 °C is especially favourable for the processes in constructed wetlands (CWs). The cleaning process is accelerated and the size of the CWs can be reduced compared to colder climates. It is estimated that ca. 0.6–2 m^2 of CWs are needed per inhabitant, dependent on the chosen engineering system (i.e. with/without aeration, with/without baffle tanks). In this way the space used for NBS can be optimised and integrated as a granular system within the existing dense urban fabric on roofs, private open spaces and specified public spaces. Existing roads, pavements and parking spaces can be transformed to accommodate NBS infrastructure. The project can also encourage property owners to invest in NBS solutions using 5% of their land mandated as Open Space Requirement (OSR) in return for "water credits" as a rebate against the property's monthly water bill. In cities such as Chennai, which receive the majority of rainwater over only 2 months a year and have a rising demand in potable water, NBS urgently need to be implemented, tested and monitored on a larger scale, in order to research their climate adaptive potential and competitive advantage over conventional infrastructure with high energy expenditures, such as desalination plants, that negate (or instantly compromise) parallel climate mitigation efforts.

Programmatic and incremental approach

To persuade all stakeholders to build NBS, the City of 1000 Tanks programme is structured to build effective, efficient and inclusive cooperation in water management. It does this through strategic coordination between an overarching, long-term, large-scale vision and concrete projects on the ground. Incremental development—from field trials to flagship projects to replication across the city through an integrated regional vision—ensures that solutions are both tailored to local communities and capable of creating large-scale change. Field trials or prototypes act as test beds to experiment with new ideas and explore how both the thinking and practical elements can be upscaled and applied to our larger scale projects and strategies. These are tools that can be used to communicate, provoke discussion and bring together diverse stakeholders around critical topics.

Financing innovative water management

With a long-term loan from an international financial institution, the Government of Tamil Nadu would be able to build NBS on public properties. In parallel, private properties can implement their own NBS with private sector financing, with the incentive of water bill rebates for aquifer recharge. In order to efficiently and effectively monitor water table levels and individual recharge contributions, remote monitors need to be installed.

CMWWSB or 'Metrowater', the local public water supply department, can increase its revenue through increased water sales. Part of Metrowater's increased revenue stream can pay for contracted local NBS operators to manage different areas of the city. Another part could go towards incentivising private property recharge. The measures can also lead to savings for other government departments through reduced public property damage risk, reduced health costs, and various other budgetary expenses. Over time the government could also receive additional tax revenue from benefits that affect society at large.

4.4 Interventions

An integrated regional vision, four flagship projects and a field trail have been developed to initiate an institutional shift from purely engineering-based solutions towards a sustainable groundwater management approach, by including the soundness of environmental and societal systems. Through an incremental approach, trust is built at the local level to ensure ownership and appropriation by the people. This can contribute to the crosscutting areas of governance and sustainability. The pilot project demonstrates the various NBS techniques and community engagement strategies, the findings of which can be later implemented in the flagship projects. The four flagship projects represent four urban types, namely Heritage (historic tank areas), Disaster Resilient Housing (disaster prone low-income housing areas), Green Industry (logistics and industrial hubs) and Smart Waterways (canals, rivers and channels). Under these types, they are scaled up to similar areas and watersheds across the city. This scaling up and working together of all projects to achieve the overall goal of a water balanced region is guided by the regional vision.

Regional vision: A Metropolitan Landscape structure

In order to solve the problems with flooding in an adequate way, it is necessary to consider the city of Chennai in its regional context. The high water events in the city are not only caused by rain storms in the city itself, but largely also by stormwater from the hills and lowlands around the city, running through the city which finds itself at the lowest level of the region. The runoff of stormwater from the hills and farmlands has increased during the last decades because of the reduction of water retention, due to deforestation, drainage and canalisation, and petrification of large amounts of the regional territory.

This process can be expected to get worse in the next decades, when the population is expected to increase from 10 million today to 14 million in 2035, resulting in an explosive urbanisation of the periphery and farmlands around the city. The task at the regional scale is the substantial increase of the capacity in the hills and farmland for the absorption and retention of stormwater, in combination with a controlled process of urbanisation. This goal can be reached by the layout of a metropolitan landscape structure, composed as a regional green-blue framework (Fig. 14). It consists of NBS elements to collect and treat runoff before discharge, combining water retention basins with recreational areas. This approach builds on experiences in comparable

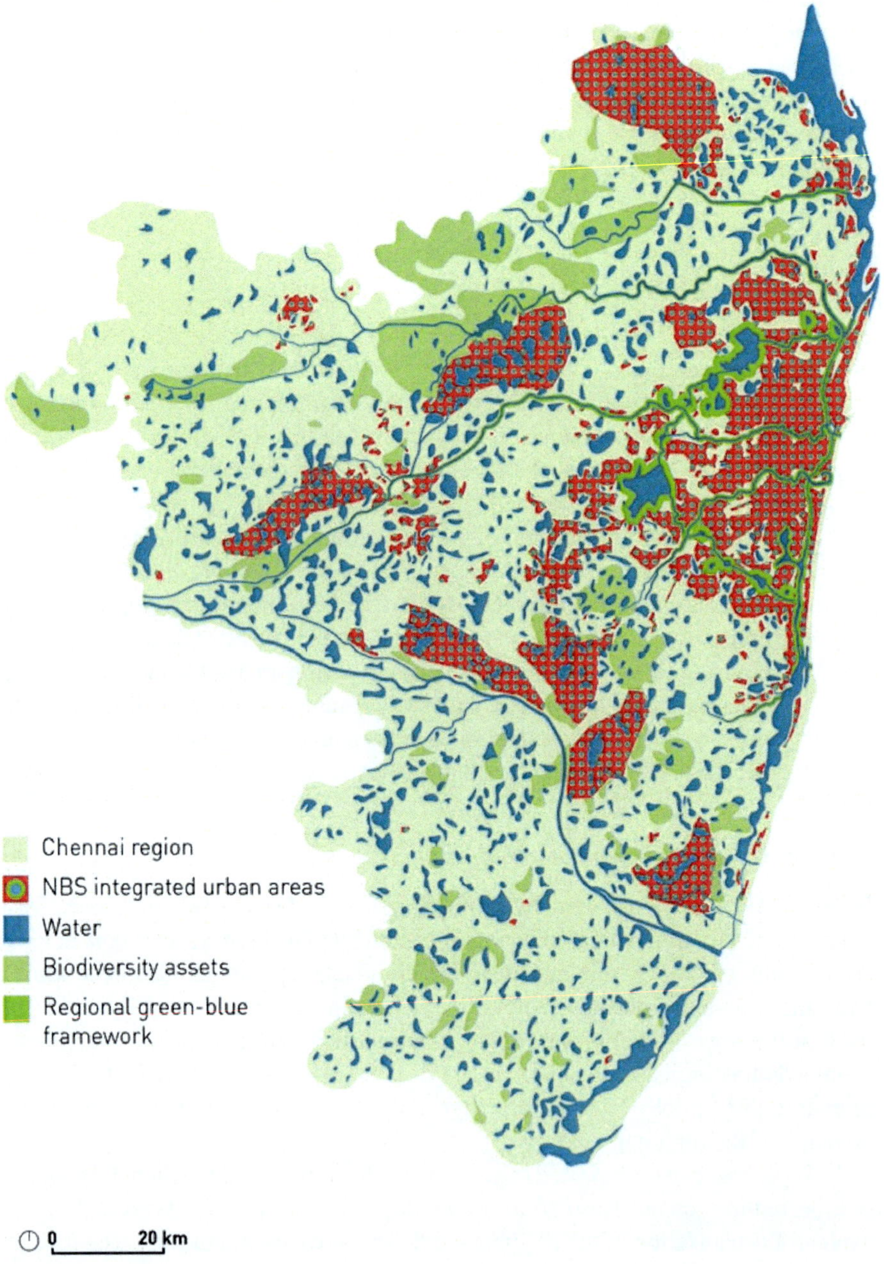

Fig. 14 Regional strategy for urban development and green-blue structures (image: OOZE Architects & Urbanists)

situations, such as the famous late nineteenth-century projects 'Emerald Necklace' for Boston by F.L. Olmsted (Schuyler 1986) and the 'Water project' for Rotterdam by W.N. Rose (Hooimeijer 2014), or more recently the Greater New Orleans Urban Water Plan (Waggonner 2013).

The composition of the green-blue framework of Metropolitan Chennai is based on a series of radial long lines from the surrounding hills towards the city, and a circular belt of parks and waters around the city, which is able to collect the stormwater from the countryside and to discharge it into the sea via the north and south flanks of the city.

Flagship Project—Mylapore, the ancient smart city

Mylapore was originally designed as a smart city with a resilient water infrastructure capable of retaining water in the monsoons for use in the dry season. It is one of the oldest parts of the city with its origins dating back to the First Century BCE. Modern development patterns in this historic core of Chennai have erased the functionality and knowledge of its temple tanks. "Mylapore Trail" seeks to restore and enhance Chennai's important water heritage, making it a proven resilient model for the rest of the city and as a sustainable tourist attraction (Fig. 15). Two historic temple tanks that no longer function are to be restored by collective efforts in increasing groundwater recharge, which eventually restore water levels in the tanks. Public spaces are designed with NBS as new dynamic tanks that treat and recharge rainwater

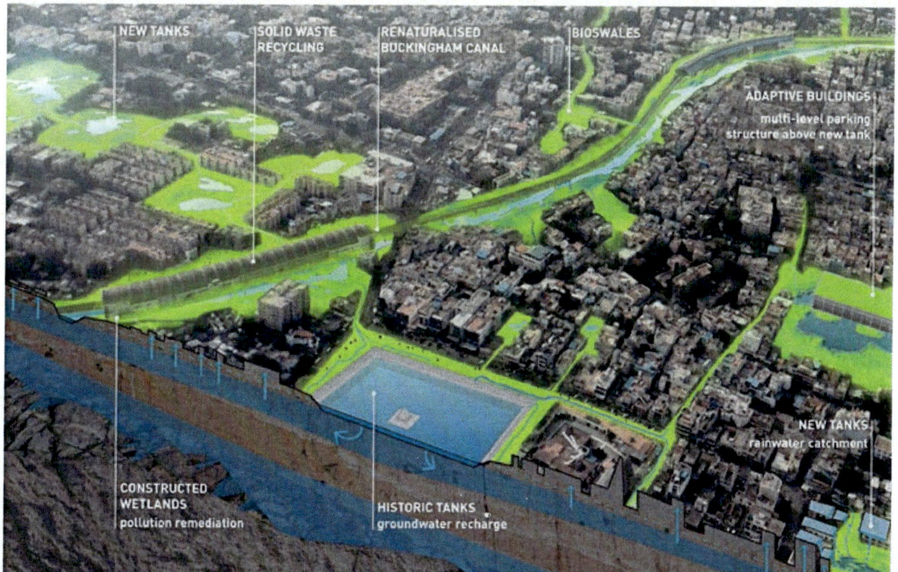

Fig. 15 Sectional view of the Mylapore Flagship project showing various recharge efforts through existing and new areas integrated with NBS (image: OOZE Architects & Urbanists)

runoff, while private properties, both residential and commercial, treat and recharge greywater through decentralised NBS. An integrated network of bioswales connects the new recharge tanks to make a resilient and highly visible system that increases awareness, highlights historic areas and improves the overall attractiveness of the neighbourhood for the benefit of tourism, commerce and liveability. The project has strong backing from the local elected representative, residents' groups and local institutions and can benefit 61,000 residents.

Field trial—School pilot as live classroom for NBS demonstration

The field trial at the PS Higher Secondary School playground and the adjacent HR&CE Temple grounds demonstrate the use of NBS in public spaces. New retention tanks that collect rainwater runoff and treated wastewater, recharging both to the aquifer to increase groundwater levels, are connected to an educational and community program for awareness and capacity building. The field trial contains three sub-projects: The first consists of the intervention on the school playground in which constructed wetlands treat wastewater from 1200 students and multiple bioswales that direct this treated water together with stormwater into designated detention and recharge zones. The second consists of an eco-literacy centre, which integrates learning by doing and interactive demonstrations of the NBS on the school playground. The third is the intervention on the adjacent temple grounds, which is redesigned as a retention park with model Erys and a flower garden sustained through recharge efforts. The entire project is accompanied by a dedicated engagement program, including public water festivals, cultural events, training programmes for technocrats and citizens, information sessions for the vulnerable and at risk, eco-literacy/water education program and social media campaign. The field trial can build capacity on the local level, empowering citizens with the knowledge to adapt and adopt NBS on a city level. The PS Higher Secondary School pilot is proposed as a working model that creates a 'live classroom' to understand and interact with the water cycle and can be developed into a city-wide or nation-wide school program.

4.5 Cascading Semarang

The island of Java is part of the pacific fire belt, a series of volcanic mountains surrounding the Pacific Ocean. About 90% of the world's earthquakes and 81% of the world's largest earthquakes occur along this belt, putting Java Island at high risk in terms of natural disasters like earthquakes and tsunamis.

Semarang (Fig. 16) is the biggest city in Central Java. Besides being part of one of the largest urban agglomerations in Indonesia, the city has the 5th largest metropolitan area and is the capital of Central Java province. In recent decades it has seen dynamic population growth—from 1 million to 1.8 million inhabitants in 20 years. This expansion not only resulted in a strong economic development and a substantial growth in wealth, but it also amplified a number of issues related to water.

Fig. 16 The map of Semarang shows main water bodies, infrastructure, and the urban area of the city (image: Defacto Urbanism)

Semarang is situated in a tropical, maritime climate zone with an average temperature of 26.7 degree Celsius and an annual precipitation of 2182 mm. The mount Ungaran, located 20 km south of the city centre of Semarang, is part of a huge mountain ridge. This contributes to a strong topography around Semarang, with a steep slope from the hilly inland to the coastal lowlands. Thanks to these conditions, the natural landscape was crisscrossed by a fine network of streams accumulating into rivers, that carried the water to the sea, before urbanisation processes took place. The soil generally consists of alluvial and fluvial clay in the coastal plains and along the rivers, Mediterranean soil in central Semarang and volcanic rocks and tuff towards the mountains. These soils have a limited infiltration capacity and cause difficult drainage conditions for the city.

In Semarang's urban history, the construction of an elaborate system of drainage canals was completed to cope with these conditions. The biggest historical canal is the Kanal Semarang, constructed by the Dutch colonial occupation. Currently the two biggest canals are the Banjir Kanal Timur situated in the east and the Banjir Kanal Barat in the west of the old City Centre. The municipality covers nine catchment areas, which discharge approximately 999 mm/m^2 rainfall per year into the Java Sea.

4.5.1 Multidisciplinary Team

The Cascading Semarang team consists of FABRICations and MLA + (core team and design lead), Deltares (water and business case development lead), Witteveen + Bos (water and business case development), Diponegoro University (UNDIP) (community development, water and stakeholder involvement), UNISSULA—Universitas Islam Sultan Agung (community development and water) and the Liveable Cities platform (community development).

4.5.2 Challenges Addressed: Rising Sea Levels, Sinking City and Drinking Water Scarcity

Continuous groundwater extraction for multiple purposes causes land subsidence in Semarang. The coastal areas sink with a rate recorded up to 17 cm per year (Murdohardono and Tobing 2001a, b; Abidin et al. 2004). In the kampungs (Indonesian village like settlements) located in the northeast area of Semarang, streets are raised every few years, severely impacting daily life, as it has widely been reported by local stakeholders. At the same time, Semarang is experiencing an increase in surface runoff due to the increase of sealed areas as the city expands. This regularly leads to flash flood.

Simultaneously, the city continues to grow into the hilly hinterland, occupying more and more territory with increased landslide risks. Additionally, these expansions also reduce the water entry into the different aquifer levels beneath the city. Drought is affecting Semarang due to a clear seasonal shift in rainfall. Insufficient water supply infrastructure has led to the proliferation of wells built by the industry, but also for commercial and domestic usage. PDAM (the drinking water company) only provides 60% of the city with clean water. Lack of available water also results in problems in the agricultural sector, affecting the livelihood of local farmers.

In relation to these issues, Semarang is facing a number of self-reinforcing mechanisms or vicious cycles that drive a process of exponential growth of vulnerability of the city as it develops. Firstly, population growth and urban expansion cause the expansion of the urban tissue and an increase in paved and built area upstream of Semarang, which decreases the infiltration rate, causing higher discharge peaks during extreme precipitation events. A higher frequency of floods generates more attractiveness to moving upstream, reinforcing the initial reduction in infiltration rates.

Secondly, an increase in flood frequency increases the Climate Value at Risk for the private (damage of assets and foregone revenues during business interruptions) as well as for the public sector (damage to public assets such as infrastructure networks). The impact on the private sector erodes the tax base of Semarang city. A lower tax base reduces the availability of resources for public spending on water infrastructure development and maintenance; impacting negatively the capacity of the city to deal with extreme events. An increased Value at Risk of public assets leads to higher expected damages. Less resources for maintenance in combination with higher expected damages leads to deteriorated maintenance levels of water infrastructure over time, reinforcing the initially increased Value at Risk of public and private assets through higher flood risk.

Thirdly, socio-economic activities in Semarang are heavily reliant on groundwater resources. Current abstraction rates are unsustainable, causing aquifer depletion and land subsidence. Land subsidence increases pluvial and coastal flood risk due to a relatively larger low-lying coastal/downstream area over time. Increased flood risk has so far led to more canalization of the city for a faster discharge of water in the ocean; negatively affecting the availability of surface water in the watershed and the infiltration of surface water, both contributing to an increase in the rate of depletion of groundwater resources.

The way the city is expanding, and the way citizens of Semarang still prefer groundwater over surface water, is leading to an exponentially increasing rate of subsidence, and together with the increasing frequency of extreme events driven by climate change, it is resulting in more frequent and damaging floods.

4.5.3 Developed Vision/Strategy

The Cascading Semarang team decided to focus on rainwater capturing and storing, thus managing and mitigating the environmental impact since rainwater has a huge potential in solving many of the water challenges of Semarang. Through a "Theory of Change"—a methodology often used in NGO's to promote social change by defining long-term goals and then map backward to identify necessary actions (Center for Theory of Change, n.d.)—the Cascading Semarang developed five concepts that aim to set in motion a systemic change. Following the watershed from upstream to downstream, opportunities have been defined and captured in five concepts named: Spongy Mountain Terrace, Rechannelling the City (Fig. 17), Feeding the Industry, Micro interventions and Recharging the Aquifer. In every concept the view is shifted

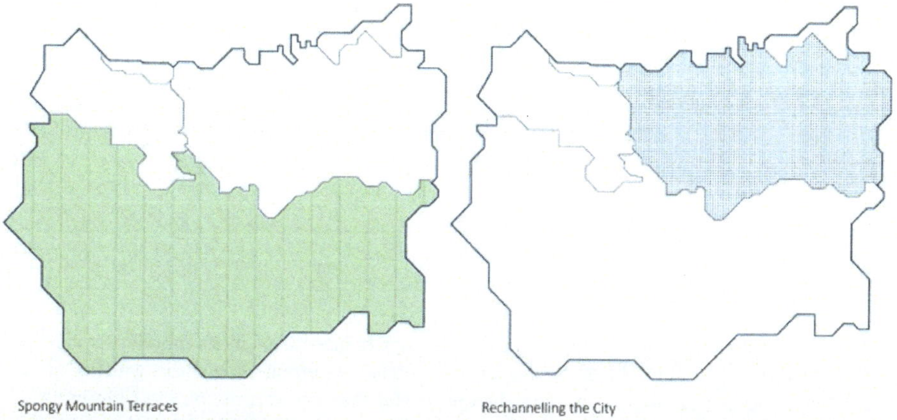

Fig. 17 Complementing projects for different parts of the city (image: FABRICations & MLA+)

from water as a threat, towards water as beneficial asset. By adding the idea of 'cascading', the five concepts completely transform the current water management system into a series of elements of storage and utilisation. Each concept has been developed in order to transform current vicious into virtuous cycles, resulting in inclusive and resilient economic growth in the long term. Thus, creating a paradigm shift from "All the water out" towards "Not a drop of water gets lost".

4.5.4 Interventions: Spongy Mountain Terrace & Rechannelling the City

Spongy Mountain Terrace

Spongy Mountain Terrace aims at the upstream areas where the future urban expansion is expected. Natural preservation, agricultural zones, new residential development, and water management infrastructure are the main realm of interventions (Fig. 18).

Additional storage and infiltration capacity can reduce increasing runoff rates caused by urban development. Urban development coupled with spongy mountain terrace structures should at least achieve a stabilisation of runoff rates, which reduces flood risk downstream. This reduces expected public and private physical damage stemming from floods. Meanwhile, the way these new neighbourhoods are designed results in better water supply for population, agriculture and industrial use. Increased water retention rates mitigate drought risk, especially for surrounding agriculture (Fig. 19).

The stabilisation techniques that can be applied, combined with better infiltration rates and stabilisation of runoffs, can decrease the risk of landslides. Through the reduction of all these different water risks, business continuity can be safeguarded,

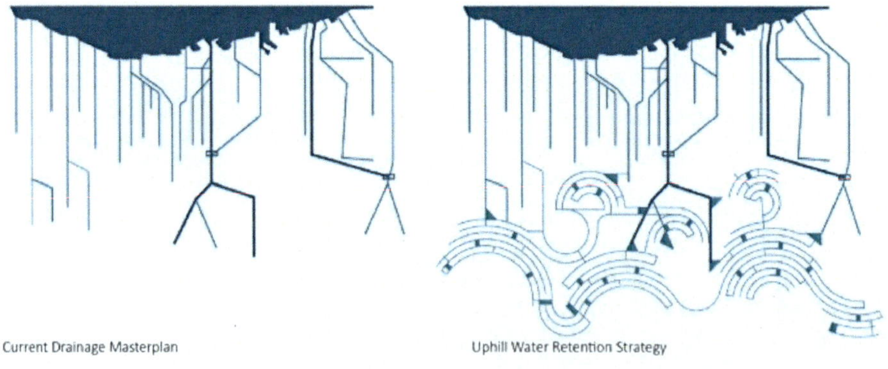

Fig. 18 Conceptual scheme for the proposed water management interventions for the Spongy Mountain Terraces. The scheme on the left represents the current situation in which all rivers in the Semarang area drain all rainwater directly to the sea. The scheme on the right indicates an extension of the current drainage system with a system of terraces in the hilly areas of Semarang (image: FABRICations & MLA+)

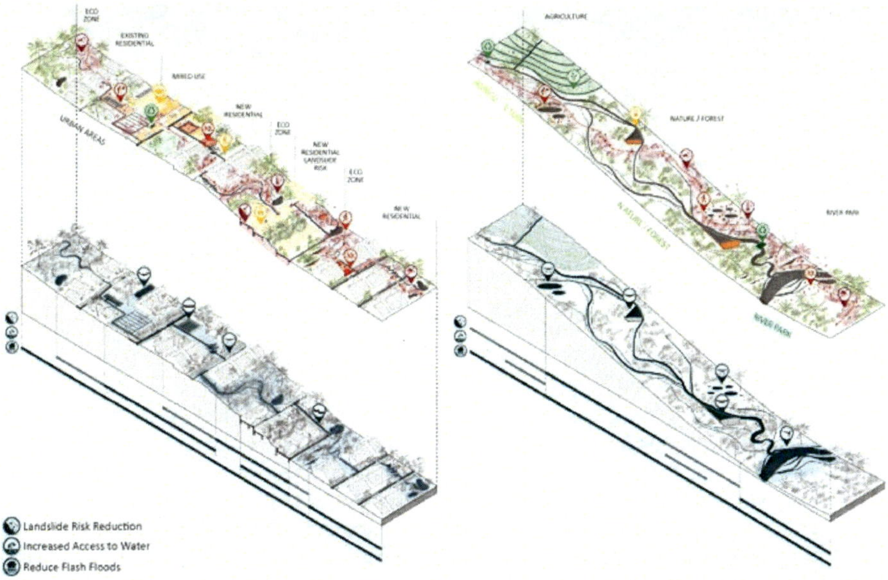

Fig. 19 The applied water management concepts of introducing terrace-like water ways and local storage ponds creates opportunities to both reduce land slide risk and at the same time introduce possibilities to better use the water on its way down, create additional benefits for the communities (image: FABRICations & MLA+)

creating a more favourable investment climate. Besides these three direct benefits, there is a variety of co-benefits as open and green spaces make part of the urban tissue.

The implementation of the Spongy Mountain concept requires customised regulatory mechanisms, including adjustments of regulations (obligatory policy) and/or providing developers with reductions and allowances to secure this expected effect. For landslide risk areas, water infrastructure like friction piles, deep foundation, retention walls and grated drainage ditches can be applied. Biotechnical slope stabilisation techniques are extra measures to ensure the safety. For areas without landslide risks, water storage and shallow aquifer recharge measures can be implemented. Water stored can also be locally filtered, using for instance helophyte filters, cleaned and transported to either local consumption or supply the industries.

Rechannelling the City

Rechannelling the City focuses on the existing urban fabric of Semarang, where opportunities lie in (re)connecting the existing streams and canals to the more fine-grain water network (Fig. 20). By carefully choosing the locations for these improved water networks within the overall water system, the highest flood risk mitigation can be achieved. The local context dictates scale and opportunities. Reduced flash flood risk decreases expected damage and economic disruption. All these effects combined, especially better water quality, are known to increase economic productivity and

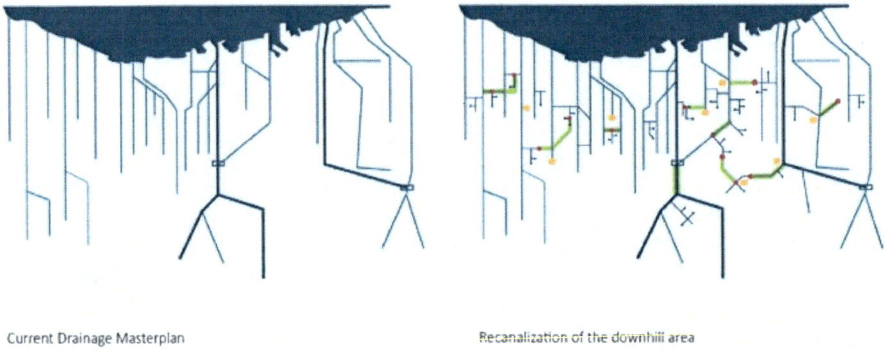

Fig. 20 Conceptual scheme for the proposed water management interventions for Rechannelling the City. The scheme on the left represents the current situation in which all rivers in the Semarang area drain all rainwater directly to the sea. The scheme on the right indicates an elaboration of the current system in the flat and densely urbanised areas of Semarang by introducing and upgrading many east-west running channels and local storage and purification facilities. In the new situation the system will still be able to drain the rainwater from the city in times of abundance. At the same time, this system has the capacity to not let all water escape to the sea but store it for when it is needed (image: FABRICations & MLA+)

decrease public health costs. Since this concept focusses on the existing urban fabric and established communities, the participation of the local communities is key.

The implementation of Rechannelling the City improves the urban water network by integrated measures aiming at water storage, conveyance and cleansing (Fig. 21). It stores stormwater locally and slowly discharges it after the storm, which reduces pluvial and fluvial flood risk. At the same time this program creates synergetic community hubs for additional beneficial programs like solid waste management, harvesting materials and energy, and social public facilities. The infrastructural projects also give an opportunity to improve and create new public space.

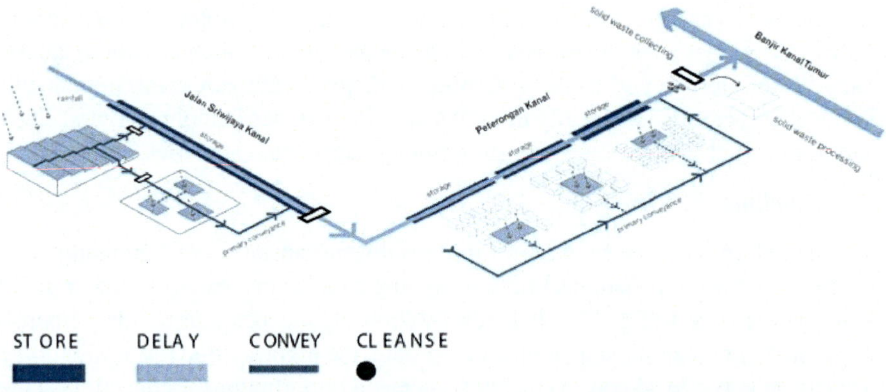

Fig. 21 System diagram of interventions to storage, convey and cleanse water (image: FABRICations & MLA+)

5 Conclusions: Reflection on Design Projects

We see striking similarities in terms of approaches, challenges and outcomes of the three design studies for Khulna, Chennai and Semarang. With regards to the challenges, we see that in all three cities the current strategy of draining rainwater when there is a surplus during the rainy season and pumping water when needed for industry and drinking (and also for agriculture during dry periods), is an unsustainable model with many negative systemic effects, such as land subsidence and groundwater depletion.

When looking at the approaches of the different teams, it becomes evident that they aim to proactively address water-related problems and not only react to the symptoms. For example, none of the teams proposes solely grey drainage measures to address flooding; instead, to prevent floods from occurring, a series of coherent measures is proposed that prevent or delay stormwater runoff.

5.1 More Balanced Water Cycles

All projects analysed the water system on a regional scale in order to better understand its essence and the relation of local phenomena within the water system. This results in awareness of the interlinkage of different aspects of urbanisation, water management interventions, water availability and quality. As a result, the strategies also address a wide spectrum of different water management issues.

We see a focus on creating a more sustainable water cycle and treating rainwater as a valuable water source instead of draining it as fast as possible. The strategies all try to address the current and future water demand (caused by urban growth, industrialisation, a mismatch between crops and water demand, pollution and salinisation) by storing, filtering, infiltrating and reusing as much rainwater as possible. This, however, requires substantial space, as demonstrated by the rainwater model developed by the Khulna team, and coinciding measures for water purification.

5.1.1 Storing, Reusing and Infiltrating Rainwater

In all three projects, the required space is both addressed on a regional as well as on a local scale. On the regional scale we see green blue networks being proposed that, next to water storage, provide benefits for mitigating urban heat stress, urban quality, ecology, water purification and recreational networks. On the local scale there are some variations: in the Khulna and Chennai cases the teams propose the protection and restoration of historic water containers in the shape of currently diminishing ponds (Khulna) and the reuse of currently unused tanks (Chennai). In the Semarang and Khulna cases we see a differentiation in strategies for different types of urban areas in which new building typologies and building regulations are proposed for new development areas, that aim to capture and infiltrate as much water as possible, avoiding to simply passing on challenges to other areas of the city.

In general, we see different approaches for the existing built areas (protect, restore and where possible extend storage capacity, draining only what is necessary) and the newly developed areas (create a fully closed water balance while developing these).

5.1.2 Water Storage and Its Quality

When storing large amounts of water, as proposed for example in public spaces, the quality of water becomes an important aspect. It is for a reason that the currently polluted water is preferably drained as quickly as possible. In the different proposals, we do therefore see a combination of storage and purification. Next to measures for sewerage systems, (industrial) wastewater treatment plants and solid waste collection, we also see NBS proposed as a natural filtering system. This multifaceted approach requires many simultaneous interventions and investments, which is a major challenge regarding the implementation of the proposed strategic projects.

5.2 Aiming for Additional Benefits

All projects aim to create additional benefits besides those related to the water management system, which also strongly relates to the system analysis of the water cycle and its linkages with other sectors.

For Khulna and Chennai for instance, the proposed building typologies and regulations not only reduce flood risks, but also reduce damages on assets and protect vital infrastructure in the case of extreme events, to reduce interruptions to the economy and daily life.

All teams propose NBS as (part of) their solutions, because the additional benefits are much wider than with solely engineered solutions. They can add urban quality and recreational spaces while reducing the urban heat island effects, like the Mayur river park in Khulna or bioswales in public spaces in Chennai; or the rechannelling in Semarang which connects to community hubs to promote solid waste management connected to social functions.

Engineering solutions are often very complex, cost-intensive and mono-functional structures. In the light of climate change, they show very limited adaptability to changing conditions and in case of failure, the effects on people and property can be dramatic. This leads to a paradigm shift towards NBS, which are by definition more adaptable to changing conditions and can to some extent regulate and recover themselves. Furthermore, NBS are associated with many additional benefits, such as reduction of the heat island effect, increase of biodiversity and improved air quality (Browder et al. 2019), and are therefore better suited to address urban challenges than engineering solutions.

NBS do require significant surfaces, which make them easier to be implemented in new development areas where they can be included in an integral design proposal, than in an existing high-density urban environment. In existing urban areas, NBS can be combined with existing engineering infrastructure to increase the adaptability to climate change and generate additional benefits for local communities.

5.3 *Need for a Coordinated Implementation*

The necessity of the proposals to be this inclusive (store water, while also improving the quality by providing sewerage, water purification and solid waste management) creates challenges in executing them. In order to prevent ending up with an area that stores and infiltrates water that is polluted, an inherent overall approach is needed, that includes projects on different scale levels and includes a range of physical as well as governance and capacity building interventions. The teams' integral approach led to integrated multi-faceted design strategies that also require integral governance and financial structures to implement them. The integral design teams are now equipped to formulate strategies for urban water-related challenges in an innovative and inclusive way, as attention is being paid to these aspects during the design process. The next challenge is to develop governance approaches for implementing and financing these strategies within the case studies contexts, and mechanisms to enforce regulations in highly unregulated urban areas such as informal settlements.

References

Abidin HZ, Djaja R, Andreas H, Gamal M, Hirose K, Maryama Y (2004). Capabilities and constraints of geodetic techniques for monitoring land subsidence in the urban areas of Indonesia. In: Third FIG regional conference for Asia and the Pacific, Jakarta, Indonesia, 3–7 October

Ahmed I, Mahmud KA, Islam S (2014) A study of historical transformation of the urban integration core of Khulna City, Bangladesh. Int J Innov Appl Stud 8(4):1410–1417

Asian Development Bank (ADB) (2011) Adapting to climate change, strengthening the climate resilience of the water sector Infrastructure in Khulna, Bangladesh

Browder GSO, Rehberger Bescos I, Gartner T, Lange G-M (2019) Integrating green and gray: creating next generation infrastructure. World Bank and World Resources Institute, Washington, DC

CCC&AR and TNSCCC (2015) Climate Change Projection (Rainfall) for Chennai. In: District-Wise Climate Change Information for the State of Tamil Nadu. Centre for Climate Change and Adaptation Research (CCC&AR), Anna University and Tamil Nadu State Climate Change Cell (TNSCCC), Department of Environment (DoE), Government of Tamil Nadu, Chennai, Tamil Nadu, India. Available at URL. www.tnsccc.in

Center for Theory of Chane (n.d.). How does theory of change work? https://www.theoryofchange.org/what-is-theory-of-change/how-does-theory-of-change-work/. Accessed 18 July 2020

Delta Committee (2008) Working together with water, a living land builds for its future: findings of the Delta Committee 2008. Delta Committee, The Hague

Ganesan M (2008) The temple tanks of Madras, India: rehabilitation of an ancient technique for multipurpose water storage. Ind J Sci Technol 1(7):1–8

General Economics Division (GED) Bangladesh (2018) Bangladesh Delta Plan 2100. General Economics Division, Dhaka

Hooimeijer F (2014) The making of polder cities. A Fine Dutch Tradition. JapSam Books, Prinsenbeek

Hore SK, Sarker MH, Ferdous R, Ahsan M, Hasan I (2013) Study of the off-take dynamics for restoring the Gorai river. In: 4th international conference on water & flood management (ICWFM-2013)

IHE Delft (2020) Strategic delta planning. Strengthening strategic delta planning processes in Bangladesh, the Netherlands, Vietnam and beyond. https://strategic-delta-planning.un-ihe.org/highlights. Accessed 18 July 2020

Khulna University of Engineering & Technology (Khulna University) (2015) A study on saving the Mayur River and its connected Canals of Khulna, Bangladesh

Kothuis B, Brand N, Sebastian A, Nillesen AL, Jonkman B (eds) (2015) Delft Delta Design. Houston Galveston Bay Region, Texas, USA. Delft University Publishers, Delft

Lerner J (2014) Urban Acupuncture. Island Press, Washington

Meyer H, Morris D, Waggonner D, Dircke P (2009) Dutch dialogues: New orleans - netherlands, common challenges in urbanized deltas (Dutch urbanism today). SUN, Amsterdam

Meyer H, Nijhuis S (eds) (2014) Urbanized deltas in transition. Techne Press, Amsterdam

Murdohardono D, Tobing (2001) Engineering geology evaluation of land subsidence in Semarang. Unpublished report, Geological Agency, Directorate of Environment of Geology and Mining, Bandung, Indonesia

Mukhopadhyay A, Payo A, Chanda A, Ghosh T, Chowdhury SM, Hazra S (2018) Dynamics of the Sundarbans Mangroves in Bangladesh Under Climate Change. In: Ecosystem services for well-being in deltas, pp 489–503. Palgrave Macmillan, Cham

Murdohardono D, Tobing (2001b) Engineering geology evaluation of landsubsidence in Semarang. Unpublished report, Geological Agency, Directorate of Environment of Geology and Mining, Bandung, Indonesia

Netherlands Enterprise Agency. Water as Leverage for Resilient Cities Asia. https://english.rvo.nl/subsidies-programmes/water-leverage. Accessed 11 Jan 2020

Nishat B (2019) Landscape narrative of the Sundarban: towards collaborative management by Bangladesh and India. World Bank Group, Washington, D.C. http://documents.worldbank.org/curated/en/539771546853079693/Landscape-Narrative-of-the-Sundarban-Towards-Collaborative-Management-by-Bangladesh-and-India

Ovink H, Boeijenga J (2018) Too big: rebuild by design: a transformative approach to climate change. Rotterdam: nai010

Prakash A, Singh S (eds) (2013) Water Security in Peri-urban South Asia. Adapting to climate change & urbanisation. SaciWATERs, India

Revi A, Satterthwaite DE, Aragón-Durand F, Corfee-Morlot J, Kiunsi RBR, Pelling M, Roberts DC, Solecki W (2014) Urban areas. In: Field CB, Barros VR, Dokken DJ, Mach KJ, Mastrandrea MD, Bilir TE, Chatterjee M, Ebi KL, Estrada YO, Genova RC, Girma B, Kissel ES, Levy AN, MacCracken S, Mastrandrea PR, White LL (eds) Climate Change 2014: impacts, adaptation, and vulnerability. Part A: global and sectoral aspects. Contribution of Working Group II to the Fifth Assessment Report of the Intergovernmental Panel on Climate Change. Cambridge University Press, Cambridge and New York, pp 535–612

Roy K, Karim R, Akter F, Islam S, Ahmed K, Rahman M, Datta DK, Khan DA (2018) Hydrochemistry, water quality and land use signatures in an ephemeral tidal river: implications in water management in the southwestern coastal region of Bangladesh. Appl Water Sci 8:78

Schuyler D (1986) The New Urban landscape. The Redefinition of City Form in Nineteenth-Century America. John Hopkins University Press, Baltimore

Seijger C, Datta DK, Douven W, van Halsema G, Khan MF (2019) Rethinking sediments, tidal rivers and delta livelihoods: tidal river management as a strategic innovation in Bangladesh. Water Policy 21:108–126. doi: https://doi.org/10.2166/wp.2018.212

Seneviratne SI, Nicholls N, Easterling D, Goodess CM, Kanae S, Kossin J, Luo Y, Marengo J, McInnes K, Rahimi M, Reichstein M, Sorteberg A, Vera C, Zhang X (2012) Changes in climate extremes and their impacts on the natural physical environment. In: Field CB, Barros V, Stocker TF, Qin D, Dokken DJ, Ebi KL, Mastrandrea MD, Mach KJ, Plattner G-K, Allen SK, Tignor M, Midgley PM (eds) Managing the risks of extreme events and disasters to advance climate change adaptation. A Special Report of Working Groups I and II of the Intergovernmental Panel on Climate Change (IPCC). Cambridge University Press, Cambridge, UK, and New York, NY, USA, pp 109–230

UN Habitat (2006) State of the World's Cities 2006/7: the millennium development goals and urban sustainability: 30 years of shaping the Habitat Agenda. UN Habitat, Nairobi, Kenya

van Dijk J, Schreijer M, Verhallen JM (2002) Drietrapsstrategie. Uitwerking naar doelstelling en gebied voor de stroomgebiedsvisie Noorderkwartier. Het Waterschap 14:638–643

Waggonner D (ed) (2013) Greater new Orleans urban water plan. Digital publication on www.livingwithwater.com

Water as Leverage (2018) Setting the scene for a call for action. Netherlands: Netherlands Enterprise Agency. https://watersleverage.org/file/download/57979535/watersleverage-settingthesceneforacallforaction.pdf. Accessed 18 Feb 2020

Water as Leverage (2019) The challenge in Asia. https://watersleverage.org/cms/view/57979211/the-challenge-asia. Accessed 22 June 2020

Wetlands International (n.d.). Building with Nature Indonesia—reaching scale for coastal resilience. Retrieved from: https://www.ecoshape.org/uploads/sites/2/2016/07/building-with-nature-in-indonesia-1.pdf. Accessed 18 July 2020

Winterwerp JC, Giardino A (2012) Assessment of increasing freshwater input on salinity and sedimentation in the Gorai river system. Deltares, Delft

Dr. Anne Loes Nillesen is founding director of the urban design firm Defacto Urbanism in Rotterdam, The Netherlands. Defacto specializes in urban research and design in the domain of water and flood risk management. With her office, Anne Loes worked on complex, large-scale urban design and flood risk management assignments such as the Dutch Delta Programme and Bangladesh Deltaplan, and several regional and local scale adaptation projects, amongst which a barrier design for Houston and a regional strategy for Khulna, Bangladesh. Current projects include an NBS study for water management in Kigali, Rwanda and 'Resilience by Design' for the Amsterdam Metropolitan Area, the Netherlands. Besides running Defacto, Anne Loes founded the Climate Adaptation Lab and Delta Interventions integral MSc graduate studios at Delft University of Technology (TU Delft), Faculty of Architecture. Her PhD research focused on the relation between urban design and flood risk strategies. She graduated with honors as an Architect and Urban designer from TU Delft and undertook postgraduate studies in Landscape Architecture and Urbanism at the Berlage Institute in Rotterdam. Anne Loes published books such as 'Amphibious Housing in the Netherlands', 'Delta Interventions: Design and Engineering in Urban Water Landscapes' and developed the 'Climate Adaptation Game', a multi-player board game.

Mona zum Felde holds a Master degree in Urbanism from Delft University of Technology. She received an honourable mention for her graduation project, where she investigated the opportunities for the integration of urban water management and urban development to reduce climate change impacts in unplanned settlements in Dar es Salaam, Tanzania. In her current function as urban designer at Defacto Urbanism, she works on complex research by design projects related to water management, climate change resilience and urban development. Current projects include a study on spatial implications of accelerated sea level rise in the Netherlands and a study on NBS for water management in Kigali, Rwanda.

Eva Pfannes Based in Rotterdam, Eva Pfannes is an architect and urban designer with over 15 years of experience in Europe, India, UK, and Brazil. She co-founded the international design practice OOZE in 2003 with Sylvain Hartenberg, working across scales, from regional strategies to architecture and research, combining an elaborate understanding of natural, ecological processes, with technological expertise and deep insights into the social-cultural behaviour of users of the built environment. The cyclic closed-loop processes found in nature are the foundation for each intervention and integrate the human scale within a comprehensive urban strategy. She is currently lead designer for the IABR Atelier Rotterdam, working on a local energy action plan (LEAP) to decarbonize a local neighbourhood and team lead for City of 1000 Tanks project in Chennai, part of the global Water as Leverage programme of the Dutch Government. Her work was widely published including the Materials Book by Ruby Press, DOMUS, AJ and The Guardian and exhibited at the Venice Biennial, Studio-X-Rio de Janeiro and the Sao Paulo Architecture Biennial. Awards include the 2017 Landscape Institute Award UK for 'Best Design for Temporary Landscape' and the LafargeHolcim Bronze Award-Latin America 2017.

Prof. Dr.Ir. Han Meyer is emeritus Professor of Urban Design at Delft University of Technology (TU Delft). He graduated as an urban designer from TU Delft and subsequently worked at the Department of Physical Planning and Urban Renewal in Rotterdam for ten years. He has been working at Delft University of Technology since 1990, obtaining his doctorate with the thesis De stad en de haven (City and Port) in 1997. He was professor of urban design - theory & methods from 2001–2016 and is now active as an advisor for national and international projects. His research and expertise focuses on two main topics. 'Delta Urbanism', which pays special attention to the search of a new balance between urbanization processes and climate change in vulnerable deltaic territories. 'City and Port' addresses the evolution of port cities in the past, presence and future, and the mutual influence of port- and urban developments. He published many articles and books on 'Delta Urbanism'. Recent books include 'Urbanizing Deltas in Transition'(co-edited with Steffen Nijhuis Amsterdam Techne Press, 2014) and 'The State of the Delta. Engineering, urban development and nation building in the Netherlands'(VanTilt 2017).

Olv Klijn is Principal at FABRICations—architecture, urbanism, regional strategies and Assistant Professor at the chair of Dwelling at the Technical University Delft. He studied architecture at the Eindhoven University of Technology and graduated Cum Laude. As an artist in residence Klijn studied the impact of current urbanization patterns at the American West Coast stayed at the Banff Centre in Canada. Klijn is (co)author of several books such as 'VMX Agenda', '10 x Den Bosch', 'Station Centraal', 'Architect by accident' and 'The making of …' and published in various journals. In 2007 he founded FABRIC together with Eric Frijters and as of then involved in the design and research of architecture, urbanism and regional strategies. In 2010 they won the Prix de Rome, the price for architects up to 35 years. In 2011 Klijn was recognized as one of the 40 emerging European architects under 40 years old. One year later he was nominated for the Iakov Chernikhov International Architecture Prize 'for designers, that stand in the field experiment with innovative architecture, education and research in architecture and urbanism'.

Some Medical Issues Related to Human-Water Interaction: A Brief Introduction

Eugen Bogdan Petcu

Abstract This chapter aims to present a brief account on potential medical health issues which may be determined by living close to a water source as well as consumption of improper water. Regardless, the type of architecture, any future projects construction project aiming at developing a sustainable habitat must comply with the United Nations-Sustainable Development Goals (UN-SDG),2030 goals for water accessibility and quality. Also, additional clues for principles and directions of an efficient futuristic architectural development SeaCities could be identified in the history of human development and medicine. The current chapter will summarize the most common medical problems related to human-water interaction regardless the type of habitat. The importance of general sanitation is presented as well as the most important aspects related to the therapy and pathogenesis of various conditions. Various diseases are presented separately under several categories: recreational water exposure, food poisoning, seafood toxicity, fish related histamine toxicity, and other seafood poisoning including tetrodotoxin poisoning. A short review of intoxications determined by ingestion of water contaminated with lead, mercury, and arsenic is also presented

Abbreviations/Glossary

ETEC	Enterotoxigenic *Escherichia Coli*
EHEC	Enterohemorrhagic *Escherichia Coli*
UN-SDG	United Nations-Sustainable Development Goals

E. B. Petcu (✉)
New York Institute of Technology, Northern Boulevard, Old Westbury, NY 11568-8000, USA
e-mail: e.petcu@griffith.edu.au

School of Medicine, Menzies Health Institute of Queensland, Griffith University, Gold Coast Campus, Brisbane, QLD 4222, Australia

J. Baumeister et al. (eds.), *SeaCities*, Cities Research Series,
https://doi.org/10.1007/978-981-15-8748-1_7

1 Introduction

The water is an essential element maintaining human vital functions. From another perspective, it is unquestionable that water represents the "*sine qua non*" condition of human development, including farming, which allowed numerous cities to flourish all over the world. In parallel, the maritime transport had the same role as the modern Internet providing the transport of information and goods. Therefore, numerous cities were built in proximity of oceans and seas. The cradle of Western civilization (Thornton 2002), the Greek-Hellenistic culture was made possible not only by a mild sub-tropical climate but also by the water itself, the Mediterranean Sea. For example, Piraeus was founded around 2500 BC, and subsequently, in the fifth century BC became a dynamic commercial seaport (Hadjimanolakis 2020), allowing Athens to become the cultural capital of the world promoting arts and culture all over the region while transforming the Mediterranean Sea into a Greek Sea. In France, the modern city of Marseille was developed on the structure provided by the old Greek commercial colony of Massalia which was founded more than 2500 years ago by Greek travelers from Asia Minor (Boucheron et al. 2019).

We could mention also the Phoenicians which built a flourishing society linked to sea travel. Stephen Krensky (1991) indicates that around 1000 BC the Phoenicians settlers created Cadiz, one of the oldest European cities (Krensky 1991). However, other human dwellings were built away from the sea but close to a freshwater source such as Jericho, a city located in the Palestinian West bank which was erected around 9000 BC in the vicinity of the Jordan River (Gates 2011).

Regardless the location of the human settlements, people have always realized that the quality of water is of paramount importance for health, and efficient solutions for storage and supply have been developed by various communities. Archeological ruins of water collection systems such as stone water basins may be seen nowadays in Middle East and Turkey (Juuti et al. 2020) while Roman aqueducts are still seen in all countries bordering the Mediterranean Sea. The Roman aqueduct is an esthetically imposing complex structure consisting of tunnels, pipes, and bridges that had the role of supplying the water to urban communities where this was scarce. Today, one of these aqueducts, Aqua Virgo which was built by the Emperor Agrippa more than 2000 year ago in Rome, is still in use.

Later, the principles of the Roman-style water supply system were further fine tuned and improved in Medieval Europe, where a significant number of cities and towns were developed near the large rivers and seashores. Subsequently, the industrial revolution of the nineteenth century paralleled by an intensive development of farming, determined a significant increase in water consumption as the urban population exploded. In sixteenth century, Sir John Harington created the first flushable toilet using a water basin mounted above a waterproof receptacle. This system became widely used only in the second part of the nineteenth century with the advent of Thomas Crapper's device for basin/tank water filling.

However, poor management of water sewage remained common during this time and was identified as the cause of a variety of diseases, including cholera outbreaks all over the Europe. The first to formulate the water-borne theory of cholera transmission was Dr. John Snow who in 1854 described in detail the London cholera outbreak (McLeod 2000). Further improvements were made possible by filtration and chlorination of the water which overall contributed to a sustained development of the cities. During the twentieth century, the population grew considerably and while in general, cities all around the world adopted efficient water systems and at least some norms of sanitation, health issues determined by consumption of improper water become more complex.

At the present time, toxic industrial by-products can be identified, unfortunately, all too frequently, representing an important public health issue. Current development of towns and cities, including building and architectural concepts, should evaluate in detail all aspects related to public health including water supply.

2 General Aspects Related to Sanitation of Water

In Antiquity, the first topic related to sanitation of water was illustrated by Vitruvius in his "De Architectura," where he debates the issue of using lead pipes. Other Roman authors such as Palladius, Columella, and Plinius also discussed the same topic (Juuti et al. 2020). As mentioned above, the Mediterranean Sea and its climate made possible the flourishing of numerous cities and became the principal route of transports of goods and people. Obviously, it also facilitated not only the mass movement of populations, but also the spread of various infectious diseases, such as malaria which is spread by mosquitoes breeding in large numbers in marshlands, close to a source of water, in warm climates. Grmek (1994) suggests that malaria was already known in Greece in the Mesolithic times, with significant morbidity during the peak of the Hellenistic period and then again during the Ottoman invasion.

Interestingly, Angeletti (1995) indicates that, quite early on, the Hippocratic disciples have focused their attention on the quality of water. One of their principles was water observation as a method of explaining the probable causes of disease in a community living in a town or city. This principle was re-discovered only later in the eighteenth–nineteenth centuries when incipient public health ideas were implemented in Europe and North America (Angeletti 1995). Around this time in France, Antoine-Augustin Parmentier, a pharmacist who promoted the potato as a reliable source of food for population, an advocate of Jenner's vaccination and a pioneer of food chemistry, studied the quality of water and its impact on public health. His research focused on river Seine and its impact on Parisian population, creating the first Committee of Sanitation of Seine (Lafont 2014). Meanwhile, there were numerous cholera outbreaks in England and United States. In 1832, the first recorded cholera epidemic in New York caused the New York Board of Health to issue a public act implementing rules and regulations of public health importance related to water sanitation and prevention of disease (Kim 2015).

All these suggest a direct relationship between water sanitation and human development. Essex et al. (2020) suggest that a minimum of 50 L per day for each person is required on average for personal and food hygiene and cleaning, while 7.5 L of water per day is necessary for drinking and cooking. However, considering the agricultural needs and food consumption, the overall indirect daily water use for every EU citizen reaches a staggering 3000 L. In addition, water is required for industrial activities. Therefore, the supply of good quality water remains even nowadays a serious problem in numerous parts of the world, amplified by a populational growth explosion, as well as by chemical pollution of waters.

In this context, the United Nations-Sustainable Development Goals (UN-SDG) related to water indicate that by 2030 a universal and equitable access to affordable and safe water must be implemented as well as facilitating an efficient access to adequate and equitable water sanitation and hygiene, significantly decreasing water pollution while improving sustainable water consumption, developing an integrated international water management method and restoring the water-depleted ecosystems (Essex et al. 2020).

The most important pathogens related to water exposure and consumption are summarized in Table 1.

3 Water-Borne Disease: Recreational Water Exposure and Food Poisoning

Water-borne diseases appear after exposure to recreational water or drinking water contaminated with various pathogens [e.g., enteric viruses such as Rotavirus, Norovirus and Adenovirus; bacteria such as *Campylobacter jejuni,* Salmonella spp. and *Escherichia coli;* and protozoa such as *Cryptosporidium* spp. and *Giardia* spp. (Chhipi-Shrestha et al. 2017)]. Hlavsa et al. (2018) have reported in the United States during 2004–2015 nearly 500 outbreaks of water-borne diseases determined as a result of exposure to contaminated recreational water. Most of these outbreaks were determined by Cryptosporidium associated with gastroenteritis (approximately 60%). Also, Legionella species produced Legionnaires's pneumonia in 16% of these outbreaks while Pseudomonas outbreaks causing the so-called hot tub rash—folliculitis and otitis externa were reported in 13% of the epidemies. Overall, there were more than 27,000 cases including eight deaths that could be identified with certainty after exposure to recreational water (Hlavsa et al. 2018).

It is quite interesting to note that members of the Bacillus group such as Paenibacillus, Brevibacillus, and Pseudomonas have become resistant to chlorine and peracetic acid which are used in the preventive treatment of water (Gu et al. 2019). This explains the previous findings of Hlavsa et al. (2018) which have suggested that the Model Aquatic Health Code by the Centers for Disease Control and Prevention (CDC) has no use in predicting the outbreak case counts (Hlavsa et al. 2018). On the other hand, O'Flaherty et al. (2019) have reported that humans could be exposed to antibiotic resistant Escherichia coli if bathing in areas located near wastewater treatment plants (O'Flaherty et al. 2019).

Table 1 Summary of the most important pathogens related to water exposure and consumption

Agent/disease	Transmission	Notes/clinical characteristics
Travelers diarrhea a. Water-borne disease: Rotavirus, Norovirus, Adenovirus, Campylobacter, Salmonella, E.Coli, Cryptosporidium, Giardia b. Food: ETEC, EHEC	a. Recreational exposure b. Eating meats, fish, uncooked eggs	• USA: 500 outbreaks between 2004–2015 • Fever, abdominal pain • Diarrhea: watery, mucoid
Vibrio cholera	Water	• Haiti: 2018–812 586 cases including 9606 deaths • Fever, abdominal pain, watery diarrhea • Antibiotic resistant strains
Giardia lamblia	Water	• Norway: 2007 outbreak • Fever, abdominal pain, diarrhea • Widespread duodenal inflammation
Seafood toxicity	Consumption of fish infested with Anisakis nematode	• Allergic reaction–IgE mediated reaction
Fish-related histamine toxicity- scombroid syndrome	Consumption of mahi-mahi, tuna inappropriately cooled	• Muscular L-histidine is transformed into histamine • Diarrhea, flushing, urticaria, hypotension
Ciguatera toxicity	Consumption of brown marbled grouper infested with dinoflagellate microalgae	• Ciguatoxins which opened the sodium channels • Paresthesia, bradycardia, hypotension, dyspnea
Shellfish poisoning	Consumption of mollusks infested with dinoflagellates	• Various gastrointestinal symptoms
Paralytic shellfish poisoning	Consumption of mollusks infested with Alexandrium	• Paresthesia, headaches, respiratory insufficiency
Amnestic shellfish poisoning	Consumption of mollusks infested with Pseudonitzschia	• Various gastrointestinal symptoms • Short-term memory loss • Seizure, coma
Tetrodotoxin poisoning	Consumption of contaminated puffer fish	• Neurotoxin opens the sodium channels: ataxia, paresthesia, ascending paralysis, vomiting, coma
Lead poisoning	Water	• Anemia, hypertension, fertility issues, immunity issues, developmental problems dental and skeletal malformations.
Mercury poisoning	Water	• Metallic taste, vomiting, abdominal pain, acute renal failure

(continued)

Table 1 (continued)

Agent/disease	Transmission	Notes/clinical characteristics
Arsenic poisoning	Water	• Cancers: bladder, kidney, lung • Pancreas damage resulting in diabetes, cardiovascular disease, immunological issues

However, drinking contaminated water with various microbiological forms or eating food which was prepared with contaminated water determines food poisoning. This clinical entity could be included also into the bigger category of Traveler's Diarrhea which apart from been caused by consumption of improper water is also determined by eating uncooked eggs, meats, or fish or drinking unpasteurized milk (Giddings et al. 2016). Food poisoning includes an array of specific clinical features such as fever, abdominal pain of various intensity, diarrhea which may appear with an increased amount of mucous or blood. Often the young patients present with reactive arthritis, especially after food poisoning with Salmonella species (Arnedo-Pena et al. 2010; Rohekar et al. 2008).

In terms of mechanisms of diarrhea, this may be caused by a non-inflammatory enterotoxin which acts upon the intestinal mucosa to produce large fluid stools. In rare cases, patients may lose large amounts of fluids with subsequent electrolytic imbalances (Satchell 2003). In contrast, inflammatory bloody/mucoid diarrhea is the result of a cytotoxic by-product synthetized by the microbial strain which damages the intestinal mucosa. While the stools are smaller and less fluid than in the non-inflammatory diarrhea, they contain leukocytes. Overall, the patients present with severe pain and fever (Hale et al. 1996).

There are several clinical forms of bacterial, viral, and protozoan induced food poisonings. While a detailed description of all of these is beyond the scope of this chapter, the most important aspects with potential therapeutic and pathogenic importance will be covered below.

3.1 Enterotoxigenic Escherichia Coli (ETEC) Food Poisoning

ETEC represents a major cause of infant mortality and morbidity in the general population. Unfortunately, numerous strains are antibiotic resistant in which case only supportive therapy may be available (Roussel et al. 2017). The presence of ETEC suggests fecal contamination of the water which represents an issue of public health concern (Barbosa et al. 2016). Von Mentzer (2020) suggests that the most important virulence factor for ETEC is represented by its capacity to bind to the glycosphingolipid receptors present on the small bowel mucosa by a structural component labeled as a colonization factor CS30. This is subsequently followed by hypersecretion as a result of activation of guanylate cyclase. In this context, it seems that the best treatment would the administration of inert analogs factors, which would bind the glycosphingolipid receptors preventing the adhesion of ETEC on colonic mucosa. However, these are yet to be used on a large scale, and more research evaluating these inert analogs ligands are required at the present time.

3.2 Enterohemorrhagic **Escherichia Coli** *(EHEC) Serotype O157:H7 Food Poisoning*

Eppinger and Cebula (2015) indicated in a recent paper that EHEC serotype O157:H7 is one of the most aggressive types characterized by massive production of shiga-toxin (Stx) which produces wide spread endothelial damage and micro-vascular thrombosis in the colonic mucosa associated with bloody diarrhea, which usually lasts for more than one week. Often, the hemorrhagic colitis is followed by Hemolytic Uremic Syndrome, which represents an important cause of renal failure in children. It seems that the key element in the pathogenesis of this type of food poisoning is represented by adhesion of the bacterium to the endothelium, which is modulated by parallel translocation of Tir receptor and over expression of intimin by the pathogen. In addition, this process seems to be dependent on special fimbrial adhesins (the Sakai strain containing 16 fimbrial gene clusters) (Eppinger and Cebula 2015; Elpers and Hensel 2020). While the diagnosis is confirmed after stool cultures, no specific antibiotic medication is recommended, and the treatment is supportive.

3.3 Vibrio Cholera Food Poisoning

Vibrio cholera may still determine occasional outbreaks. Mavian et al. (2020) reiterate that person to person is the most important method of transmission during epidemies; however, Vibrio cholera grows and survives for a very long period in natural water sources. One of the most important epidemies that started from a natural source of water was recorded in Haiti. In July 2018, there were 812,586 cholera patients and 9606 deaths caused by Vibrio cholera O1, suggesting new bacterial strains determined by possible mutations related to integration in a system represented by water (Mavian et al. 2020). Recent research has revealed that the transfer between aquatic systems, and humans is achieved by regulatory systems such as the EnvZ/OmpR complex activated by a drop in pH as seen in some bacterial forms. Interestingly, an alkaline pH inhibits genes which control the synthesis of virulence factors. However, it may be possible that the same alkaline pH would modulate also the transfer between human and aquatic ecosystems (Kunkle et al. 2020). In terms of pathogenicity, recent research shows that Vibrio cholera has a type VI secretion system (T6SS) which releases the actual toxin in the tissue, producing significant damage with widespread inflammation (Crisan and Hammer 2020). In 2018 in Santiago, Chile, an outbreak of Vibrio cholera was reported, and the strain was labeled non-O1/non-139 Vibrio. These forms are less virulent than the O1 and O139, lacking cholera toxin and toxin-coregulated pilus but carrying genomic islands controlling expression of type III and VI secretions systems as well as antibiotic resistance, a very important aspect of paramount importance for this type of patients (Arteaga et al. 2018).

4 Viral Gastroenteritis

Despite modern requirements and policies for sanitation and improved quality water, including treatment with chlorine, viral gastroenteritis represents a significant source of morbidity all around the world, in both adults as well as children. The literatures indicate that the most important viral agents responsible for viral gastroenteritis are the following: astrovirus, norovirus, rotavirus, sapovirus, and enterovirus (Scarcella et al. 2009; Zbinden 2019; Inns et al. 2019). However, the correct confirmation of the etiological factor requires molecular analysis of the stool (Zbinden 2019). In the United States, after the introduction of routine immunization in infants, the rotavirus season became delayed, shorter, and of lower magnitude compared with the pre-vaccination period (Tate et al. 2009). Currently, two live attenuated rotavirus vaccines, RotarixRTM, RotaTeqRTM, are available in numerous countries (Desselberger 2017).

Regarding the mechanisms of disease in viral gastroenteritis, Lorrot and Vasseur (2007) have indicated that some viruses such as rotavirus inhibit intestinal disaccharidases as well as the sodium symports systems which are related with water transport. As a result of this, the carbohydrates are not digested and accumulated in parallel with malabsorption of nutrients and water retention in the lumen of the intestinal tract which results in diarrhea. In addition, the NSP4 enterotoxin modulates a phospholipase C-dependent calcium signaling pathway, which in turn promotes chloride secretion by the intestinal villi contributing to the mechanisms of diarrhea.

Viral gastroenteritis is particularly very dangerous in infants and children less than five years of age. The American Academy of Pediatrics recommends rapid oral rehydration in those children with mild to moderate dehydration as a result of diarrhea in the context of viral gastroenteritis. In severe cases, intravenous rehydration is preferred. It seems that early administration of milk or even food after initial rehydration is quite beneficial in reducing the duration of diarrhea. However, the use of anti-diarrheal medication should be avoided in infants, toddlers, and young children (American Academy of Pediatrics 1996).

It is also reported that viral gastroenteritis represents a serious problem for elderlies, especially those that live in homes for elderlies. A study conducted in North East England between 2016 and 2018 has revealed there were 566 gastroenteritis cases, with a majority having a viral cause. In this setting, norovirus was identified in more than 60% of the outbreaks while sapovirus and rotavirus were noted in 13 and 11% of outbreaks, respectively, (Inns et al. 2019).

5 Giardia Lamblia

Giardia lamblia induced gastroenteritis is a very common cause of protozoan diarrheal disease related to poor sanitation of water, seen in both developed and underdeveloped countries. In October 2007, an outbreak of infectious diarrhea that was traced back to water infestation was reported in Oslo, Norway. Apart from bacteriological findings, Giardia cysts were also identified in water. As a result, more than 460,000 citizens of Oslo were requested by the local public health authorities to use for their daily needs boiled water for five days (Robertson et al. 2009). Also, in 2007, another outbreak of intestinal giardiasis was recorded in New Hampshire, USA. Drinking tap water was identified as a risk for contracting Giardia. Although only 31 patients developed clinical features of gastroenteritis, it is considered the most significant giardiasis recorded in recent times in the United States (Daly et al. 2010).

This protozoan is very motile due to its complex micro-tubule cytoskeleton which also modulates transition between the cyst and trophozoite developmental stage with flagella promoting also attachment and cell division (Hagen et al. 2020). It is not known why, but Giardia can escape the immunological system. Experimental studies conducted in a murine model have shown that in macrophages this organism can inhibit the nitric oxide (NO) synthase and cyclooxygenase-2 via activation of a set of three proteases which degrade a subunit of nuclear factor-kappa B (NF-κB). This explains the immunological escape seen in patients infected through poor quality water with Giardia lamblia (Desselberger 2017). Other studies conducted by Fink et al. (2019) have revealed that Giardia promotes accumulation of macrophages in the small bowel wall which are active and continue to proliferate, but while this would indicate some involvement of the immunological system; the real significance of this event is poorly understood (Faria et al. 2020).

Usually, dehydration due to diarrhea is treated with oral rehydration, while, to neutralize the protozoan, the patient is given metronidazole. However, a significant number of these type of patients may develop chronic disease or may become carriers. In Bergen region, Norway, over a period of 15 months, 124 patients with giardiasis were diagnosed during an outbreak. They were treated with metronidazole. However, the intestinal infection persisted in more than 30% of cases. Moreover, histopathological evidence of inflammation was present in 87.2% of those that still tested positive for Giardia after therapy and recovery. Also, histological evaluation of duodenal biopsies showed inflammation in 28% of those that recovered and tested negative for giardiasis (Fink et al. 2019).

6 Seafood Toxicity and Fish-Related Histamine Toxicity-Scombroid Syndrome

Seafood toxicity and fish-related histamine toxicity-scombroid syndrome can occur after consumption of fish, shellfish, and other products that grow in the water. While the signs and symptoms appear after a variable interval of time, in some patients the symptomatology presents as an allergic reaction. This varies from a mild localized urticarial reaction to a potentially fatal anaphylactic reaction. *Anisakis larvae* is a nematode which infests various species of fish. If this fish is eaten by people that produce increased amount of IgE in contact with the nematodes, an IgE-like mediated reaction takes place. However, it is very important to note that sensitization to seafood allergens follows not only after consumption, but also after inhalation or even only skin contact. In this context, tropomyosin represents the most important allergen from shellfish (Hanevik et al. 2007).

Clinical symptomatology seen in seafood toxicity often is similar with clinical features of the so-called fish-related histamine toxicity or scombroid syndrome, which is occurs after eating inappropriately cooled or handled fish from the *Scombridae* family (mahi-mahi, tuna). The cause of this clinical syndrome is represented by transformation of l-histidine normally seen in the fish muscle, into histamine by spoilage bacteria which subsequently determines a variety of symptoms from diarrhea to headaches, facial flushing, urticaria in milder cases to hypotension in patients with severe forms (Prester 2016; Hungerford 2010; Knope et al. 2014). The therapy in mild and moderate cases includes antihistamines such as diphenhydramine and cimetidine or adrenaline administered subcutaneous in those with severe form of disease associated with bronchospasm. In Australia, between 2001 and 2013, there were 57 outbreaks of histamine fish poisoning, and overall, 187 cases were recorded. However, 14% of the patients required hospital care but no fatalities were recorded (Hungerford 2010). Remarkably, the number of cases per outbreak is very small compared with other types infectious agents. There were between 1 and 22 cases for every outbreak, and nearly 90% of outbreaks included less than 5 patients (Hungerford 2010).

7 Ciguatera Toxicity

Ciguatera toxicity has been reported in patients living in the Pacific Ocean as well as Caribbean region and Mediterranean Sea, potentially affecting up to 500,000 people every year. This condition is determined by the metabolic transformation of gambiertoxins to ciguatoxins. Gambiertoxins are initially synthesized by dinoflagellate microalgae such as *Fukuyoa or Gambierdiscus* which are eaten by fish such as *Epinephelus fuscoguttatus*, the brown marbled grouper which is consumed by humans. It seems that fish that grows slower may accumulate higher amounts of ciguatoxins (Colombo et al. 2018; Soliño and Costa 2018; Chan 2014). Ciguatoxins open sodium channels, which subsequently will impact the overall cellular metabolism with subsequent increased intestinal secretion and diarrhea starting within hours after ingestion of infested fish. The patients present also with vomiting and pruritus. Chan (2014) reports also neurological signs, including paresthesia of all limbs and cardiovascular symptoms such as sinus bradycardia and hypotension requiring dopamine and atropine therapy as well as intravenous fluid replacement. Some patients developed also dyspnea and ventricular ectopic beats. Remarkably, the cardiovascular side-effects were present at higher incidence in those that consumed alcohol (Colombo et al. 2018; Clausing et al. 2018; Chan and Kwok 2001). Traylor (2020) suggests that initially activated charcoal should be administered after presentation, as this may inhibit further absorption of the toxin. Also, antihistamines should be administered for pruritus and IV rehydration for those with severe symptomatology. Calcium channel blockers could be administered for those patients with headache while amitriptyline, a tricyclic anti-depressant, could be administered for those with paresthesia (Pierre 2018). In addition, those with neurological symptoms may respond well to an intravenous infusion of mannitol solution (an osmotic diuretic) 20% (200 ml) slowly administered over 1 h (Clausing et al. 2018; Traylor and Singhal 2020).

8 Other Seafood Poisonings

The scientific literature mentions other types of poisoning caused by improper consumption of fish and mollusks. Shellfish poisoning is encountered in the context of harmful algal blooms, which offer optimal development conditions for dinoflagellates such as *Gymnodium, Pseudonitzschia, Dinophysis, and Alexandrium,* which are subsequently eaten and metabolized by shellfish as well as mussels, clams, and oysters, into domoic acid, brevetoxin, okadic acid, azaspiracid, and saxitoxin (Mullins and Hoffman 2017; Farabegoli et al. 2018). In addition, Arnich et al. (2020) have recently identified a new group of marine biotoxins called pinnatoxins which are synthesized by *Vulcanodinium rugosum*, a dinoflagellate encountered in the Mediterranean Coast in the South of France as well as on the Atlantic shores of France. However, this toxin seems to be harmless in humans, while a lethal effect has been described only experimentally in a murine model (Centers for Disease Control and Prevention 2017).

Paralytic shellfish poisoning represents a severe form of paresthesia, characterized by an onset within minutes, initially perioral which extends to include the limbs, severe headache, peripheral nerves paralysis, which may include the respiratory muscles with subsequent respiratory insufficiency and arrest. There is no specific therapy, but those with severe respiratory distress require intubation. The toxin is in fact an alkaloid produced by dinoflagellates as *Alexandrium* which represents the nutrition of mussels such as *Mytilius galloprovincialis* a bivalve mollusk. In some settings, the diagnosis could be confirmed by liquid chromatography or by bio-enzymatic assays which detect the toxin (Arnich et al. 2020; Bazzoni et al. 2020).

Amnestic shellfish poisoning is caused by domoic acid which is synthesized by members of the diatom genus *Pseudonitzschia* which contaminate the shellfish used for human consumption. Previously conducted studies have indicated that domoic acid is a glutamate acid agonist with neuroexcitatory action in the central nervous system. The first outbreak of amnestic shellfish poisoning was reported in 1987 in Prince Edward Island, Canada, in 107 diagnosed patients after consumption of cultivated Mytilus edulis mussels. It represented a major public health issues as the disease was not previously recognized. The most important clinical diagnostic criteria were the appearance of gastrointestinal symptoms within 24 h and neurological symptoms within 48 h of eating mussels. Most patients presented also with vomiting, abdominal pain, and diarrhea. Loss of short-term memory was present in 25% of patients and 12 patients developed seizures and become comatose. There were three fatal cases and the epidemiological investigation established the link between these cases and infested mussels cultivated in three riverbeds on Prince Edward Island (Raposo et al. 2020). No other cases of amnestic shellfish poisonings have been reported yet, but domoic acid poisoning have been documented in marine mammals and birds. However, at the present time, we do not understand the human health implications for low-level exposure to domoic acid (Perl et al. 1990; Lefebvre and Robertson 2010).

9 Tetrodotoxin Poisoning

This type of poisoning occurs sometimes after eating contaminated puffer fish, which belongs to the family of *Tetraondontidae*, considered an exquisite culinary experience in parts of South East Asia and Japan. However, it was also described after eating contaminated frogs, Australian blue-ringed octopuses, snails, and crabs. Numerous cases of tetrodotoxin poisoning have been also reported in Atlantic, European, and Pacific waters. Clinical features appear within minute after consumption. As tetrodotoxin is a strong neurotoxin actin upon sodium channels, the patients develop vomiting, ataxia, paresthesia, ascending paralysis, and coma. Often the prognosis is fatal as to date, we do not have a specific antidote against tetrodotoxin (Pulido 2008; Bane et al. 2014; Madejska et al. 2019)

10 Heavy Metals Toxicities

One of the oldest issues related to quality of water is related to the presence of lead. Prioreschi (1998) mentions that lead was widely used in Roman times in Roman aqueducts. However, Vitruvius identifies an unusual taste of the water which was supplied to the cities through lead pipes (Pinto et al. 2019). In England, before the eighteenth century, the so-called saturnine gout was produced after chronic ingestion of large amount of wine which was sweetened with lead acetate (Prioreschi 1998). More than 180 papers were published in English before the beginning of the XX century describing health issues related to lead toxicity. Jean-Martin Charcot (1825–1893), the founder of Neurology, is the first to identify neurological complications of lead poisoning (Eknoyan and Charcot 2011). Common symptoms of lead poisoning such as abdominal pain associated with constipation, irritability, memory issues were already known (Couper 2006). Even nowadays lead ends up in the drinking water from lead plumbing as this material degrades, especially if the supplied water has high acidity or minerals that corrode lead pipes, faucets, and other kitchen and bathroom fixtures (Couper 2006; Jonasson and Afshari 2018).

In April 2014, in Flint, Michigan a change in the municipal water source which was inappropriately tested determined a significant contamination with lead of water supplied to the population. While this was addressed at that time and health outcomes improved in the following years, the citizens of Flint experienced significant amount of psychological stress until 2018 (Sneed et al. 2020).

Currently, it is known that in adults chronic lead poisoning results in anemia in association with hypertension, fertility, and immune problems, whereas in children in addition there is abnormal deciduous dental development in association with delayed skeletal maturation. The molecular mechanisms responsible for all these clinical features are poorly understood. However, at the present time, the literature abounds with articles reporting as potential causes of clinical symptomatology a variety of mechanisms such as: single nucleotide polymorphisms, epigenetic changes, miRNAs. However, populational studies are needed to validate these theories (Mitra et al. 2017).

Mercury is a very toxic compound inducing neurologic, renal, and gastrointestinal problems, regardless of its form, methylmercury or as an organic compound. Acute mercury poisoning presents with metallic taste in the mouth, vomiting, and abdominal pain of variable quality. Ultimately, the patients will develop acute renal failure. This type of patients should be managed by chelation therapy with dimercaprol, which binds to toxic compound and charcoal to prevent its absorption. It is already notorious the May 1956 incident when on Kyushu Island, Japan, in Minamata Bay, harvested fish, shellfish, and other biological sea products indicated a high methylmercury contamination between 5.61 and 35.7 ppm, which was dumped in the sea by a local chemical plant. Moreover, the hair of inhabitants showed extremely high levels of mercury of up to 700 ppm, while they presented with severe neurological symptoms such as ataxia, glove and stocking sensory loss, reduction of visual fields, auditory disturbance, and dysarthria suggesting massive neuronal destruction in the cerebellum and cortex. Until 1995, 2,252 Minamata disease patients were recognized while 1,043 have since passed away (Harada 1995). Sadly, in 1965, the Agano River located in Niigata, Japan was also contaminated with methylmercury (Shimohata et al. 2015). Remarkably, even the fetuses of those with Niigata Minamata disease showed increased level of methylmercury, suggesting that this toxic compound may cross through the placenta. Currently, the typical patient with Minamata disease is an individual born during maximal contamination of Minamata Bay with methylmercury, between 1955 and 1959. During their biological development, they presented with postural abnormalities in head control and sitting as well as walking. These patients presented also with mental retardation (Sakamoto et al. 2017).

11 Arsenic Poisoning

In case of accidental poisoning with arsenic, the symptoms are described within minutes to one hour. The patients will develop vomiting, abdominal pain with diarrhea, and muscular contractions. Some patients will develop encephalopathy. Anyway, the therapy consists of sequestrants (dimercaprol, dimercaptosuccinic acid) and charcoal to prevent further absorption of arsenic. Hemodialysis may be used in some patients. However, arsenic is found naturally in groundwaters. Huq et al. (2020) have reported that in Bangladesh, groundwaters located in two counties contain arsenic levels significantly above the World Health Organization (WHO) requirements for drinking water, most likely of Himalayan origin (Hamann et al. 2014). However, about 150 million people all over the continents are affected by arsenic contamination. The most important outcomes to chronic exposure to high levels of arsenic are represented by development of a variety of cancers affecting the bladder, kidney, and lung. Also, arsenic exposure determines pancreas damage which results in diabetes (Mori et al. 2018). Luna et al. (2010) have shown that arsenic has the potential to interfere with immunological system. In a selected group of 87 children, arsenic had the ability to determine oxidative stress in blood cells activation oxidative pathways in the cells (Luna et al. 2010). Other studies have showed that chronic exposure to high levels of arsenic significantly increases the incidence of cardiovascular disease. A prospective study conducted over 19 years in 3,575 American Indian men in which the inorganic arsenic and methylated arsenic was measured in urine as an indicator or toxin exposure and showed that arsenic levels of more than 15.7 micro grams/g creatinine increased the risk for cardiovascular disease including coronary heart disease. However, while these findings are suggesting a serious public health issue affecting American Indians, the study is limited in interpretation as the level of arsenic in the drinking water was not measured (Moon et al. 2013). The cardiotoxic potential of arsenic is a very important aspect in oncology as several arsenic-based drugs are used in various therapeutic protocols. Arsenic trioxide is used in patients with acute promyelocytic leukemia but some of these patients have developed fatal arrhythmias. The risk of fatal cardiac events during arsenic trioxide therapy increases substantially if the patient presents with concomitant hypomagnesemia and hypokalemia (Raghu et al. 2009). On the skin, it determines acral keratosis, squamous cell carcinoma as well as melanosis and depigmentation (Mori et al. 2018). Last but not least, it seems that arsenic may have a role in Alzheimer's disease, promoting hyperphosphorylation of tau protein, which destabilizes the neuronal axons. Also, it may be involved in the formation of the amyloid precursor protein, which is the most important component of amyloid plaques characteristic for Alzheimer's. In addition, arsenic may release free oxygen radicals which would increase the neuronal damage (Gong and O'Bryant 2010).

12 Discussion and Conclusions

In 1993, in Milwaukee, an outbreak of Cryptosporidium parvum resulting from defective handling at the local water filtration plant resulted in more than 400,000 people (25% of the total population of the city) infected with this parasite. Overall, more than 96 million dollars were spent in medical costs. The most important factor leading to this outbreak was due to water transmission processes allowing the infectious people to discharge the parasite into the water sewage, which after transport to Lake Michigan was returned very soon to the water treatment plant. It was revealed that the distance between sewage effluent and drinking water supply influent was inappropriately short (Eisenberg et al. 2005). Therefore, while the architectural esthetics in designing a water treatment plant are not that important, the distance between the sewage water effluent and the drinking water influent is crucial for the general public health system. As there are numerous variables involved, a detailed evaluation is necessary prior to commencing any relevant project. However, current modeling methodology allows in some cases a separation between a wastewater effluent and the drinking water source to less than 40 m (Moore et al. 2020).

In 2000, the water supply in Walkerton, Ontario was contaminated with Escherichia coli O157:H7 and Campylobacter duodenales infecting more than 2300 people. Unfortunately, seven people died. The causes of this bacterial contamination outbreak were multiple from wrong practices, utility fraud, privatization of water testing facility, and non-existence of criteria for water quality and sanitation. In addition, there were provincial government financial cuts related to water supply to the population. This economic decision was taken without a real assessment from the public health specialists, although the politicians were warned about potential health implications (Salvadori et al. 2009).

We could conclude that the management of waters represents a complex issue with political, economic, cultural, and most importantly, health implications. Therefore, multiple partners must be involved and preferably management and monitoring agreements operational over multiple countries or regions should be implemented. Any type of architectural construction development project should fit within the United Nations-Sustainable Development Goals (UN-SDG) 2030 goal, providing the access to a sustainable source of water for all the residents while maintaining the water sanitation and avoid chemical pollution. Therefore, a dynamic inter-disciplinary dialog is required between all the parties involved, such as architects, builders, politicians, and health professionals.

References

American Academy of Pediatrics (1996) Practice parameter: the management of acute gastroenteritis in young children. Provisional Committee on Quality Improvement, Subcommittee on Acute Gastroenteritis. Pediatrics 97(3):424–35. ISSN: 0031-4005

Angeletti LR (1995) Environment and political institutions between antiquity and contemporary medicine. Med Secoli 7(3):415–423

Arnedo-Pena A, Beltrán-Fabregat J, Vila-Pastor B, Tirado-Balaguer MD, Herrero-Carot C, Bellido-Blasco JB, Romeu-Garcia MA, Safont-Adsuara L, Pac-Sa MR, Guillen-Grima F (2010) Reactive arthritis and other musculoskeletal sequelae following an outbreak of Salmonella hadar in Castellon. Spain J Rheumatol 37(8):1735–1742. https://doi.org/10.3899/jrheum.091250 Epub 2010 Jun 15

Arnich N, Abadie E, Delcourt N, Fessard V, Fremy JM, Hort V, Lagrange E, Maignien T, Molgó J, Peyrat MB, Vernoux JP, Mattei C (2020) Health risk assessment related to pinnatoxins in French shellfish. Toxicon 1(180):1–10. https://doi.org/10.1016/j.toxicon.2020.03.007

Arteaga M, Velasco J, Rodriguez S, Vidal M, Arellano C, Silva F, Carreño LJ, Vidal R, Montero DA (2018) Genomic characterization of the non-O1/non-O139 Vibrio cholerae strain that caused a gastroenteritis outbreak in Santiago, Chile, 2018. Microb Genom 6(3). https://doi.org/10.1099/mgen.0.000340

Bane V, Lehane M, Dikshit M, O'Riordan A, Furey A (2014) Tetrodotoxin: chemistry, toxicity, source, distribution and detection. Toxins (Basel) 6(2):693–755. https://doi.org/10.3390/toxins6020693

Barbosa LJ, Ribeiro LF, Lavezzo LF, Barbosa MM, Rossi GA, do Amaral LA (2016) Detection of pathogenic *Escherichia coli* and microbiological quality of chilled shrimp sold in street markets. Lett Appl Microbiol 62(5):372–8. https://doi.org/10.1111/lam.12562. Epub 2016 Mar 30

Bazzoni AM, Cangini M, Mudadu AG, Lorenzoni G, Arras I, Sanna G, Pino F, Milandri A, Virgilio S (2020) Recent findings of paralytic shellfish toxins linked to the genus Alexandrium Halim in Mediterranean mollusc production areas. Toxicon 30(174):48–56. https://doi.org/10.1016/j.toxicon.2019.12.157 Epub 2019 Dec 24

Boucheron P et al (eds) (2019) France in the world: a new global history, pp 30–35

Centers for Disease Control and Prevention. Harmful Algal Bloom (HAB)-Associated Illness. CDC.gov. Available at https://www.cdc.gov/habs/illness-symptoms-marine.html. December 13, 2017; Accessed 18 Apr 2020

Chan TY, Kwok TC (2001) Chronicity of neurological features in ciguatera fish poisoning. Hum Exp Toxicol 20(8):426–428

Chan TY (2014) Large outbreaks of ciguatera after consumption of brown marbled grouper. Toxins (Basel) 6(7):2041–2049. https://doi.org/10.3390/toxins6072041

Chhipi-Shrestha G, Hewage K, Sadiq R (2017) Microbial quality of reclaimed water for urban reuses: probabilistic risk-based investigation and recommendations. Sci Total Environ 576:738–751. https://doi.org/10.1016/j.scitotenv.2016.10.105

Clausing RJ, Losen B, Oberhaensli FR, Darius HT, Sibat M, Hess P, Swarzenski PW, Chinain M, Dechraoui Bottein MY (2018) Experimental evidence of dietary ciguatoxin accumulation in an herbivorous coral reef fish. Aquat Toxicol 200:257–265. doi: 0.1016/j.aquatox.2018.05.007. Epub 2018 May 26

Colombo FM, Cattaneo P, Confalonieri E, Bernardi C (2018) Histamine food poisonings: a systematic review and meta-analysis. Crit Rev Food Sci Nutr 58(7):1131–1151. https://doi.org/10.1080/10408398.2016.1242476 Epub 2017 Jun 2

Couper RT (2006) The severe gout of Emperor Charles V. N Engl J Med 355(18):1935–6; author reply 1936. Erratum in: N Engl J Med. 355(25):2712

Crisan CV, Hammer BK (2020) The Vibrio cholerae type VI secretion system: toxins, regulators and consequences. Environ Microbiol. https://doi.org/10.1111/1462-2920.14976

Daly ER, Roy SJ, Blaney DD, Manning JS, Hill VR, Xiao L, Stull JW (2010) Outbreak of giardiasis associated with a community drinking-water source. Epidemiol Infect 138(4):491–500 (ISSN: 1469-4409)

Desselberger U (2017) Viral gastroenteritis medicine (Abingdon). 45(11):690–694. https://doi.org/10.1016/j.mpmed.2017.08.005. Epub 2017 Oct 10

Eisenberg JN, Lei X, Hubbard AH, Brookhart MA, Colford JM Jr (2005) The role of disease transmission and conferred immunity in outbreaks: analysis of the 1993 Cryptosporidium outbreak in Milwaukee, Wisconsin. Am J Epidemiol 161(1):62–72

Eknoyan G, Charcot J-M (2011) Neurologist by avocation, nephrologist by yearning. J Nephrol 24(17):S4–11. https://doi.org/10.5301/jn.2011.6453

Elpers L, Hensel M (2020) Expression and functional characterization of various chaperon-usher fimbriae, Curli Fimbriae, and Type 4 Pili of Enterohemorrhagic *Escherichia coli* O157:H7 Sakai. Front Microbiol 11:378. https://doi.org/10.3389/fmicb.2020.00378

Eppinger M, Cebula TA (2015) Future perspectives, applications and challenges of genomic epidemiology studies for food-borne pathogens: a case study of Enterohemorrhagic Escherichia coli (EHEC) of the O157:H7 serotype. Gut Microbes 6(3):194–201. https://doi.org/10.4161/19490976.2014.969979 Epub 2014 Sep 1

Essex B, Koop SHA, Van Leeuwen CJ (2020) Proposal for a national blueprint framework to monitor progress on water-related sustainable development goals in Europe. Environ Manage 65(1):1–18. https://doi.org/10.1007/s00267-019-01231-1

Farabegoli F, Blanco L, Rodríguez LP, Vieites JM, Cabado AG (2018) Phycotoxins in marine shellfish: origin, occurrence and effects on humans. Mar Drugs 16(6). pii: E188. https://doi.org/10.3390/md16060188

Faria CP, Neves BM, Lourenço Á, Cruz MT, Martins JD, Silva A, Pereira S, Sousa MDC (2020) Giardia lamblia decreases NF-κB p65RelA protein levels and modulates LPS-induced pro-inflammatory response in macrophages. Sci Rep 10(1):6234. https://doi.org/10.1038/s41598-020-63231-0

Fink MY, Maloney J, Keselman A, Li E, Menegas S, Staniorski C, Singer SM (2019) Proliferation of resident macrophages is dispensable for protection during giardia duodenalis infections. Immunohorizons 3(8):412–421. https://doi.org/10.4049/immunohorizons.1900041

Gates C (2011) Ancient Cities: the archaeology of urban life in the ancient near east and Egypt, Greece and Rome, 2nd edn. Routledge. ISBN-10: 0415498643, ISBN-13: 978-0415498647

Giddings SL, Stevens AM, Leung DT (2016) Traveler's Diarrhea. Med Clin North Am 100(2):317–330. https://doi.org/10.1016/j.mcna.2015.08.017 Review. PubMed PMID: 26900116

Gong G, O'Bryant SE (2010) The arsenic exposure hypothesis for Alzheimer disease. Alzheimer Dis Assoc Disord 24(4):311–6. ISSN: 1546-4156

Grmek MD (1994) Malaria in the eastern Mediterranean in prehistory and antiquity. Parassitologia 36(1–2):1–6

Gu G, Ottesen A, Bolten S, Luo Y, Rideout S, Nou X (2020) Microbiome convergence following sanitizer treatment and identification of sanitizer resistant species from spinach and lettuce rinse water. Int J Food Microbiol 318:108458. https://doi.org/10.1016/j.ijfoodmicro.2019.108458. Epub 2019 Nov 23

Hadjimanolakis YE (2020) The port of piraeus through the ages. http://www.greece.org/poseidon/work/sea-ports/piraeus.html

Hagen KD, McInally SG, Hilton ND, Dawson SC (2020) Microtubule organelles in Giardia. Adv Parasitol 107:25–96. https://doi.org/10.1016/bs.apar.2019.11.001 Epub 2020 Feb 5

Hale TL, Keusch GT (1996) Shigella. In: Baron S (ed) Medical microbiology, 4th edn. University of Texas Medical Branch at Galveston, Galveston (TX)

Hamann CR, Boonchai W, Wen L, Sakanashi EN, Chu CY, Hamann K, Hamann CP, Sinniah K, Hamann D (2014) Spectrometric analysis of mercury content in 549 skin-lightening products: is mercury toxicity a hidden global health hazard? J Am Acad Dermatol 70(2):281–7.e3. ISSN: 1097-6787

Hanevik K, Hausken T, Morken MH, Strand EA, Mørch K, Coll P, Helgeland L, Langeland N (2007) Persisting symptoms and duodenal inflammation related to Giardia duodenalis infection. J Infect 55(6):524–530 Epub 2007 Oct 26

Harada M (1995) Minamata disease: methylmercury poisoning in Japan caused by environmental pollution. Crit Rev Toxicol 25(1):1–24. ISSN: 1040-8444

Hlavsa MC, Cikesh BL, Roberts VA, Kahler AM, Vigar M, Hilborn ED, Wade TJ, Roellig DM, Murphy JL, Xiao L, Yates KM, Kunz JM, Arduino MJ, Reddy SC, Fullerton KE, Cooley LA, Beach MJ, Hill VR, Yoder JS (2018) Outbreaks associated with treated recreational water—United States, 2000–2014. Morbidity and Mortality Weekly Report 67(19):547–551

Hungerford JM (2010) Scombroid poisoning: a review. Toxicon 56(2):231–243. https://doi.org/10.1016/j.toxicon.2010.02.006 Epub 2010 Feb 10

Huq ME, Fahad S, Shao Z, Sarven MS, Khan IA, Alam M, Saeed M, Ullah H, Adnan M, Saud S, Cheng Q, Ali S, Wahid F, Zamin M, Raza MA, Saeed B, Riaz M, Khan WU (2020) Arsenic in a groundwater environment in Bangladesh: Occurrence and mobilization. J Environ Manage 262:110318. https://doi.org/10.1016/j.jenvman.2020.110318

Inns T, Wilson D, Manley P, Harris JP, O'Brien SJ, Vivancos R (2019) What proportion of care home outbreaks are caused by norovirus? An analysis of viral causes of gastroenteritis outbreaks in care homes, North East England, 2016–2018. BMC Infect Dis 20(1):2. https://doi.org/10.1186/s12879-019-4726-4

Jonasson ME, Afshari R (2018) Historical documentation of lead toxicity prior to the 20th century in English literature. Hum Exp Toxicol 37(8):775–788. https://doi.org/10.1177/0960327117737146 Epub 2017 Oct 27

Juuti PS, Katko TS, Vuorinen HS. A brief history of water and health from ancient civilizations to modern times. https://www.iwapublishing.com/news/brief-history-water-and-health-ancient-civilizations-modern-times Accessed on 19 Apr 2020

Kim S (2015) Epidemic Cholera and American reform movements in the 19th century. Uisahak 24(3):783–814. https://doi.org/10.13081/kjmh.2015.24.783

Knope K, Sloan-Gardner TS, Stafford RJ (2014) Histamine fish poisoning in Australia, 2001 to 2013. Commun Dis Intell Q Rep 38(4):E285–E293

Krensky S (1991) Who really discovered America? p 30. ISBN 0-590-40854-2. Hastings House Book Publishers

Kunkle DE, Bina XR, Bina JE (2020) Vibrio cholerae OmpR contributes to virulence repression and fitness at alkaline pH. Infect Immun pii: IAI.00141-20. https://doi.org/10.1128/iai.00141-20

Lafont O (2014) Parmentier hygiene and public health. Ann Pharm Fr 72(3):194–201. https://doi.org/10.1016/j.pharma.2013.12.007 Epub 2014 Jan 21

Lefebvre KA, Robertson A (2010) Domoic acid and human exposure risks: a review. Toxicon 56(2):218–30. ISSN: 1879-3150

Lorrot M, Vasseur M (2007) How do the rotavirus NSP4 and bacterial enterotoxins lead differently to diarrhea? Virol J 4:31. ISSN: 1743-422X

Luna AL, Acosta-Saavedra LC, Lopez-Carrillo L, Conde P, Vera E, De Vizcaya-Ruiz A, Bastida M, Cebrian ME, Calderon-Aranda ES (2010) Arsenic alters monocyte superoxide anion and nitric oxide production in environmentally exposed children. Toxicol Appl Pharmacol 245(2):244–51. ISSN: 1096-0333

Madejska A, Michalski M, Osek J (2019) Marine tetrodotoxin as a risk for human health. J Vet Res 63(4):579–586. https://doi.org/10.2478/jvetres-2019-0060. eCollection 2019 Dec

Mavian C, Paisie TK, Alam MT, Browne C, Beau De Rochars VM, Nembrini S, Cash MN, Nelson EJ, Azarian T, Ali A, Morris JG Jr, Salemi M (2020) Toxigenic Vibrio cholerae evolution and establishment of reservoirs in aquatic ecosystems. Proc Natl Acad Sci U S A. 117(14):7897–7904. https://doi.org/10.1073/pnas.1918763117. Epub 2020 Mar 30

McLeod KS (2000) Our sense of snow: the myth of John Snow in medical geography. Soc Sci Med 50(7–8):923–935. https://doi.org/10.1016/s0277-9536(99)00345-7

Mitra P, Sharma S, Purohit P, Sharma P (2017) Clinical and molecular aspects of lead toxicity: an update. Crit Rev Clin Lab Sci 54(7–8):506–528. https://doi.org/10.1080/10408363.2017.1408562. Epub 2017 Dec 7

Moon KA, Guallar E, Umans JG, Devereux RB, Best LG, Francesconi KA, Goessler W, Pollak J, Silbergeld EK, Howard BV, Navas-Acien A (2013) Association between exposure to low to moderate arsenic levels and incident cardiovascular disease. A prospective cohort study. Ann Intern Med 159(10):649–59

Moore C, Nokes C, Loe B, Close M, Pang L, Smith V, Osbaldiston S (2020) Keeping waters apart—new separation distance guidelines for onsite wastewater systems and wells. https://www.waternz.org.nz/Attachment?Action=Download&Attachment_id=1135

Mori S, Lowenstein EJ, Steffen C (2018) The largest mass poisoning in history: arsenic contamination of well water in Bangladesh. Skinmed 16(4):265–267. eCollection 2018

Mullins ME, Hoffman RS (2017) Is mannitol the treatment of choice for patients with ciguatera fish poisoning? Clin Toxicol (Phila) 55(9):947–955

O'Flaherty E, Solimini A, Pantanella F, Cummins E (2019) The potential human exposure to antibiotic resistant—*Escherichia coli* through recreational water. Sci Total Environ 650(Pt 1):786–795. https://doi.org/10.1016/j.scitotenv.2018.09.018 Epub 2018 Sep 6

Perl TM, Bédard L, Kosatsky T, Hockin JC, Todd EC, Remis RS (1990) An outbreak of toxic encephalopathy caused by eating mussels contaminated with domoic acid. N Engl J Med 322(25):1775–80. ISSN: 0028-4793

Pierre O, Misery L, Talagas M, Le Garrec R (2018) Immune effects of the neurotoxins ciguatoxins and brevetoxins. Toxicon 149:6–19. https://doi.org/10.1016/j.toxicon.2018.01.002 Epub 2018 Jan 31

Pinto EP, Rodrigues SM, Gouveia N, Timóteo V, Costa PR (2019) Tetrodotoxin and saxitoxin in two native species of puffer fish, Sphoeroides marmoratus and Lagocephalus lagocephalus, from NE Atlantic Ocean (Madeira Island, Portugal). Mar Environ Res 151:104780. https://doi.org/10.1016/j.marenvres.2019.104780 Epub 2019 Sep 3

Prester L (2016) Seafood allergy, toxicity, and intolerance: a review. J Am Coll Nutr 35(3):271–83. ISSN: 1541-1087

Prioreschi P (1998) A history of medicine, volume 3 of roman medicine. Horatius Press, p 279. ISBN 978-1-888456-03-5

Pulido OM (2008) Domoic acid toxicologic pathology: a review. Mar Drugs 6(2):180–219. https://doi.org/10.3390/md20080010

Raghu KG, Yadav GK, Singh R, Prathapan A, Sharma S, Bhadauria S (2009) Evaluation of adverse cardiac effects induced by arsenic trioxide, a potent anti-APL drug. J Environ Pathol Toxicol Oncol 28(3):241–52. ISSN: 2162-6537

Raposo M, Botelho MJ, Costa ST, Gomes MTSR, Rudnitskaya AA (2020) Carbamoylase-based bioassay for the detection of paralytic shellfish poisoning toxins. Sensors (Basel). 20(2):E507. https://doi.org/10.3390/s20020507

Robertson L, Gjerde B, Hansen EF, Stachurska-Hagen T (2009) A water contamination incident in Oslo, Norway during October 2007; a basis for discussion of boil-water notices and the potential for post-treatment contamination of drinking water supplies. J Water Health 7(1):55–66. ISSN: 1477-8920

Rohekar S, Tsui FW, Tsui HW, Xi N, Riarh R, Bilotta R, Inman RD (2008) Symptomatic acute reactive arthritis after an outbreak of salmonella. J Rheumatol 35(8):1599–1602 Epub 2008 Jun 1

Roussel C, Sivignon A, de Wiele TV, Blanquet-Diot S (2017) Foodborne enterotoxigenic *Escherichia coli*: from gut pathogenesis to new preventive strategies involving probiotics. Future Microbiol 12:73–93. https://doi.org/10.2217/fmb-2016-0101 Epub 2016 Dec 16

Sakamoto M, Itai T, Murata K (2017) Effects of prenatal methylmercury exposure: from minamata disease to environmental health studies. Nihon Eiseigaku Zasshi 72(3):140–148. https://doi.org/10.1265/jjh.72.140

Salvadori MI, Sontrop JM, Garg AX, Moist LM, Suri RS, Clark WF (2009) Factors that led to the Walkerton tragedy. Kidney Int Suppl 112:S33–S34. https://doi.org/10.1038/ki.2008.616

Satchell KJ (2003) Activation and suppression of the proinflammatory immune response by Vibrio cholerae toxins. Microbes Infect 5(13):1241–1247. https://doi.org/10.1016/j.micinf.2003.08.007 PMID: 14623020

Scarcella C, Carasi S, Cadoria F, Macchi L, Pavan A, Salamana M, Alborali GL, Losio MM, Boni P, Lavazza A, Seyler T (2009) An outbreak of viral gastroenteritis linked to municipal water supply, Lombardy, Italy, June 2009. Euro Surveill 14(29). ISSN: 1560-7917

Shimohata T, Hirota K, Takahashi H, Nishizawa M (2015) Clinical aspects of the Niigata Minamata disease. Brain Nerve 67(1):31–38. https://doi.org/10.11477/mf.1416200084

Sneed RS, Dotson K, Brewer A, Pugh P, Johnson-Lawrence V (2020) Behavioral health concerns during the flint water crisis, 2016–2018. Community Ment Health J 56(5):793–803. https://doi.org/10.1007/s10597-019-00520-7

Soliño L, Costa PR (2018) Differential toxin profiles of ciguatoxins in marine organisms: chemistry, fate and global distribution. Toxicon 150:124–143. https://doi.org/10.1016/j.toxicon.2018.05.005 Epub 2018 May 17

Tate JE, Panozzo CA, Payne DC, Patel MM, Cortese MM, Fowlkes AL, Parashar UD (2009) Decline and change in seasonality of US rotavirus activity after the introduction of rotavirus vaccine. Pediatrics 124(2):465–71. ISSN: 1098-4275

Thornton BS (2002) Greek ways: how the Greeks created western civilization, Revised edn. Encounter Books. ISBN-10: 1893554570, ISBN-13: 978-1893554573

Traylor J, Singhal M (2020) Ciguatera toxicity. StatPearls [Internet]. Treasure Island (FL). StatPearls Publishing, 2019 Oct 18

Von Mentzer A, Zalem D, Chrienova Z, Teneberg S (2020) Colonization factor CS30 from enterotoxigenic *Escherichia coli* binds to sulfatide in human and porcine small intestine. Virulence 11(1):381–390. https://doi.org/10.1080/21505594.2020.1749497

Zbinden A (2019) Acute viral gastroenteritis: viruses other than norovirus. Praxis (Bern 1994) 108(5):335–339. https://doi.org/10.1024/1661-8157/a003182

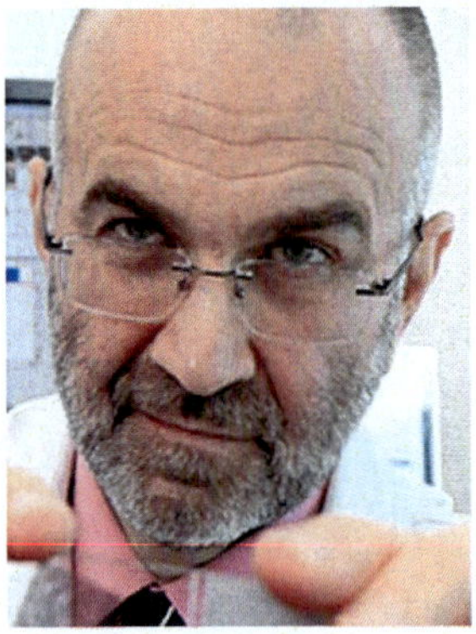

Dr Eugen Bogdan Petcu MD, PhD is an academic pathologist with interests in angiogenesis and regenerative medicine research including 3-D additive bio-manufacturing. Also, he enjoys neuroesthetics and artistic anatomy and has published a significant number of articles and monographs in both medical and humanities journals including Leonardo.

The Relation Between Coastal Flood Risk and Ecosystem Services Affecting Coastal Tourism: A Review of Recent Assessments

Carlotta Quagliolo, Alessandro Pezzoli, Elena Comino, and Marco Bagliani

Abstract It is widely accepted by the scientific community that coastal cities are considered the most vulnerable urban areas. Future climate change scenarios and impacts are expected to exacerbate coastal flood risk with ecological and economic influence on the ecosystem services and human activities as coastal tourism. In this context, the aim of this chapter, reviewing recent research, is to explore the state of the art on the vulnerability of coastal zones, mainly due to the risk of "great floods", and the potential adaptation of coastal tourism to address this issue in the face of climate change. The perspective of this review analysis is to deepen the interrelations among climate change and flood risk related to coastal ecosystem services and their impacts on coastal tourism. Particularly, by conducting a systematic review focused on Australian and European regions, the present study examines the emergent themes about these topics while addressing the need to deal with an integrated approach. We searched two academic literary databases and identified 16 studies that met the selection criteria for this review. Findings highlighted that relatively few studies had analysed the interrelationships between these issues, showing a research gap both from a regional geographic coverage and from a strategy perspective. Emergent themes of the recent studies reviewed here could help to discover critical features and future research axes for more resilient coastal urban communities.

C. Quagliolo · A. Pezzoli · E. Comino (✉)
Politecnico di Torino, Turin, Italy
e-mail: elena.comino@polito.it

C. Quagliolo
e-mail: carlotta.quagliolo@polito.it

A. Pezzoli
e-mail: alessandro.pezzoli@polito.it

M. Bagliani
University of Torino, Turin, Italy
e-mail: marco.bagliani@unito.it

J. Baumeister et al. (eds.), *SeaCities*, Cities Research Series,
https://doi.org/10.1007/978-981-15-8748-1_8

1 Introduction

Coastal zones have always been crucial and attractive for humans due to their abundant supply of resources, logistical reasons, trade and transport activities as well as recreational and cultural ones. Indeed, the oceans and particularly coastal zones contribute more than 60% of the total economic value of the biosphere, while providing ecosystem services with considerable environmental and economic value (Costanza et al. 1997). Critical infrastructures as ports are key elements for the economic growth of both coastal and inland areas enhancing land–sea connection. Tourism, as the main industry globally, has the potential to modify local economies by providing multibenefits to the coastal communities, that are mainly composed of microenterprises. Since tourism has become the main economic engine of coastal areas mostly during the summer season, the majority of coastal and maritime tourism is based on nautical activities and the sun-and-sand model. As a consequence, this sector depends highly on good environmental conditions of coastal zones (i.e. the water and beach quality) (European Commission 2019). In this context, coastal tourism can be considered both as providing cultural services and as one of the main drivers of ecosystem variation, and thus strictly linked to regulating and provisioning services (Drius et al. 2019).

Although coastal areas cover only 4% of the Earth's total land area, coastal population densities are about three times higher than the global average of inland areas (Sanchez-Reaza and Carletto 2013). Moreover, the future population and urbanisation rates are projected to increase rapidly in the low-elevation coastal zone—LECZ (coastal areas below 10 m of elevation), specifically by 50–71% by 2050, compared to the year 2000; this is equivalent to 630 million people (McGranahan et al. 2007; Lichter et al. 2011; Merkens et al. 2016). The attraction of living near the coast is so strong that, as a result, 15 out the 20 present-day world's megacities are located in LECZ (Neumann et al. 2015; Mehvar et al. 2018). These coastal cities are particularly exposed to natural hazards, such as periodical saltwater intrusion, coastal erosion, flooding and storms, increasing risk to key infrastructures. Besides, they also deal with typical problems of large urban centres, such as water, air and noise pollution, as well as social issues (such as inequality and gentrification) and rapid increase in tourism activities, which represent a significant environmental-socio-economic burden (Oppenheimer et al. 2019).

The risk of "great floods" (i.e. the floods with discharges exceeding 100-year levels) has substantially increased throughout the past century, due to coastal ecosystems degraded by human activities (Milly et al. 2002). This trend is expected to continue in the future, showing that, under climate change, flood frequency will rise in 42% of the land area, more likely due to the increased frequency of extremes weather events (Hirabayashi et al. 2013). Even though direct consequences of climate change remain limited to temperature extremes and sea-level rise, change in flooding features and other hydro-meteorological hazards are associated with climate change processes, leading to higher probability and severity of these events (IPCC 2019).

Globally, changes in precipitation patterns from climate change increase the sensitivity of urban areas, due to their high level of impervious surfaces (Scholz 2013). In this context, urban flooding is mainly caused by local high-intensity precipitation (pluvial flood), high flows in river-prone systems (fluvial flood) and high sea levels (coastal flood). Simultaneous occurrence of these flood types exacerbates the effects of a single event and increases the risk of coastal hazards (Berndtsson et al. 2019; IPCC 2019).

The global direct economic loss due to flooding was estimated to be over $ 1 trillion, across the period 1980–2013, with a high probability of even larger damages by the end of the century (Winsemius et al. 2016). About 70% of this escalation can be attributed to economic growth in flood-prone areas, posing a critical problem to policymakers (Berndtsson et al. 2019). Due to the severity and frequency of flood events, interest in flood risk has grown rapidly in both research and policy-making communities. In recent years, many studies have conducted an integrated vulnerability assessment and cost–benefit analysis at different scales, for flood protection in coastal and inland areas, aiming at supporting mitigation and adaptation measures to reduce flood risk impacts (Haer et al. 2017; Aroca-Jiménez et al. 2018; Rehan 2018; de Ruig et al. 2019; Garrote et al. 2019; Aitali et al. 2020; Fang et al. 2020; Ogato et al. 2020). Other studies have estimated the potential impacts and losses associated to flood risk under climate change scenarios in different coastal contexts and discussed the underling need for non-stationary risk frameworks (Nicholls 2002; Graff et al. 2019; Kundzewicz et al. 2019; Reed et al. 2020; Toimil et al. 2020). Furthermore, floods and sea-level rise have been analysed from microeconomic perspectives considering not only direct and indirect damage costs, but also loss of future potential economic growth through dynamic consumption and investment decisions (Grames et al. 2016).

There is a need to establish the current state of knowledge about risks and opportunities for the local economy in coastal areas also based on a recent assessment, i.e. the Special Report on the Ocean and Cryosphere in a Changing Climate (SROCC), where the Intergovernmental Panel on Climate Change (IPCC) identified various regional hazards (that cause physical changes) and impacts (positive and negative on the various systems) on oceans.

The present study addresses these needs by presenting a coherent literature review on latest research focused on this topic, especially by considering the sea-level rise and coastal flood risk as natural hazards. The SCROCC from IPCC has shown that the observed sea-level is projected to increase for all geographical areas. Therefore, we investigated the link among climate change and flood risk related to coastal ecosystem services and their impacts on a specific aspect of the local economy, that is tourism. This perspective represents the novelty of the present review. Indeed, researches have explored the effect of tourism at a local level (Richins and Pearce 2000; Moscardo et al. 2001); nevertheless, the literature lacks comparative reviews between two different regional areas selected for this review, i.e. Europe and Australia.

Figure 1, which is the result of our elaboration, based on the SROCC assessment in relation to variables of interest for this review analysis, illustrates various regional physical changes (increase/decrease or both) in the ocean regions and local

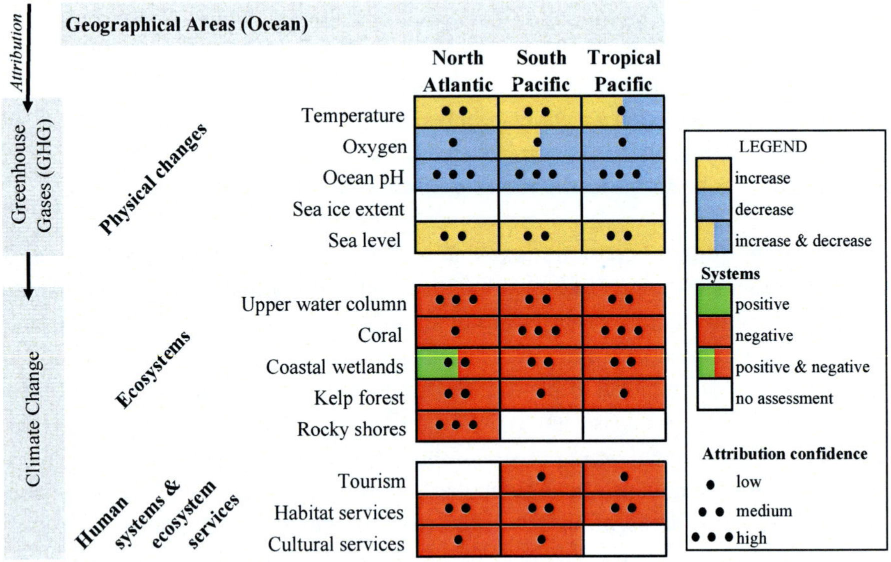

Fig. 1 Observed regional hazards and impacts in ocean regions. Authors' elaboration based on Figure SPM.2 of the SROCC (IPCC 2019)

climate change impacts (positive/negative or both) on ecosystems, human systems and ecosystem services. For each ocean regions, the confidence level refers to the confidence in attributing observed changes in response to variation in greenhouse gas forcing for physical changes and to climate change ecosystems, human systems and ecosystem services. What emerges from Fig. 1 concerns the lack of assessment (i.e. not assessed at a regional scale or insufficient evidence in this area) of climate change impact on coastal tourism in the North Atlantic's ocean region.

The main contribution of this literature review would be the identification of regional knowledge gaps by revealing the spatial pattern of climate change and coastal flood risk related to coastal ecosystem services and impacts on coastal tourism. In doing so, it identifies "hotspots" that should be prioritised for further analysis while evidencing recommendations for addressing these gaps. Specifically, this review can serve as an excellent foundation to develop more considerable advancements in this field by giving an overview of emergent themes from literature. In doing so, first, we present the method used to conduct this review analysis (Sect. 2). Secondly, we describe and analyse the papers under multiple classification criteria (Sect. 3.1) to present the major findings as emergent themes from each study organised under three topics of interest: coastal flood risk, coastal ecosystem services and coastal tourism (Sect. 3.2). Finally, we provide a brief discussion and conclusion (Sect. 4–5).

2 Research Methodology

A literature review refers to the examination of recent or current literature which has been subjected to a peer-review process. This typology differs from rapid or scoping reviews which provide preliminary assessments on the topic interested (Grant and Booth 2009).

This review was conducted based on a research methodology consisting of two subsequent stages. The initial phase of research was based on the definition of the search strategy and the identification of inclusion criteria needed for creating the inventory of articles (Sect. 2.1). Subsequently, the selected studies were reviewed through the codification of classification criteria to organise notions of each study in categories. This process helps to conduct an analysis systematically while obtaining a first overview of the results (Sect. 2.2).

2.1 *Research Strategy and Inclusion Criteria*

For the identification of the studies in this research, we adopted a three-step procedure illustrated in Fig. 2. In the first step, we search papers indexed in two online literary databases—Science Direct (www.sciencedirect.com) and Web of Science (www.webofknowledge.com), being a common strategy among researchers (Thompson et al. 2017; Sadiq et al. 2019). The use of two literature database allowed to include more articles, but paid attention to exclude the repeated papers. Since we need to conduct a careful search analysis, meaning that we initially defined inclusion criteria,

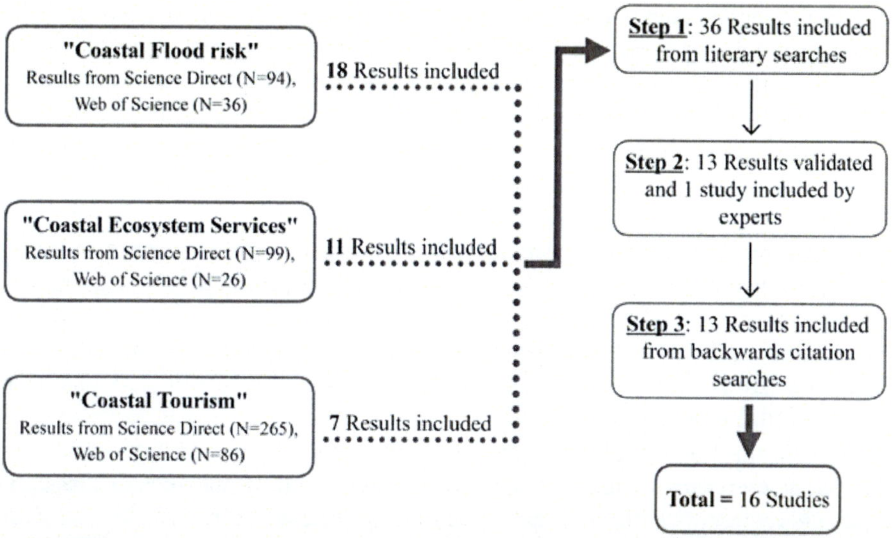

Fig. 2 Diagram of the search strategy and selected studies

we decided not to use the Google Scholar database because it does not adapt to our research. Indeed, this database does not allow "advanced searches form" allowing to specify one or more pieces of information (e.g. keywords, dates, authors, etc.) to find related studies (i.e. explicit query using multiple Boolean operators and nested clauses).

The inclusion criteria for this first step were (1) papers written in English; (2) only peer-reviewed journal articles; and (3) temporal coverage of the last three years (2018–2019–2020). The reason why we decided to define a short and very recent timeframe as search criterion is due to the nature of this review article, since our literature review also included review articles of older relevant studies. This literature review does not represent an exhaustive inventory; still, it wants to show the recent trend of literature with the scope limited to the coastal flood risk impacts on the coastal ecosystem services, which are both attractive for coastal tourism and vulnerable at the same time. In this context, this literature research takes into consideration the multifaceted interactions between ecosystem services and the human activities connected to the tourism sector.

We began our investigation using the keywords search "Coastal Flood risk". This keywords search yielded 94 studies from Science Direct and 36 from Web of Science. To refine the analysis assuring to include other useful studies, we did another search on two online databases using the following keywords: "Coastal Ecosystem Services" and "Coastal Tourism". The "Coastal Ecosystem Services" keyword search yielded 99 studies from Science Direct and 26 from Web of Science. The "Coastal Tourism" keyword search returned 265 studies from Science Direct and 86 from Web of Science. Although the keywords searches ended, we used Web of Science Alerts to integrate the research with studies uploaded to Web of Science that contained the pre-identified search terms up to the beginning of March 2020. Nevertheless, no additional remarkable studies for our research were found.

In the second step, we reviewed the title, abstract, keyword and type of journal for every paper generated by each keyword search of the first stage (36 results), to assess if it met the criteria for inclusion. Specifically, the primary method for determining the reason for exclusion was whether the study's focus was on Europe, Australia, or at a global level. The second method to validate our list was to assume that the coastal flood risk influences the coastal ecosystem services, which have a potential impact on coastal tourism. A total of 13 studies met the inclusion criteria. Through the involvement of an experts' focus group with experience in the field representing varied perspectives from academia and private sector (e.g. industry), it was recommended the inclusion of an important article review at global scale published on 2013. Even if this article is out of the search criterion of timeframe (range of 2018–2020), it was added to the total 13 studies, as considered fundamental basis for this review.

In the third and final step, we conducted a backwards citations search going through references of all 14 studies that resulted during the second phase. This process represents a further improvement of the articles' search analysis to identify the missing studies from the first step eventually. Indeed, this backwards citation process allowed to identify an additional two new studies that met the inclusion criteria. Finally, in total, 16 studies were included in this review. The adopted strategy may have excluded some articles considering, for example, the wide range of disciplines analysing the coastal flood risk impacts, the coastal ecosystem services and coastal tourism.

2.2 Classification Criteria

Two of the authors reviewed the 16 studies included in this review though systematic codification of the information coming from these articles following the classification criteria by covering the study area, year of publication, article type, approach, "climate change issue" and descriptors (variables). Table 1 in Appendix presents a synthesis of all the criteria used to organise the notion of each of the 16 eligible studies. At the same time, the following sections illustrate the meaning of some classification criteria.

Research objectives to conduct the analysis were characterised by reviewing the research question, purpose and method, and describing the main interesting findings of each study and then identifying the common aspects among the 16 studies. Due to the complexity of this review due to exploring the link among four macro-argument (climate change-coastal flood risk-coastal ecosystem services-coastal tourism) and consequently, the wide range of disciplines and approaches used by authors, we describe the path under multiple, different perspectives. We also structured the results of this review around the emergent issues that need to be explored. Using this method, we systematically identified the significant current knowledge gaps both regarding the geographical coverage of studies and the connection among our keywords by comparing all the reviews.

2.2.1 Study Area and Geographical Scale

To determine the geographic scope and thus, the spatial level of each study (e.g. global, national, regional, local), we extracted the information on the study area. Three studies included in this review conducted a global-scale analysis derived by the results of different local and regional case studies (Scott et al. 2019; Van Coppenolle and Temmerman 2019, 2020). One study describes the status quo of coastal and marine ecosystem services research at a global level (Liquete 2013). A higher number of studies were conducted in Europe, especially at a regional or local scale, analysing specific coasts and coastlines of some countries (Roebeling et al. 2013; Hérivaux et al. 2018; Pueyo-Ros et al. 2018; Toimil et al. 2018; Drius et al. 2019; López-Dóriga et al. 2019; Cooper et al. 2020) even if one study focused at the same time on both the regional and national scale for the Faroe Islands (Plieninger et al. 2018). One looked at coastal and marine ecosystem services at a local level in Australia (Sangha et al. 2019), while two studies did not specify its geographic scope (Caro et al. 2018; Debele et al. 2019). The locations of the local- and regional-scale

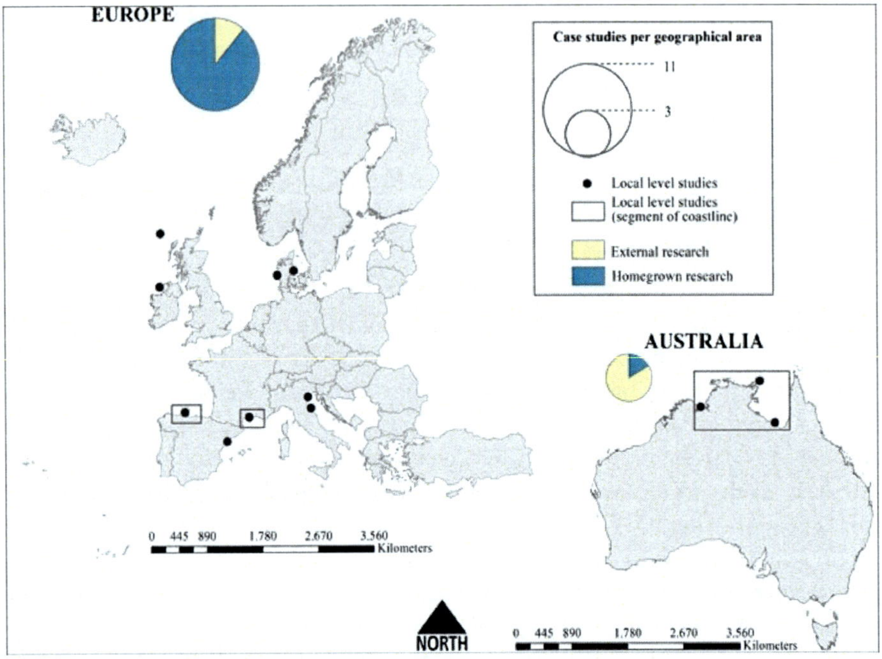

Fig. 3 Location of case studies and the affiliation of the first author

case studies are illustrated in Fig. 3. Of those studies, we also investigated the correspondence between the affiliation of the first author and the case study, in line with the methodology developed by Liquete (2013). In the map, the pie size represents the number of studies per each geographical areas (Australia and Europe). Points describe the location of the local-level case studies, while the rectangle depicts the local-level studies extended along a segment of coastline. The colours within the pie show the proportion of the studies conducted by researchers from the same geographical region (light yellow) or different geographic areas (blue). Most of the selected papers were led by the institution situated in the same country of the case, although in two cases the first author is, respectively, a Canadian and American researcher.

2.2.2 Article Type and Approach of the Study

We coded the typology of the article by classifying it as either a review or research article. It is important to note that four studies are review articles (Liquete 2013; Caro et al. 2018; Mora et al. 2018; Debele et al. 2019). What emerges though is that no articles review on coastal tourism has been recently conducted.

The approach of each study is coded as qualitative or quantitative, based on whether the study uses qualitative or quantitative data. Of the total studies reviewed, there are significantly more quantitative ($N = 10$) than qualitative ($N = 5$) ones. In one case (i.e. the research on the Faroe Islands), the study includes both qualitative and quantitative information (Plieninger et al. 2018). Some of them are subcategorised by whether the qualitative/quantitative approach is mapping oriented or uses a conceptual framework.

2.2.3 Climate Change Issue

We introduced the codification of the "climate change issue" by specifying how each study deals with climate change, mainly if this issue has been analysed marginally or extensively. Since the present study aims to understand the connection between climate change and coastal flood risk, we thus explored the intensity of this link. Precisely, we dug deeper through this literature review identifying whether each study had cited the climate change phenomenon as a boundary condition, coding it "Contextual", or used some data provided by climate change analysis, defining it "Analytical", or focused on climate change analysis representing it "Pivotal". We verified that our sample resulted in the same portion of studies for "Contextual" ($N = 7$) and "Analytical" approaches ($N = 7$). Only two studies categorised as "Pivotal" have demonstrated their objective of analysing the "climate change" (see Table 1—Appendix A).

3 Results

3.1 Overview Analysis

Figure 4 shows the descriptive statistics where pies illustrate, respectively: (a) the percentage of studies characterised as review or research articles; (b) the proportion of studies that use a qualitative, quantitative or mixed (qualitative and quantitative) approach; (c) the proportion of studies which tackle the climate change issue through "Contextual", "Analytical" or "Pivotal" approach. Finally, the column chart (d) shows the focus on the typology of analysis adopted by each qualitative, quantitative or mixed approach. The percentages within columns represent the proportion of studies for each type of analysis: mapping, framework or not defined. Generally, the propensity of researchers to study this topic in recent years is to structure the article type in research studies (75% of analysed cases compared to 25% of review articles—Fig. 4a). Since our analysis is based on a good portion of review articles, our choice to conduct a recent literature review was confirmed by the fact that this work wants to exemplify the previous literature, and thus indirectly we cite other references.

In relation to the approach adopted by each article, 63% is quantitative, the 31% qualitative and 6% mixed (qualitative and quantitative). Moreover, within the quantitative approach, the majority of studies tends not to define the type of analysis (60%), while 20% employs the mapping approach and the same portion uses quantitative

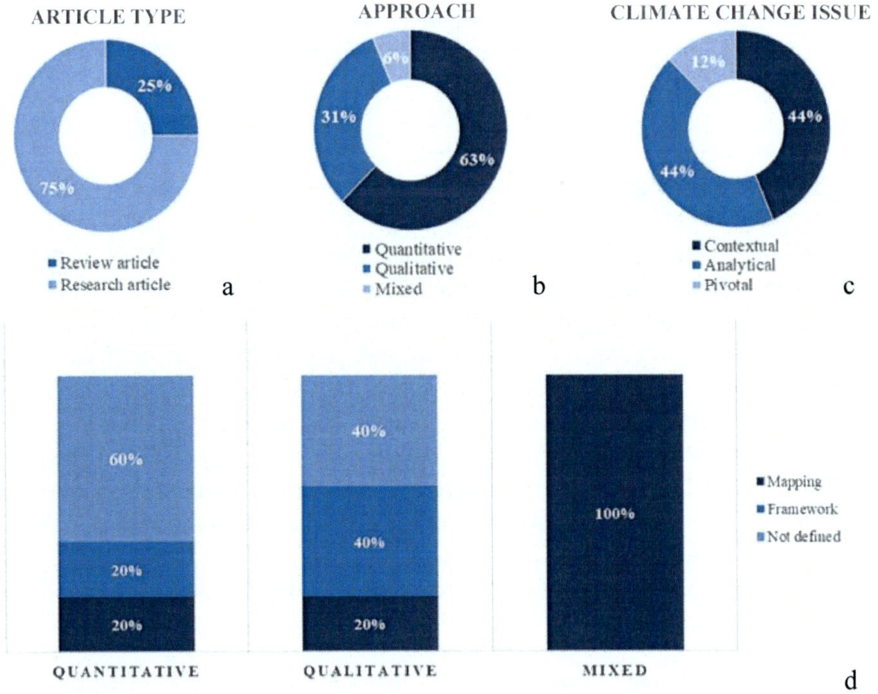

Fig. 4 Descriptive statistics of analysed articles

frameworks. For qualitative studies, the same percentage of 40% emerges for ones that do not specify the type of analysis and for mapping ones (Fig. 4d). The "climate change issue" outcome indicates that the same number of studies deal with climate change through the citations as a boundary condition or analysis of some data (Fig. 4c).

The following three Sects. (3.2–3.3–3.4) describe the major emergent themes from recent literature organised in key issues: coastal floods, coastal ecosystem services and coastal tourism.

3.2 *Emergent Theme: Coastal Floods*

The review article of Debele et al. (2019) highlighted the need for classifying extreme events into hydro-meteorological hazards to have a better interpretation of risks prediction. Usually, the hydro-meteorological hazards are categorised based on their driving mechanisms and their statistical behaviour (i.e. duration and magnitude of extreme events, the potential of occurrence). Among hydro-meteorological hazards, floods, storm surges, landslides, drought and heatwaves are recognised as the critical natural events. Anomalies trigger these events in the atmosphere and the hydrological cycle as weather-, climate- and hydrological-related events, which are projected to increase in their frequency and magnitude due to climate change impacts (Debele et al. 2019). In this context, many previous relevant studies showed that the trend in climate model projections on flood risk is expected to increase within mid- and long-term across Europe. Meanwhile, other studies evidenced projected growing patterns for storm surge risk to be one of the primary triggers with sea-level rise for coastal flood (Alfieri et al. 2015a; b; Vormoor et al. 2016). What emerges about coastal floods from selected literature is that the leading cause is a severe storm (like a hurricane). A coastal flood can be induced by stormwater surge along the coast due to high tides combined with low atmospheric pressures and strong winds. Storm surge can have an impact for several kilometres upstream causing further flooding (Debele et al. 2019). Storm surge is predicted to exceed by 30% the relative sea-level rise along the European coast under the "Business as Usual" IPCC scenario (Vousdoukas et al. 2016).

Particularly, the results from a regional analysis on sea-level rise projections conducted in the Languedoc-Roussillon coastline (France) by Hérivaux et al. (2018) showed that, overall, around 39,000 ha of the coastal area might be exposed to coastal flooding in 2100 compared to 15,000 ha in a scenario without sea-level rise. Precisely, this area is likely to be affected mainly by extreme flooding, causing a flooding area 55% greater by 2100 (Hérivaux et al. 2018).

3.3 *Emergent Theme: Coastal Ecosystem Services*

The review article of Liquete (2013) aiming at defining the status quo of coastal and marine ecosystem services research framed a conceptual framework based on a cascade model for contextualising the link between the natural provision of ecosystem services and the impacts on human well-being. This cascade model begins with the biophysical structure and process of the ecosystem that determines its functions defined as the capacity of ecosystems to provide services. These functions are considered as a flow of ecosystem services because they contribute to human well-being. Therefore, the flow may be translated into societal benefits and often allows allocating monetary or alternative values to these benefits (illustration of the cascade model in Figure 2 page 4 of the review article (Liquete et al. 2013)).

Caro et al. (2018) evidenced the crucial role of mapping coastal and marine ecosystem services to enhance sustainable management, since they are increasingly subjected to anthropogenic stressors such as related to tourism. In Europe, some efforts to facilitate this process of ecosystem services' mapping has been developed by the European Group of Mapping and Assessment of Ecosystems and their Services (MAES) to monitor the environmental policies implementation. Generally, the recent review of Caro et al. (2018) recalled similar findings observed by a number of authors, concerning the insufficient use of quantitative and qualitative information for describing marine and coastal ecosystem services (Caro et al 2018). In particular, one of the main issues seems to be related to the lack of quantitative studies about this topic, especially in an explicit spatial matter and with enough resolution (Lavorel et al. 2017; Caro et al 2018).

Sangha et al. (2019) underlined the essential role of local-scale economic assessment of coastal and marine ecosystem services, by adopting a framework for measuring each type of ecosystem services (provisioning, regulating and cultural) including its economic impact (as direct and indirect market value) and economic value (as of market and non-market values for contributing to social welfare) applied in the Northern Territory (Australia). A total amount of about \$ 1.3 million per year of economic impact and \$ 1.4 million per year of economic contribution to the local economy were estimated, with additional > 6000 jobs as economic value. Among the selected ecosystem services, the economic value and impact of the cultural ones (including tourism) resulted to be particularly high. The authors highlighted the contribution of this assessment, considering non-marketable services that are vital to support the local community (Sangha et al. 2019). In another study, Roebeling et al. (2013) argued the importance of European Union (EU) member states' strategy in dealing with coastal erosion impacts on the economic and ecological assessment of their coastal areas. For this reason, they conducted research assessing to what extent historical (1975–2006) and projected (2006–2050) coastal erosion patterns lead to ecosystem service value losses. The estimation resulted in almost a 15% decrease in coastal ecosystem service values as compared to 1975, leading to losses up to €2.9 billion per year by 2050 (Roebeling et al. 2013).

Despite coastal flood and erosion are often cited as posing a risk to the infrastructures and causing human and economic losses, coastal erosion can also provide many societal benefits contributing to coastal ecosystem services. For example, periodic mild flood events supply wetlands (e.g. saltmarshes) with fresh nutrients and sediments. In this view, coastal defences compromise natural processes associated to coastal ecosystem services, as seen in the case of Northern Ireland coast, where construction of sea defences took little account of environmental and ecological implications for the coast (Cooper et al. 2020). Globally, the one-dimensional anthropic view of coastal protection (i.e. for infrastructures' defence) is causing severe impacts on the coastal ecosystem services and, also, compromising their ability to adapt to sea-level rise.

Since the conservation of coastal ecosystems, such as tidal wetlands, contributes to nature-based mitigation from sea-level rise through their capacity to protect coasts against flood and erosion risk, Van Coppenolle and Temmerman (2019) completed a first research study on a global scale to identify land surface areas suitable for restoring or creating tidal wetlands. This study revealed that among 135 studied towns, about 60% of the area below the mean high tide is urbanised or densely populated, and 34% is potentially available for wetland creation or restoration. For instance, many European cities presented the largest area below high tide, as the case of Hamburg (Germany) and Rotterdam (Netherlands), of which the former has the most upper potentially available area for tidal wetlands creation. Key factors influencing this potentially available area are geomorphology, population density and land use (Van Coppenolle and Temmerman 2019). In line with these results, another study at a global scale assessed that about 30% of the flood-exposed low-lying coasts and 35% of the global flood-exposed population could benefit from nature-based storm surge mitigation through tidal wetlands. Generally, the coastal plains with the largest surface benefiting from storm surge mitigation are situated in densely populated deltas and estuaries, of which the majority is located in five countries (such as Netherland and Germany in North-western Europe) (Van Coppenolle and Temmerman 2020).

3.4 Emergent Theme: Coastal Tourism

Recently, tourism has been recognised as a highly climate-sensitive sector, demonstrating the growing awareness of its vulnerability to climate change (Mora et al. 2018; Scott et al. 2019). Indeed, Mora et al. (2018) conducted a comprehensive review of how ten types of climate change hazards are impacting on humanity, considering six different aspects of anthropogenic systems (health, food, water, infrastructure, economy and security). The results highlighted that tourism-based economies are considered particularly sensitive, being the only sector of "economy" affected by all ten types of climate impacts (Mora et al. 2018). Although the tourism vulnerability to climate change appears globally recognised, the regional implications of climate change impacts on tourism competitiveness and sustainability remain uncertain. Scott et al. (2019) presented an analysis of the Climate Change Vulnerability Index on Tourism (CVIT) in 181 countries, integrating 27 indicators to represent the internal and transnational impacts on the tourism sector. What emerges from this analysis is that climate change is likely to be most impactful in those countries where the tourism sector contributes to at least 15% of gross domestic product (GDP), that is having a greater relative importance on the national economy. However, looking at future climate change, the vulnerability increases in regions where the most considerable portion of the growth in tourism is projected. For instance, Australia evidenced a projected annual growth equal to the 2% of GDP with a very high CVIT score (Scott et al. 2019). Considering that tourism has become the main economic engine of many coastal areas, of which the Mediterranean Basin is the world's leader destination representing about 30% of international tourism, preserving the coastline quality under different climate and sea-level rise scenarios is essential (Plan Blue 2016; López-Dóriga et al. 2019). Indeed, López-Dóriga et al. (2019) conducted an assessment of the potential impacts of sea-level rise on coastal tourism for improving mitigation and adaptation activities for Catalan beaches (López-Dóriga et al. 2019). In the same country, Toimil et al. (2018) explored the effects of erosion on the recreation value of beaches in Asturias as an indicator of the tourism sector. They determined the risk of loss of beach recreation value triggered by climate change in probabilistic terms by including sea-level rise uncertainty. The results evidenced severe risks for the Asturias beaches leading to an increase of the regional "Expected Cumulative Damage (ECD)" under "Business as Usual" IPCC scenario of around 220%, implying economic losses up to 6.5% of the Asturian capital stock (Toimil et al. 2018).

The massive fluxes of tourists concentrated in small areas are resulting in a substantial environmental impact on coastal ecosystems, intensified by other human impacts. The excellent functionality of coastal ecosystem services lies at the heart of the tourist attraction as a prerequisite for coastal tourism (Drius et al. 2019). Indeed, Drius et al. (2019) provided a conceptual framework underlying, on the one hand, the benefits arising from ecosystem services with their effects on human well-being, and on the other hand, showing the negative impact of anthropological activities on the Mediterranean coastal tourism through an ecosystem perspective. The study defines coastal ecosystem services key component for enabling coastal tourism to contribute positively to socio-economic well-being and to be a positive force for nature conservation (Drius et al. 2019). For instance, coastal wetland ecosystems have great potential for tourism by offering recreational opportunities (i.e. cultural services) to local inhabitants and tourists. Since coastal wetland is one of the most affected ecosystems in the world due to human pressures as recreational activities and tourism, recreational restoration in the coastal wetlands is considered vital for preserving the ecosystem. At the same time, this practice could have potential adverse effects on tourism; therefore, Pueyo-Ros et al. (2018) developed an analysis to evaluate the socio-economic value of the ecological restoration in coastal wetlands at a mass tourism destination on Costa Brava (Spain). In a context of sun and beach tourism, the results of this evaluation based on citizens' willingness to pay (combining travel cost and contingent behaviour model) showed that the ecological restoration of areas with potentially high natural value could be compatible with some tourism activities (Pueyo-Ros et al. 2018). Thus, demonstrating potential synergies between two activities that seem opposed.

The research developed by Plieninger et al. (2018) in remote and small Northern Atlantic islands, i.e. the Faroe Islands, showed diverging viewpoints among local narratives on development activities such as tourism, renewable energy and fish farming/processing. In the light of these coexisting local preferences and conflicts, especially for tourism development that prevailed on the other activities, it is recognised the need for new economic opportunities related to the adverse effects on ecosystems. The authors explicitly confirmed the cruciality of an integrated consideration of marine, coastal and landscape spatial planning, which is usually performed separately, to manage the potential for trade-offs in the land- and seascape development (Plieninger et al. 2018).

4 Discussion

Coastal flood risk, coastal ecosystem services and coastal tourism are well connected within the climate change issue; this literature review identified a small sample size due to the complexity of studying such many interrelated factors with various perspectives, and thus in turn the paucity of recent research studies in this specific area. We systematically identify 16 studies and synthesise their findings.

Europe is the region with the greatest number of studies. This is not surprising given that in Europe, flood events are estimated to be the most damaging natural hazard during 1980–2018 (Debele et al. 2019), and especially, the coastal flood is considered the most catastrophic floods (Simmonds et al. 2019). However, also, coastal tourism is the leading industry in the Mediterranean Basin (López-Dóriga et al. 2019). On the other side, the main focus of the studies in Australia was coastal ecosystem services, especially to what concerns the role of wetlands on the sea-level rise risk reduction. Moreover, it is not unexpected that the majority of the studies adopt a quantitative approach, assessing the effects of climate change on the past, present and future. Results also indicate that there are significantly more studies using quantitative approaches compared to qualitative approaches. This finding is not surprising either considering that both coastal flood, coastal ecosystem services and coastal tourism have an economic consequence.

To what concerns coastal flood risk, the studies included in this review are mainly focused on the past and projected trends of sea-level rise. At the same time, the increasing number of severe storms events, associated with the storm surge risk, is often assessed as the main trigger for a coastal flood. Indeed, coastal floods caused by sea-level rise will be a future and essential research focus considering that the total area exposed to coastal flooding by 2100 will be more than double compared to the scenario without sea-level rise. Several studies on coastal ecosystem services were risk assessment/management oriented, demonstrating the need for the creation and restoration of ecosystem services. Little research generally assessed the economic impacts of sea-level rise on ecosystems. The existing studies are usually focused on physical effects of sea-level rise on a specific coastal ecosystem (for instance Li et al. 2015 for mangroves), and more comprehensive assessments are needed for the coastal ecosystem services evaluation (Hérivaux et al. 2018). For coastal tourism, it is quite evident how this economic sector in the coastal area is crucial in terms of local development. For this reason, a few studies analysed climate change impacts on tourism, mainly focusing on sea-level rise effects, since this is recognised as one of the main risks at a coastal level. Some of this research deepened the essential functionality of ecosystem services for attracting tourism, by underlying the need for analysis of the restoration of these coastal ecosystems.

In the light of this, what emerges from this review analysis is the need to develop more integrated research on coastal cities by exploring the opportunity to adopt more coastal management strategies that "accept" sea-level rise. This future perspective should include response practices such as the "accommodation" one, which comprises occupancy and use of vulnerable lands without trying to prevent it from being damaged by natural events, allowing migration or conservation of the ecosystem. In some cases, coastal buildings such as residential houses or small commercial shops could be elevated to protect them from erosion and floods (i.e. floating buildings). This strategy could also reveal itself as convenient from the economic point of view by providing new opportunities for inundated areas that can be used for new purposes such as tourism activities (Williams et al. 2018; Coppola 2020). In this view, facing changing climate requires the deployment of adaptive solutions such as a nature-based coastal defence to decrease the vulnerability of coastal cities, reducing ecological degradation and loss of biodiversity while enhancing activities such as nature-based tourism (Tisdell and Wilson 2012; Morris et al. 2018, 2019).

5 Conclusion

Over the recent past, researchers from diverse disciplines have tried to better understand, for instance, the climate change impacts on coastal tourism, or the effects of coastal floods on coastal ecosystem services. The present study contributes to this diverse body of literature by presenting a systematic review on recent assessments of the interrelationships between climate change-related coastal flood risk, and coastal ecosystem services and associated impacts on coastal tourism. Despite the restrictions included at the inception of searching analysis of this "brief review"—conducted based on the past three years—the general trend shows publications encouraging a call for additional research exploring even more such integrative approach.

This review also provides an opportunity to deliver a few recommendations for future research on this thematic area, believing that an increase in the number of studies along these innovative lines of research could advance this "integrated literature". For example, many studies included in this review do not highlight the critical role of biodiversity in increasing the economic value and the appreciation of coastal tourism. Similarly, previous studies focusing on coastal flood and coastal ecosystem services only cited biodiversity without deepening its role for coastal defence. The element of biodiversity could be considered as the critical component for connecting all the currently studied topics, for further analysis.

In summary, this literature review reflects geographical coverage of areas more than others within the regions considered for the analysis (Europe and Australia), which both represent the areas of most significant interest to researchers and points out the areas receiving less attention. This suggests a gap in recent literature, where a dense description or more detailed analysis regarding the link among the climate change effects on the coastal flood risk, coastal ecosystem services and coastal tourism may be lacking.

Finally, the importance of a multidisciplinary approach with the integration of qualitative and quantitative analysis should be emphasised. We encourage researchers to develop future studies based on cost–benefit analysis of ecosystem services applied to coastal flood protection in term of verifying the development of coastal areas and the relative increase in local tourism.

Appendix

See Appendix Table 1.

Table 1 Classification criteria of studies

Citation	Year of publication	Article type	Study area	Approach	Climate change issue	Descriptors (variables)
Caro, C, Pinto, R., Marques, J.C.	2018	Review article	Europe ($N = 193$)	Qualitative (mapping)	Contextual	Open source spatial database useful to map (GIS software) marine and coastal ecosystem services
Cooper, J.A.G., O'Connor, M.C., Mclvor, S.	2020	Research article	Europe (Northern Ireland coast)	Quantitative (mapping)	Contextual	Low-level oblique helicopter-based photographic survey to map in GIS the nature and extent of sea defence structures.
Debele, E.S., Kumar, P., Sahara, J., Marti-Cardona, B., Mickovski, S.B., Leo, L.S., Portcu, F., Bertini, F., Montesi, D., Vojinovic, Z., Di Sabatino, S.	2019	Review article	Europe ($N = 288$)	Qualitative	Analytical	Critically harmonisation of literature on Hydro-meteorological hazards/risk and NBS for mitigating the adverse impacts
Lopez-Doriga, U., Jimenez, J.A., Valdemoro, H.I., Nicholls, R.J.	2019	Research article	Europe (Catalan coast)	Quantitative	Analytical	Assessment of the potential impact of sea level-rise on the recreational carrying capacity of Catalan beaches and hence the influence on the tourism economic model over the coming decades under climate change scenarios

(continued)

Table 1 (continued)

Citation	Year of publication	Article type	Study area	Approach	Climate change issue	Descriptors (variables)
Drius, M., Bongiorni, L., Depellegrin, D., Menegon, S., Pugnetti, A., Stifter, S.	2018	Research article	Europe (Italian Northern Adriatic coast)	Qualitative (conceptual framework)	Contextual	Complex relationships and trade-offs among threats from coastal tourism, from other human activities and from ecosystem services (focus on cultural ones)
Herivaux, C, Rey-Valette, H., Rulleau, B., Agenais, A.L., Grisel, M., Kuhfbss, L., Maton, L., Vinchon, C.	2018	Research article	Europe (Languedoc-Rousillon coastline)	Quantitative framework	Analytical	Hazard-modelling approaches with different economic valuation methods under sea-level rise scenarios
Liquete, C, Piroddi, C, Drakou, E.G., Gurney, L., Katsanevakis, S., Charef, A., Egoh, B.	2013	Review article	Global scale	Qualitative (conceptual framework)	Contextual	Summarized the available information related to marine and coastal ecosystem services (MCES) for cataloguing indicators and classify MCES

(continued)

Table 1 (continued)

Citation	Year of publication	Article type	Study area	Approach	Climate change issue	Descriptors (variables)
Mora, C, Spirandelli, D., Franklin, E.C., Lynham, J., Kantar, M.B., Miles, W., Smith, C.Z., Freel, K., Moy, J., Louis, L.V., Barba, E.W., Bettinger, K., Frazier, AG., ColburnLX, J.F., Hanasaki, N., Hawkins, E., Hirabayashi, Y., Knorr, W., Little, C.M., Emanuel, K., Sheffield, J., Patz, J. A, Hunter, C.L.	2018	Review article	Global scale	Qualitative	Pivotal	Systematic literature search to identify observed impacts on people from climate hazards developing a global map of a cumulative index of projected changes in these hazards to evaluate the extent that humanity will be exposed to different concurrent hazards
Plieninger, T., avRana, H.A., Fagerholm, N., Ellingsgaard, G.F., Ey3finn, M., Raymond, CM., Olafsoon, AS., Verbrugge, L.N.H.	2018	Research article	Europe (Faroe Islands)	Qualitative and quantitative (mapping survey)	Contextual	Landscape values and ecosystem services analysis for eliciting different preferences and potential for land-use conflicts
Pueyo-Ros, J., Garcia, X., Ribas, A., Fraguell, R.M.	2018	Research article	Europe (Catalancoast—Costa Brava)	Quantitative	Contextual	Socioeconomic value of the ecological restoration of a coastal wetland at a mass tourism destination

(continued)

Table 1 (continued)

Citation	Year of publication	Article type	Study area	Approach	Climate change issue	Descriptors (variables)
Roebeling, P.C., Costa, L., MagalMes-Filho, L., Tekken, V.	2013	Research article	Europe (coasts of Belgium, Denmark, France, Germany, Greece, Ireland, Italy, Netherlands, Portugal, Spain)	Quantitative	Analytical	Environmental-economic (direct cost) approach to assess to what extent past and future coastal erosion combining benefit transfer valuation for Ecosystem services, historical coastal land use map and future coastal erosion projections
Sangha, K.K., Stoeckl, N., Crossman, N., Costanza, R.	2019	Research article	Australia (Northern Territory coastline)	Quantitative	Contextual	State-wide assessment of coastal and marine ecosystems services including Indigenous perspectives
Scott, D., Hall, C.M., Gossling, S.	2019	Research article	Global scale (181 Countries)	Quantitative framework	Pivotal	Climate Change Vulnerability Index of Tourism (CVIT)

(continued)

Table 1 (continued)

Citation	Year of publication	Article type	Study area	Approach	Climate change issue	Descriptors (variables)
Toimil, A., Diaz-Simal, P., Camus, P.	2018	Research article	Europe (Asturias coast)	Quantitative	Analytical	Economic assessment of loss of beach recreation value under climate change through the risk estimation of erosion as a key indicator in the tourism sector
Van Coppenolle, R., Temmerman, S.	2019	Research article	Global scale (135 Coastal cities)	Quantitative	Analytical	Analysis of the surface areas available for potential creation/restoration of tidal wetlands (salt marshes and mangrove forests)
Van Coppenolle, R., Temmerman, S.	2020	Research article	Global scale (Large deltas around the world)	Quantitative (mapping)	Analytical	Estimation of the flood routing of storm surge in coastal areas resulting in land surface area and population number that would be reached by a storm surge, thus identifying where coastal wetland conservation contribute to nature-based flood risk mitigation

References

Aitali R, Snoussi M, Kasmi S (2020) Coastal development and risks of flooding in Morocco: the cases of Tahaddart and Saidia coasts. J Afr Earth Sci 164:103771. https://doi.org/10.1016/j.jafrearsci.2020.103771

Alfieri L, Feyen L et al (2015a) Ensemble flood risk assessment in Europe under high end climate scenarios. Glob Environ Change 35:199–212. https://doi.org/10.1016/J.GLOENVCHA.2015.09.004

Alfieri L, Burek P et al (2015b) Global warming increases the frequency of river floods in Europe. Hydrol Earth Syst Sci 19(5):2247–2260. https://doi.org/10.5194/hess-19-2247-2015

Aroca-Jiménez E et al (2018) A quantitative methodology for the assessment of the regional economic vulnerability to flash floods. J Hydrol 565:386–399. https://doi.org/10.1016/J.JHYDROL.2018.08.029

Berndtsson R et al (2019) Drivers of changing urban flood risk: a framework for action. J Environ Manage 240(October 2018):47–56. https://doi.org/10.1016/j.jenvman.2019.03.094

Caro C, Pinto R, Marques JC (2018) Use and usefulness of open source spatial databases for the assessment and management of European coastal and marine ecosystem services. Ecol Indicators 95(June):41–52. https://doi.org/10.1016/j.ecolind.2018.06.070

Cooper JAG, O'Connor MC, McIvor S (2020) Coastal defences versus coastal ecosystems: a regional appraisal. Marine Policy 111:102332. https://doi.org/10.1016/j.marpol.2016.02.021

Van Coppenolle R, Temmerman S (2019) A global exploration of tidal wetland creation for nature-based flood risk mitigation in coastal cities. Estuarine Coastal and Shelf Science 226(June):106262. https://doi.org/10.1016/j.ecss.2019.106262

Van Coppenolle R, Temmerman S (2020) Identifying global hotspots where coastal wetland conservation can contribute to nature-based mitigation of coastal flood risks. Global Planet Change 187(December 2019):103125. https://doi.org/10.1016/j.gloplacha.2020.103125

Coppola T (2020) Artificial surfing reefs : the right solution for Vagueira beach? Università degli Studi di Torino

Costanza R et al (1997) The value of the world's ecosystem services and natural capital. Nature 387(6630):253–260. Available at: https://www-nature-com.ezproxy.royalroads.ca/articles/387253a0.pdf

de Ruig LT et al (2019) A micro-scale cost-benefit analysis of building-level flood risk adaptation measures in Los Angeles. Water Resour Econ 100147. doi: https://doi.org/10.1016/j.wre.2019.100147

Debele SE et al (2019) Nature-based solutions for hydro-meteorological hazards: Revised concepts, classification schemes and databases. Environ Res 179(June):108799. https://doi.org/10.1016/j.envres.2019.108799

Drius M et al (2019) Tackling challenges for Mediterranean sustainable coastal tourism: an ecosystem service perspective. Sci Total Environ 652:1302–1317. https://doi.org/10.1016/J.SCITOTENV.2018.10.121

Dwyer L, Forsyth P, Dwyer W (2020) Tourism economics and policy, 2nd edn. Channel View Publ, UK

European Commission (2019) The EU blue economy report. Luxemburg. https://doi.org/10.2771/2185

Fang J et al (2020) Coastal flood risks in China through the 21st century—an application of DIVA. Sci Total Environ 704:135311. https://doi.org/10.1016/J.SCITOTENV.2019.135311

Garrote J et al (2019) Civil engineering works versus self-protection measures for the mitigation of floods economic risk. A case study from a new classification criterion for cost-benefit analysis. Int J Disaster Risk Reduc 37:101157. https://doi.org/10.1016/j.ijdrr.2019.101157

Graff K et al (2019) Analysis and quantification of potential consequences in multirisk coastal context at different spatial scales (Normandy, France). Nat Hazards 99(2):637–664. https://doi.org/10.1007/s11069-019-03763-5

Grames J et al (2016) Modeling the interaction between flooding events and economic growth. Ecol Econ 129:193–209. https://doi.org/10.1016/j.ecolecon.2016.06.014

Grant MJ, Booth A (2009) A typology of reviews: an analysis of 14 review types and associated methodologies. Health Inf Libr J 26(2):91–108. https://doi.org/10.1111/j.1471-1842.2009.00848.x

Haer T et al (2017) Economic evaluation of climate risk adaptation strategies: cost-benefit analysis of flood protection in Tabasco, Mexico. Atmósfera 30(2):101–120. doi: https://doi.org/10.20937/ATM.2017.30.02.03

Hérivaux C et al (2018) Benefits of adapting to sea level rise: the importance of ecosystem services in the French Mediterranean sandy coastline. Reg Environ Change 18(6):1815–1828. https://doi.org/10.1007/s10113-018-1313-y

Hirabayashi Y et al (2013) Global flood risk under climate change. Nat Clim Change 3(9):816–821. https://doi.org/10.1038/nclimate1911

IPCC (2019) Summary for policymakers. In: Pörtner H-O et al (eds) IPCC Special report on the ocean and cryosphere in a changing climate. In press. doi: http://www.ipcc.ch/publications_and_data/ar4/wg2/en/spm.html

Kundzewicz ZW et al (2019) Flood risk in a range of spatial perspectives—from global to local scales. Nat Hazards Earth Syst Sci 19(7):1319–1328. https://doi.org/10.5194/nhess-19-1319-2019

Lavorel S et al (2017) Pathways to bridge the biophysical realism gap in ecosystem services mapping approaches. Ecol Ind 74:241–260. https://doi.org/10.1016/J.ECOLIND.2016.11.015

Li S et al (2015) Vulnerability assessment of the coastal mangrove ecosystems in Guangxi, China, to sea-level rise. Reg Environ Change 15(2):265–275. https://doi.org/10.1007/s10113-014-0639-3

Lichter M et al (2011) Exploring data-related uncertainties in analyses of land area and population in the "Low-Elevation Coastal Zone" (LECZ). J Coastal Res 757–768. https://doi.org/10.2112/jcoastres-d-10-00072.1

Liquete C et al (2013) Current status and future prospects for the assessment of marine and coastal ecosystem services: a systematic review. PLoS ONE 8(7). https://doi.org/10.1371/journal.pone.0067737

López-Dóriga U et al (2019) Impact of sea-level rise on the tourist-carrying capacity of Catalan beaches. Ocean Coast Manag 170(December 2018):40–50. https://doi.org/10.1016/j.ocecoaman.2018.12.028

McGranahan G, Balk D, Anderson B (2007) The rising tide: assessing the risks of climate change and human settlements in low elevation coastal zones. Environ Urban 19(1):17–37. https://doi.org/10.1177/0956247807076960

Mehvar S et al (2018) Quantifying economic value of coastal ecosystem services: a review. J Mar Sci Eng 6(1). https://doi.org/10.3390/jmse6010005

Merkens JL et al (2016) Gridded population projections for the coastal zone under the Shared Socioeconomic Pathways. Glob Planet Change 145:57–66. https://doi.org/10.1016/j.gloplacha.2016.08.009

Milly PCD et al (2002) Increasing risk of great floods in a changing climate. Nature 415(6871):514–517. https://doi.org/10.1038/415514a

Mora C et al (2018) Broad threat to humanity from cumulative climate hazards intensified by greenhouse gas emissions. Nat Clim Change 8(12):1062–1071. https://doi.org/10.1038/s41558-018-0315-6

Morris RL et al (2018) From grey to green: Efficacy of eco-engineering solutions for nature-based coastal defence. Glob Change Biol 24(5):1827–1842. https://doi.org/10.1111/gcb.14063

Morris RL et al (2019) Developing a nature-based coastal defence strategy for Australia. Aust J Civil Eng 17(2):167–176. https://doi.org/10.1080/14488353.2019.1661062

Moscardo G et al (2001) Understanding coastal and marine tourism demand from three european markets: implications for the future of ecotourism? J Sustain Tour 9(3):212–227. https://doi.org/10.1080/09669580108667399

Neumann B et al (2015) Future coastal population growth and exposure to sea-level rise and coastal flooding—a global assessment. PLOS ONE 10(3):1–34. https://doi.org/10.1371/journal.pone.0118571

Nicholls RJ (2002) Analysis of global impacts of sea-level rise: a case study of flooding. Phys Chem Earth 27(32–34):1455–1466. https://doi.org/10.1016/S1474-7065(02)00090-6

Ogato GS et al (2020) Geographic information system (GIS)-Based multicriteria analysis of flooding hazard and risk in Ambo Town and its watershed, West shoa zone, oromia regional State, Ethiopia. J Hydrol: Regional Stud 27:100659. https://doi.org/10.1016/j.ejrh.2019.100659

Oppenheimer M et al (2019) Sea level rise and implications for low lying Islands, coasts and communities. In: Pörtner H-O et al. (eds) IPCC special report on the ocean and cryosphere in a changing climate. In press, pp 126–129. https://doi.org/10.1126/science.aam6284

Plan Blue (2016) Tourism. Economic activities and sustainable development. Plan Bleu Notes, No. 32

Plieninger T et al (2018) Identifying and assessing the potential for conflict between landscape values and development preferences on the Faroe Islands. Glob Environ Change 52(May):162–180. https://doi.org/10.1016/j.gloenvcha.2018.07.006

Pueyo-Ros J et al (2018) Ecological restoration of a coastal wetland at a mass tourism destination. Will the recreational value increase or decrease?. Ecological Economics. 148(January 2017):1–14. https://doi.org/10.1016/j.ecolecon.2018.02.002

Reed D et al (2020) Modeling wetland transitions and loss in coastal Louisiana under scenarios of future relative sea-level rise. Geomorphology 352:106991. https://doi.org/10.1016/j.geomorph.2019.106991

Rehan BM (2018) Accounting public and individual flood protection measures in damage assessment: a novel approach for quantitative assessment of vulnerability and flood risk associated with local engineering adaptation options. J Hydrol 563:863–873. https://doi.org/10.1016/j.jhydrol.2018.06.061

Richins H, Pearce P (2000) Influences on tourism development decision making: coastal local government areas in eastern Australia. J Sustain Tour 8(3):207–231. https://doi.org/10.1080/09669580008667359

Roebeling PC et al (2013) Ecosystem service value losses from coastal erosion in Europe: historical trends and future projections. J Coastal Conserv 17(3):389–395. https://doi.org/10.1007/s11852-013-0235-6

Sadiq AA, Tyler J, Noonan DS (2019) A review of community flood risk management studies in the United States. Int J Disaster Risk Reduc 41(July):101327. https://doi.org/10.1016/j.ijdrr.2019.101327

Sanchez-Reaza J, Carletto A (2013) Resilient coastal cities: the economic, social and environmental dimensions of risk. Washington DC

Sangha KK et al (2019) A state-wide economic assessment of coastal and marine ecosystem services to inform sustainable development policies in the Northern Territory, Australia. Marine Policy 107(June):103595. https://doi.org/10.1016/j.marpol.2019.103595

Scholz M (2013) Water quality improvement performance of geotextiles within permeable pavement systems: a critical review. Water 5(2):462–479. https://doi.org/10.3390/w5020462

Scott D, Hall CM, Gössling S (2019) Global tourism vulnerability to climate change. Ann Tour Res 77(June):49–61. https://doi.org/10.1016/j.annals.2019.05.007

Simmonds J, Gómez JA, Ledezma A (2019) The role of agent-based modeling and multi-agent systems in flood-based hydrological problems: a brief review. J Water Clim Change. https://doi.org/10.2166/wcc.2019.108

Thompson RR, Garfin DR, Silver RC (2017) Evacuation from natural disasters: a systematic review of the literature. Risk Anal 37(4):812–839. https://doi.org/10.1111/risa.12654

Tisdell C, Wilson C (2012) Nature-based tourism and conservation. In: Elgar E (ed) Edward Elgar Publishing, Cheltenham

Toimil A et al (2018) Estimating the risk of loss of beach recreation value under climate change. Tour Manag 68(April):387–400. https://doi.org/10.1016/j.tourman.2018.03.024

Toimil A et al (2020) Addressing the challenges of climate change risks and adaptation in coastal areas: a review. Coastal Eng 156:103611. https://doi.org/10.1016/J.COASTALENG.2019.103611
Vormoor K et al (2016) Evidence for changes in the magnitude and frequency of observed rainfall vs. snowmelt driven floods in Norway. J Hydrol 538:33–48. https://doi.org/10.1016/J.JHYDROL.2016.03.066
Vousdoukas MI et al (2016) Projections of extreme storm surge levels along Europe. Clim Dyn 47(9):3171–3190. https://doi.org/10.1007/s00382-016-3019-5
Williams AT et al (2018) The management of coastal erosion. Ocean Coastal Manage 156:4–20. https://doi.org/10.1016/j.ocecoaman.2017.03.022
Winsemius HC et al (2016) Global drivers of future river flood risk. Nat Clim Change 6(4):381–385. https://doi.org/10.1038/nclimate2893

Carlotta Quagliolo is a Ph.D. student in Urban and Regional Development at Politecnico di Torino. Research topic focused on climate change impacts on coastal urban areas through the integrated spatial analysis by coupling climate variables and ecosystem services values. Master Degree in Environmental Economics and Policies (Università degli studi di Torino). Research experiences as Research fellow at the Responsible Risk Resilience Centre—R3C (Politecnico di Torino) deepening vulnerability analysis and urban resilience in cities. Experiences in climate analysis are especially in urban context. Coordination activities in development of project proposal of two H2020-calls focused on nature-based solutions in cities to improve health and well-being while enhancing climate resilience in face to hydrometeorological hazards.

Alessandro Pezzoli pursued Master Degree in Civil Engineering—Hydraulic Option (Politenico di Torino—Italy) and a PhD in Meteorology and Oceanography (Université de Toulon—France). He is Senior Lecturer (Assistant Professor) in Weather Risk Management for the MBA in Environmental Economics and in Meteo-Hydrology Risk Assessment for the MSc in Geography and Regional Development of the University of Turin. He has a large expertise in Applied Meteorology and Applied Climatology considering its research about the analysis of the adaptation's strategies to natural hazards generated by the extreme weather situations. He participated in different research among which it remembers the "Rede Litoral Project" (Analysis of the Climate Change effect in the Sao Paolo area and Santos coastal zone—Brazil), "Prometeo Project" (analysis of the Climate Change effect in the urban area of Guayaquil—Ecuador), COOPI-Chaco Paraguayan (analysis of the Climate Change effect on the drought's patterns in the Chaco Paraguayan—Paraguay), UniCoo—Etiopia (analysis of the gender effect on the water's use in the Ethiopia face to Climate Change effect), Anadia 2 in Niger (analysis of the flood security related to food security around Sirba river) and One Health in Kenya. He is Associate Fellow of the Royal Institute of Navigation (AFRIN), Fellow of the Royal Meteorological Society (FRMetS) and Fellow of the Royal Geographical Society (FRGS).

Elena Comino is Professor of Applied Ecology at Politecnico di Torino—Environmental, Territory and Infrastructure Engineering Department. After a scientific high school, she obtained a Biology Degree cum Laude. PhD in Geological and Environment Engineering and a Master Degree in Environmental Engineering, with a particular interest on the integration of phytotechnology in strategies and programs for the sustainable development. She teaches Applied Ecology and Environmental Sustainability Procedure. She is responsible for the Laboratory in Applied Ecology, and most of her work is now oriented in the fields of phytoremediation and environmental sustainability. She coordinates an interdisciplinary team group, technical specialists, young researchers and graduate students. She was visiting professor Melbourne University (Australia), EPFL (Lausanne), USP (Universitade de Sao Paulo Brasile). She is Professor in Turin Taskent Politechnica Universitate (Uz). She is scientific and economically responsible for projects on process for the biological cleaning of contaminated water and for bioenergy production. Recently, she focused all these studies even to the social and planning aspects. She is a member of Ecology Italian Society (SItE), Wetland Scientist Society (SWS) and International Phytotechnology Society (IPS).

Marco Maria Bagliani pursued Bachelor Degree in Physics and Ph.D. in Physics, University of Torino. He is Associate Professor of Economic Geography, University of Torino. His research subjects are environmental geography, environmental accounting, sustainable development and indicators of sustainability. He is referee for the following scientific journals: Ecological Economics, Journal of Environmental Management, Ecological Indicators, Italian Journal of Regional Science, Fresenius Environmental Bulletin and Sustainability. He is member of the National Account Committee of the Global Footprint Network responsible for the ecological footprint calculations at global level. He has been scientific coordinator of research projects for a total funding of about €450.000 over the last ten years. He is author of numerous international publications pertaining environmental geography and territorial sustainability.

Flood Pulse Design Principles—A Time-Based Approach to Urban Flooding

Ioana Corina Giurgiu

Abstract This chapter explores the potential of a time-based, nature-inspired approach to urban flood resilient design in the context of SeaCities. As a vast majority of existing coastal cities develop along continental shorelines and are in the proximity of rivers (accounting for freshwater supply, anchorage, etc.), coastal urban development is exposed to tidal, storm surge as well as riverine flooding. Based on historical effects of urban flooding, three key flooding hazards have been identified: damage to infrastructure, dangerous debris and loss of economic productivity. In contrast, within flood-adapted natural wetland ecosystems, the three factors which negatively impact the urban context represent positive ecosystem growth drivers within a range of hydrological regimes. By reviewing flood response mechanisms in relation to the three factors (examples from riverine floodplains, coastal salt marsh and estuarine mangrove forests), temporal correlations between local hydrological regime and ecosystem processes are identified as key drivers of flood resilience in the ecosystem. Based on the strategies identified within the ecosystem review and review of selected examples of current projects and urban flood protection strategies, the analysis concludes by speculating on how the ecological strategies and temporal correlations identified could be adapted into urban design principles. Proposed strategies are assessed in terms of mitigation of the three key flood hazards and the potential for exploiting natural flooding phenomena to generate new urban growth opportunities.

1 Introduction

While the industrial revolution strived to liberate human growth from its dependence on natural phenomena and patterns, it has, paradoxically, evolved into an urban model which seems more vulnerable to natural phenomena than its historical counterpart. As urban infrastructure, connectivity and interdependency transcend localization, the impacts of climatic changes and associated disruptive events such as floods can,

I. C. Giurgiu (✉)
SeaCities, Cities Research Institute, Griffith University, Gold Coast, QLD, Australia
e-mail: ioana.giurgiu@griffithuni.edu.au

J. Baumeister et al. (eds.), *SeaCities*, Cities Research Series,
https://doi.org/10.1007/978-981-15-8748-1_9

in some cases, result in cascading effects which are difficult to predict (Lawrence et al. 2020, Fekete 2019) and counteract.

Looking forward, if current predictions of accelerated flood/drought cycles (Trenberth 2008) and sea level rise (IPCC 2019) due to climate change prove accurate, the intrinsic vulnerability of the modern urban network to environmental factors will increase dramatically. One possible future scenario is, therefore, that of urban fabric being shaped primarily by environmental factors with political, economic and social spheres as subordinate growth drivers.

In the specific context of coastal cities, a large number of which develop along continental shorelines in the proximity of rivers, sea level rise as well as tidal, storm surge and riverine flooding pose significant risks (Woetzel et al. 2020). The notion of SeaCities as future urban interfaces that mediate the relationship between land and water emerges as a design opportunity for increased resilience, flexibility and adaptation to predicted future flooding scenarios.

In order to truly achieve adaptability and flexibility in relation to environmental factors, causal environmental relationships must be understood through their manifold interdependencies as well as via their behaviour in time. In the case of flooding, timelines are particularly important as flood types and their associated impacts can vary greatly with different seasonal frequency, flooding duration or flood water speeds.

A pertinent example of the impact of timelines is the concept of flood pulse, referring to the impact of frequency and periodicity of riverine flooding. A 2015 study (Jardine et al. 2015) into tropical riverscapes in Australia found that *"systems with rhythmic annual floods have higher fish species richness, more stable avian populations, and elevated rates of riparian forest production compared with those with arrhythmic flood pulses"*. Flood pulse is, however, just one type of timeline that may be useful in understanding the benefits and challenges of flooding.

Within the realm of architectural and urban design, the concept of time-based design has, thus far, been implemented as spatial or programmatic flexibility in the short term (e.g. multi-use/purpose spaces) or as weathering and lifespan design in the long term (e.g. design for deconstruction, planning of future development zones). However, these approaches have developed punctually (localized) rather than holistically (at the level of policy or development plans). The shifting city landscape, thus, emerges as a symptom of political trends, business agendas, crises and social migration pressures applied to a relatively static infrastructure rather than being the targeted output of a dynamic design strategy.

In the context of this paper, the term flood pulse design is used to broadly describe temporal periodic and non-periodic sequences of inundation in riverine, estuarine and coastal scenarios which could then become the base design scenarios for urban development. The aim is to identify which temporal correlations and timelines represent beneficial pulses within natural ecosystems and evaluate the potential for applying similar flood resilience strategies in new urban development models.

As described in the chapter structure (Fig. 1), the following sections will first discuss a set of selected examples of wetland ecosystems, identifying pulses and natural "time management" strategies from short-term cycadean rhythms (e.g. tidal

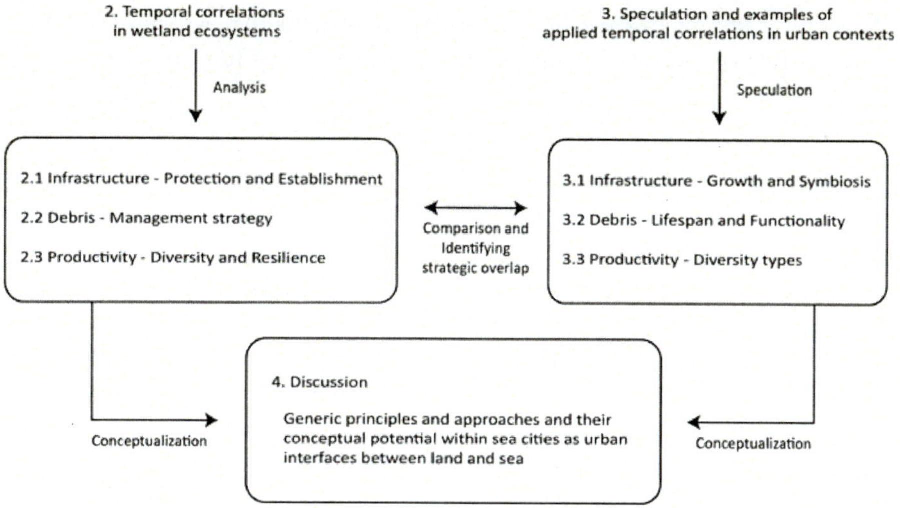

Fig. 1 Chapter structure showing relevance and aims for analysis, speculation and conceptualization sub-sections

flooding) to long-term progressive flooding (e.g. sea level rise), lifespans, cycles, flood stages and on-set patterns. Based on the natural pulses identified, we will further speculate on potential ways of integrating the flood pulse design concept as a temporal dimension of urban design which allows for targeted dynamic behaviour, flexibility and capitalizing on the beneficial effects of flooding.

In both urban and natural settings, the effects of each timeline will be assessed through the lens of the following three factors: infrastructure, debris and productivity.

In the urban setting, these factors are among the main flood-related hazards. Damage to infrastructure may cause displacement and limit the availability of food and clean drinking water. Hazardous debris such as steel and glass shards is difficult to clear out and dangerous at high velocity water flows.

In terms of productivity, the economic impacts of flooding provide clear examples of long- and short-term detrimental effects in the urban context. State of Queensland Government document "Understanding Floods: Questions and Answers" identifies floods as "the most expensive natural disasters in Australia" and highlights not only the direct costs resulting from property damage, but also associated "flow-on effects of reduced agricultural production" which "can often impact well outside the production area as food prices increase due to shortages in supply" (Queensland Government 2011).

A less obvious negative flood-related effect which derives from loss of productivity is the associated decrease in population. In the aftermath of Hurricane Katrina in 2005, more than a million people were displaced temporarily with up to 600,000 households still displaced a month after the event. Aside from displacement experienced as a direct result of flooding, due to economic loss (decrease in productivity) in the area, flooding events may also trigger population migrations. The population of

New Orleans fell from 484,674 before Katrina (April 2000) to an estimated 230,172 after Katrina (July 2006) and had only risen to 369,250 by July 2012 (Plyer 2016).

2 Temporal Correlations in Wetland Ecosystems

According to the Ramsar Convention Global Wetland Outlook (2018), wetlands "are one of the most biologically productive ecosystems" which "play a major role in the water cycle by receiving, storing and releasing water, regulating flows and supporting life" and offer ecosystem services that "far exceed those of terrestrial ecosystems".

Viewed as natural transition zones between aquatic and terrestrial landscapes, these flood-adapted ecosystems could provide excellent case studies in terms of identifying temporal correlations and flood resilient strategies that can be adapted into urban design in order to mitigate and transform potentially hazardous effects of flooding into opportunities.

The following section is structured around the three key factors and investigates three types of wetland ecosystems adapted to flood scenarios applicable in coastal development: freshwater floodplains adjacent to rivers, salt marsh ecosystems mitigating the transition between land and sea and mangrove forests populating estuaries as transitions between land, river and sea. Flood response strategies are evaluated by looking at how each ecosystem responds to short-term, seasonal and long-term flooding events.

2.1 Infrastructure—Protection and Establishment

The term infrastructure is generally defined as "the basic physical and organizational structures and facilities (e.g. buildings, roads, power supplies) needed for the operation of a society or enterprise" (Oxford Dictionary) or "the underlying foundation or basic framework of a system or organization" (Merriam-Webster Dictionary). In order to facilitate comparison between natural and urban contexts, here, the term is also used to define similar underlying frameworks for each wetland example. Although the functioning of wetland ecosystems can be influenced by any number of interlinked factors, the basic functions of nutrient and habitat provision as well as the topography and typology of each system are largely driven by the hydrological regime and characteristics of the soil substrate (Tiner 2016).

The wetland "structures" which provide for nutrient transfer, oxygenation, plant establishment and other vital wetland processes are, therefore, topographic features such as tidal creeks, floodplain channels or elevated flats. For the purpose of establishing comparisons with urban infrastructure, these features have been considered as representing wetland infrastructure. However, the relationship between topography

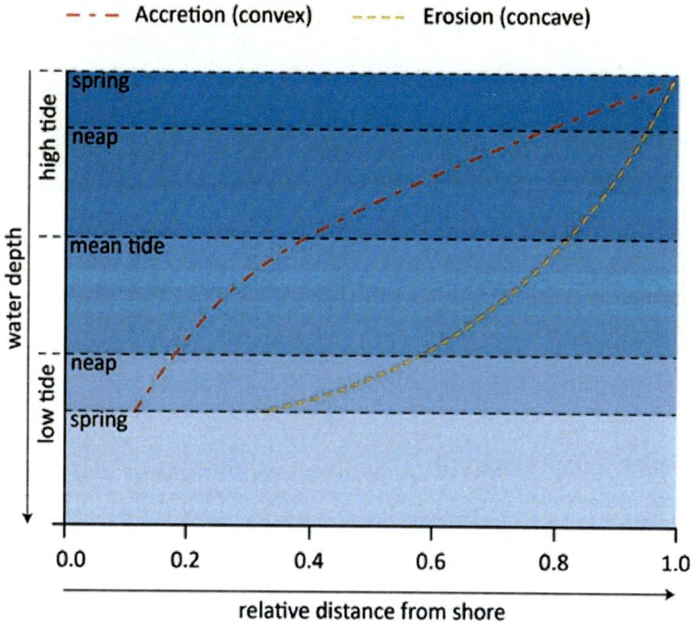

Fig. 2 Schematic diagram showing morphology of concave and convex mudflats relative to tide levels and relative shore-normal distance (Based on Adnitt et al. 2007)

and hydrological regime is a dual one as topography evolves in response to hydrological regime but can, in turn, also influence the hydrological patterns (Adnitt et al. 2007).

The evolution of wetland topography is directly linked to the hydrological regime as tidal creeks and elevated flats are primarily shaped by either sediment accretion or erosion processes. Figure 2 shows the typical morphology of concave and convex features of a salt marsh. The relative distance from shore has an impact on water speed and depth which in turn determine whether tidal action has an eroding or accreting effect. Generally, slower water flow speeds favour sediment accretion while higher flow speeds favour erosion (Adnitt et al. 2007).

The accretion/erosion processes have a marked impact on all three wetland types. In intermittently flooded zones, due to the function of plant roots in fixing soil and decreasing water speeds, the emergence of a long-term stable topography is strongly linked to plant establishment. This link between infrastructure and plant establishment leads to a strong temporal correlation within the short-term and seasonal timelines.

An example of the relationship between hydrological regime and seasonal plant behaviour is shown in Fig. 3.[1] The mean rainfall, river discharge, tidal levels (adapted from Monteiro 2017) and plant phenology graphs (adapted from Nadia et al. 2012

[1]Figure 4 used under CC BY 4.0 from Rodríguez et al. 2017/ Original Figures 3, 4 and 5 were cropped and collated to show results comparison and scenario numbering has been removed https://creativecommons.org/licenses/by/4.0/.

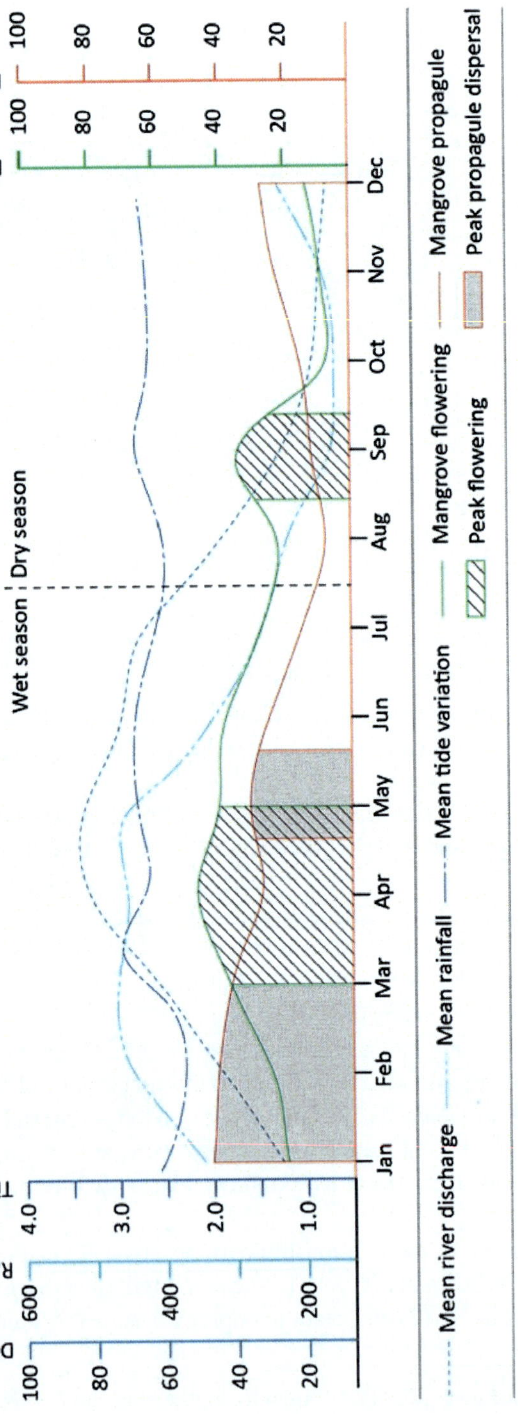

Fig. 3 Hydrological regime and red mangrove phenology-temporal correlations

and Mehlig 2006) show the average behaviour of the red mangrove populating the Caete river estuary in relation to the hydrological regime (average conditions). The Caete estuary is a tidally dominated estuary populated by a well-established red mangrove forest in Para, north-eastern Brazil.

It is especially interesting to note the correlation between periods of peak propagule dispersion and high water levels during the rainy season. Propagule dispersal peaks just before the highest river discharge and tidal range of the rainy season, a fact which has a dual effect in terms of the propagation of mangrove vegetation and implicitly the establishment of new infrastructure. On the one hand, due to higher water levels and velocities typical for the peak propagule dispersal period, propagules may be carried further afield from the parent plant decreasing local competition and potentially increasing the coverage of the mangrove forest. On the other hand, given that freshly established shoots are much more vulnerable to extreme tide and storm surges during additional root development periods (approximately 1 year to first secondary roots), it seems peak propagule dispersal is optimally timed in order to provide the highest chances of new plant establishment during the milder dry season.

The implication of the above temporal correlations between seasonal hydrological cycles and mangrove growth cycles is that, via the dependency of infrastructure stability on vegetation development, flood resilience increases with the degree of correlation between hydrological and vegetative patterns.

This phenomenon can also be observed in riverine floodplains, where due to correlations between plant growth seasons and flood timing, summer flood events are shown to have a much greater impact on plant distribution and survival rates than winter flooding events (van Eck et al. 2006).

In the short term, periodic factors such as diurnal or semidiurnal tidal action have a strong impact on the successful establishment of pioneer species and implicitly new infrastructure establishment. For example, in relation to salt marsh development, a 2019 study (Fivash et al. 2019) shows that elevated (convex) micro-topography amplifies sediment oxygenation via tidally driven hydrodynamic forcing, supporting faster vegetation establishment. As noted by the authors, "this natural phenomenon could play a central role in reducing the required disturbance-free interval for natural recruitment in the field and thereby increase the frequency at which natural recruitment occurs".

The disturbance-free interval and recruitment frequency is particularly relevant for long-term flood resilience in the context of rising sea levels via the impact of vegetation on the speed of sediment accretion. In coastal wetlands, the long-term survival of the coastal ecosystem and positive impact in terms of mitigating sea level rise is dependent on the relative speeds of sea level rise and sediment accretion. A 2017 study (Rodríguez et al. 2017) exploring a more complex hydrodynamic model (compared to commonly used bathtub models) for an Australian subtropical wetland shows that the risk of significant wetland loss over an 80-year forecast dramatically decreases in a scenario of variable sea level rise and high levels of suspended sediments (Fig. 4).

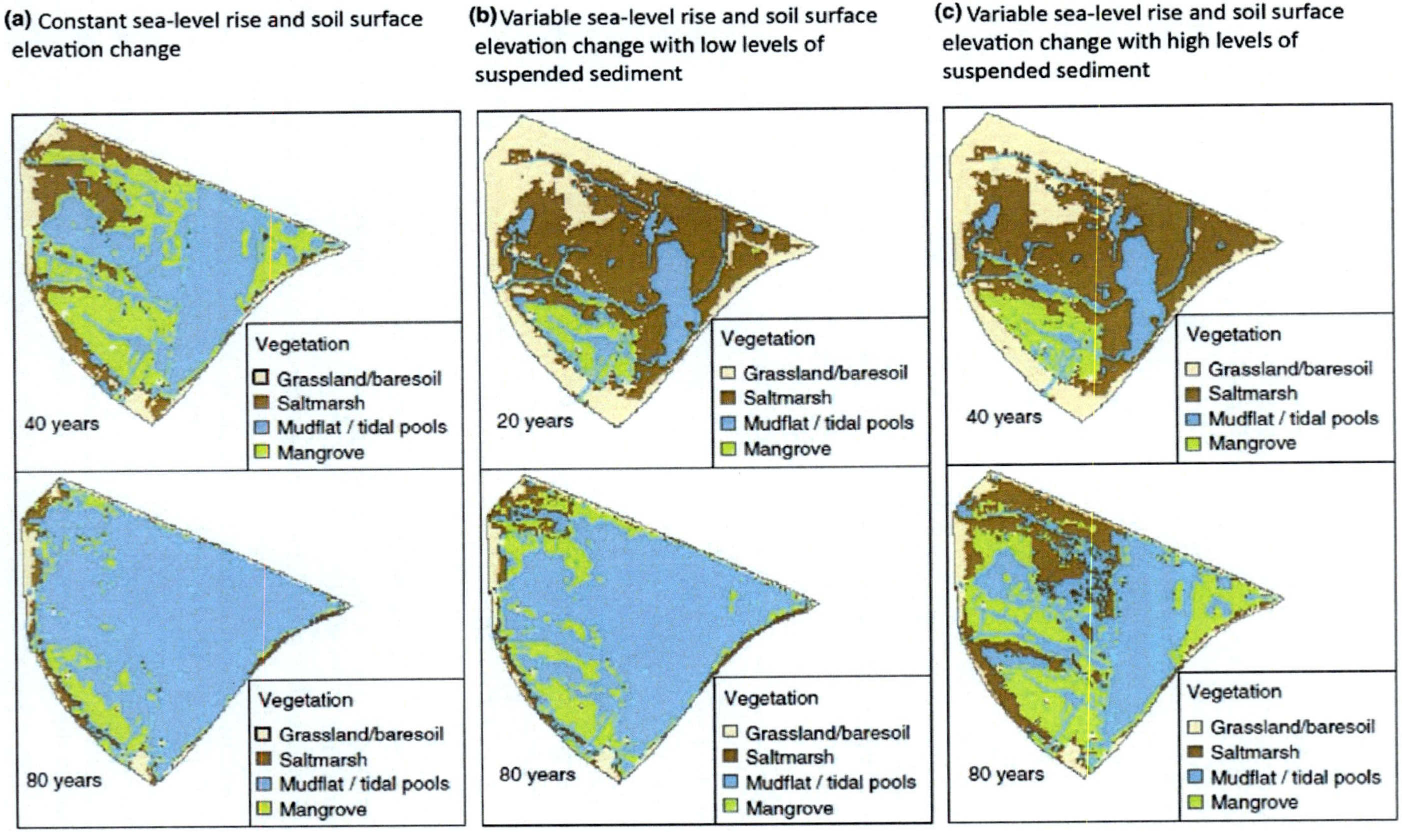

Fig. 4 (Adapted from Rodríguez et al. 2017) Hydrodynamic model outputs showing comparative vegetation changes in constant sea level rise and constant accretion speed scenario **a**, variable sea level rise and low suspended sediment availability scenario **b** and variable sea level rise and high suspended sediment availability scenario **c**

In addition to temporal correlation patterns which are either gradual (long term) or periodic (short term and seasonal), wetlands as infrastructure have a marked impact on flooding timelines and effects on adjacent ecosystems during non-periodic short-term events such as storms.

In all three ecosystem types, the wetlands have the capacity to temporarily store run-off or tidal surge waters thereby decreasing the impact on adjacent land-based ecosystems. A 2019 study (Kim et al. 2019) on the effect of antecedent soil moisture conditions on flood event type and frequency shows that "for saturated soil conditions, a 7-year precipitation event could trigger a 100-year flood", and "conversely, a 200-year precipitation event with dry soil conditions might generate only a 15-year flood event". The ability of flood plains to store water through run-off or infiltration therefore has a significant impact as it decreases the saturation speed of adjacent land, implicitly decreasing the overall impact of the flood.

In estuarine ecosystems, the delay of inundation due to the presence of intertidal wetlands influences water speeds and levels in the upper estuary, implicitly affecting the impact of storm tides on adjacent upstream land areas.

It is interesting to note the temporal correlations identified in a 2015 study (Smolders et al. 2015) relating to the role of location, elevation and size of intertidal wetlands in tide and storm tide attenuation along a confined estuary. The authors identify the "time the water enters the wetland, the duration of flow onto the wetland and the rate at which water flows onto the wetland" as important temporal factors which determine the amount of attenuation achieved. The study suggests that for a storm tide, higher elevation and location further along the estuary result in higher attenuation as this allows for longer flow duration and offsets the time when water enters the wetland, generating better synchronization between wetland water storage and storm surge evolution. The duration of flow onto the wetland is also impacted by the size of the wetland (storage capacity).

Figure 5 shows the simulated output for the impact, in terms of storm surge attenuation, of different wetland elevations on flows in and out of the wetland. On the one hand, the higher elevation scenario (P.S.4) shows the later on-set of water flowing into the wetland compared to lower elevations. On the other hand, the higher elevation scenario provides attenuation closer to the critical peak flood moment (peak water levels in the main channel), therefore providing attenuation when the flood poses the greatest risk to adjacent land. By comparison, although water is also stored within the lower elevation wetlands, storage capability decreases before the peak of the storm, thus having a negative impact by adding to the water levels in the main channel during the peak of the storm surge (main channel water level).

Overall, in terms of the natural correlations that relate to infrastructure, the main functions that wetlands perform relate to increasing resilience and fixing existing new land infrastructure through the seasonal and long-term dependency between plant growth cycles, sedimentation and hydrological cycles. The degree of correlation achieved via wetlands as transition zones therefore impact the relative extent of adjacent land and water-based areas. Additionally, in the short timeline, wetlands can help mitigate flood impacts on adjacent land areas via their ability to act as temporary water stores.

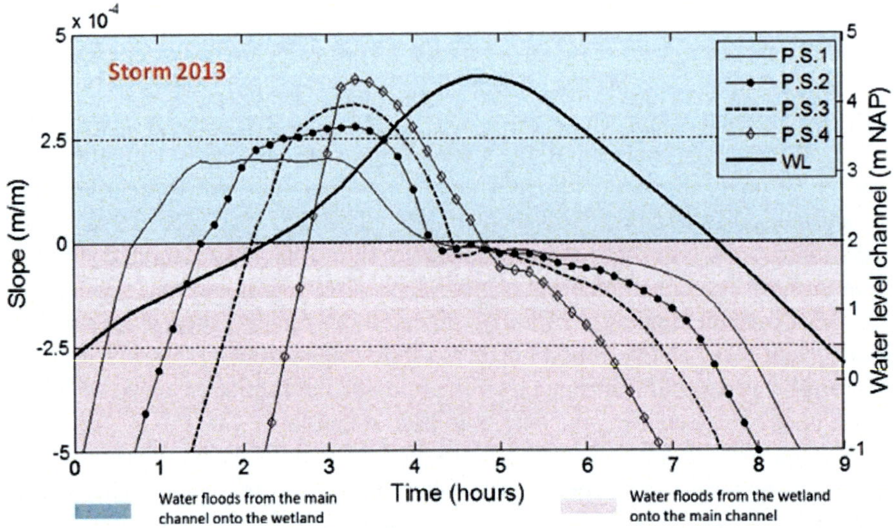

Fig. 5 (From Smolders et al. 2015) simulated wetland and main channel water levels during 2013 storm tide. Main channel water level (WL) is shown in continuous black line with scenarios P.S. 1–4 representing different elevations (P.S.1–0.85 m; P.S.2–1.4 m; P.S.3–1.85 m; P.S.4–2.5 m). P.S 1–3 elevations taken from measured site elevations in 1931, 1963 and 2009

The following section discusses similar correlations and their impact in terms of the natural functions of debris and their impact on wetland resilience and development.

2.2 *Debris—Management Strategy*

In the urban context, flood-related debris is defined as "broken or torn pieces of something larger" (Cambridge Dictionary) and represents one of the main hazards encountered in the aftermath of flood events (WHO 2007). Whether in relation to organic material or other types of debris, urban flood generated debris falls almost exclusively under the heading of waste. In the natural world, however, flood-related debris can have an important role in wetland development. In this context, flood-related debris mainly consists of organic matter such as plant litter, woody debris and mud.

Riverine floodplains and estuarine mangrove ecosystems are exposed to a large range of naturally occurring debris types such as mud, silt, plant litter and large woody debris. In relation to how debris is incorporated into the wetland ecosystem, organic decomposition rates, sediment accretion processes as well as flora and fauna lifecycles play an important role in positively utilizing debris to further the development of the ecosystem (Rothero et al. 2016).

The functional value of each debris type (over time) can be correlated to short-term, seasonal and long-term wetland processes. The decomposition rate of each debris type is a particularly important factor as it directly impacts the length of time over which each specific function may be performed. In other words, an important temporal correlation is the proportional relationship between the debris lifespan, its associated decomposition rates and its functionality. For example, under the same environmental conditions, woody debris decomposition rates can be correlated with wood density (Barbosa et al. 2017).

Currently, the functional aspect of flood generated debris is mostly studied and recognized in the field of wetland and water management. The Floodplain Meadows technical manual 2016 (Rothero et al. 2016) relating to the functions and management of floodplain meadows in the UK notes organic matter deposited via flooding is "important as a store of carbon and as a modifier of the soil's chemical and physical properties. Chemically, it has the ability to bind nutrients and thus to increase the fertility of the soil. Physically, it increases the stability of the structural units and makes the soil more resilient to compaction". The quality of the soil in turn impacts plant life as well as soil water storage capacity. The same manual highlights the fact that, where karstic soils are encountered upstream of a floodplain area, flood-deposited calcium (Ca) has the effect of maintaining a healthy soil pH by counteracting soil acidification usually due to acid rain and other anthropogenic factors.

At the seasonal and short-term timescales, it is interesting to note the relationship between nitrogen (N) and phosphorus (P) availability and decomposition processes which have been shown to accelerate as the availability of N and P increases (Rothero et al. 2016). This phenomenon is important in terms of wetland resilience to short-term extreme flooding events. Effectively, decomposition rates increase in response to the intensity of flooding via the direct connection between amount of deposited debris and available N and P levels. This temporal correlation has a twofold effect as higher decomposition rates speed up removal of excess debris and also provide more nutrients for plants and animals, thus also speeding up regeneration processes.

Similar to the correlations observed for infrastructure development, seasonal synchronization between hydrological cycles and ecosystem growth is determined by plant development regimes. In addition to the sediment accretion processes discussed in the previous section, debris of medium density such as plant litter (leaves, twigs, etc.) also plays important roles in aiding the establishment of pioneer species (Rothero et al. 2016).

However, in terms of nutrient release as a debris function, peak decomposition rates and implicitly peak fungi and other animal activity in the vicinity of plant litter deposits are usually experienced in the first year after deposition onto the wetland. For example, a 2015 study (Gingerich et al. 2015) on the role of fungi and invertebrates in litter decomposition notes that "biomass for most functional feeding groups peaked prior to 224 days and then decreased and levelled off for the remainder of the study" based on measured decomposition rates across six wetlands in West Virginia USA.

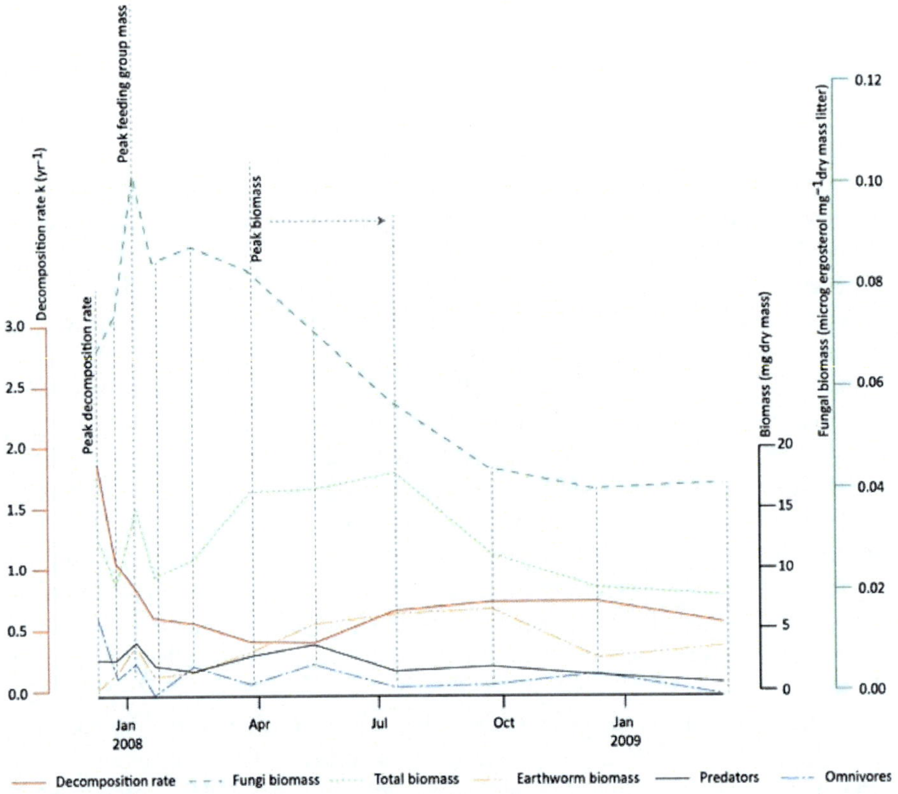

Fig. 6 Decomposition rates versus functional feeding group biomass (Based on Gingerich et al. 2015)

Figure 6,[2] based on the same study, shows a clear correlation between activity of nearly all functional feeding groups and decomposition rates. Functional feeding groups ensure nutrient release and continuity within wetland specific trophic chains. Seasonal replenishment via flood-deposited debris of this type is therefore a critical way in which flooding and, more specifically, debris have a positive role in wetland development.

In the short and seasonal timelines, the availability of nutrients is linked to either plant litter or silt, while in the long term, large woody debris which decomposes much slower plays a vital role in decreasing water flow speeds, filtering water, forming new infrastructure and providing nutrients and habitat for a variety of terrestrial and aquatic species.

The European Commissions' guidance on Individual Natural Water Retention Measures (2013) identifies a series of scenarios where the management and inclusion

[2]Figure 5 used under CC BY 3.0 from Smolders et al. 2015/Original figure 6 was colour coded to clarify negative and positive slopes shown https://creativecommons.org/licenses/by/3.0/.

of coarse woody debris have highly beneficial biophysical impacts and can enhance ecosystem services.

The document highlights that coarse woody debris is most effective in smaller rivers (0.1–1 km^2 upstream catchment area) but that it is possible for their effectiveness to be extended up to 100 km^2 upstream catchment areas. In terms of beneficial effects, the document lists highly positive biophysical impacts in terms of reduction of river water speeds, high and moderate beneficial impact in providing aquatic and terrestrial habitat, respectively. These factors in turn have a positive effect on enhancing ecosystem services relating to fish stocks and recruiting, flood risk reduction and creating new recreational opportunities.

A 2001 review (Maschhoff et al. 2001) also highlights a series of vital functions performed by coarse woody debris in wooded wetland ecosystems (Fig. 7). The article reviews the beneficial effects of reintroducing woody debris into managed wetland ecosystems, highlighting their importance in natural, undisturbed wetlands.

Depending on wood density, course woody debris, which may decay over periods as long as 20 years (Barbosa et al. 2017), can aid the formation of new periodically inundated floodplain areas. A 2017 study (Zen et al. 2017) exploring the role of debris in meander formations in the Tagliamento River in Italy found that the lateral evolution of meanders and implicitly the formation of new vegetated flood plain areas are linked to the deposition of uprooted trees via flooding. Large woody debris impacts the way sediment is trapped in the surrounding bend area and provides a base for pioneer vegetation growth, ultimately leading to the formation of new persistently vegetated zones. The study also highlights the fact that, between 1980 and 2016, the frequency of flooding which deposited this type of debris occurred with a quasi-decadal frequency for the Tagliamento River. Interestingly, the hydrological frequency seems to correlate with average decomposition timelines for higher density woody debris (Fig. 8).

In overview, the key principle identified, especially within short and seasonal timelines, is the correlation between the useful function of debris and its rate of decomposition which ultimately eliminates waste. In other words, the re-use of waste material (debris) can be viewed as an alternative strategy that reinforces the overall resilience of the ecosystem by providing alternative ways of infrastructure formation and thus alternative options for shelter and nutrient provision at the level of the whole system.

Zooming out to the bigger picture and long-term resilience to changing climatic patterns, this diversification of adaptation strategies plays an important role for wetland survival. The following section explores the impact of interlinked, diverse strategies through the notion of productivity as an indicator of resilience.

2.3 Productivity—Diversity and Resilience

In ecology, the term productivity is associated with the amount of biomass produced by an ecosystem in a given time (usually over the course of one year). Oxford

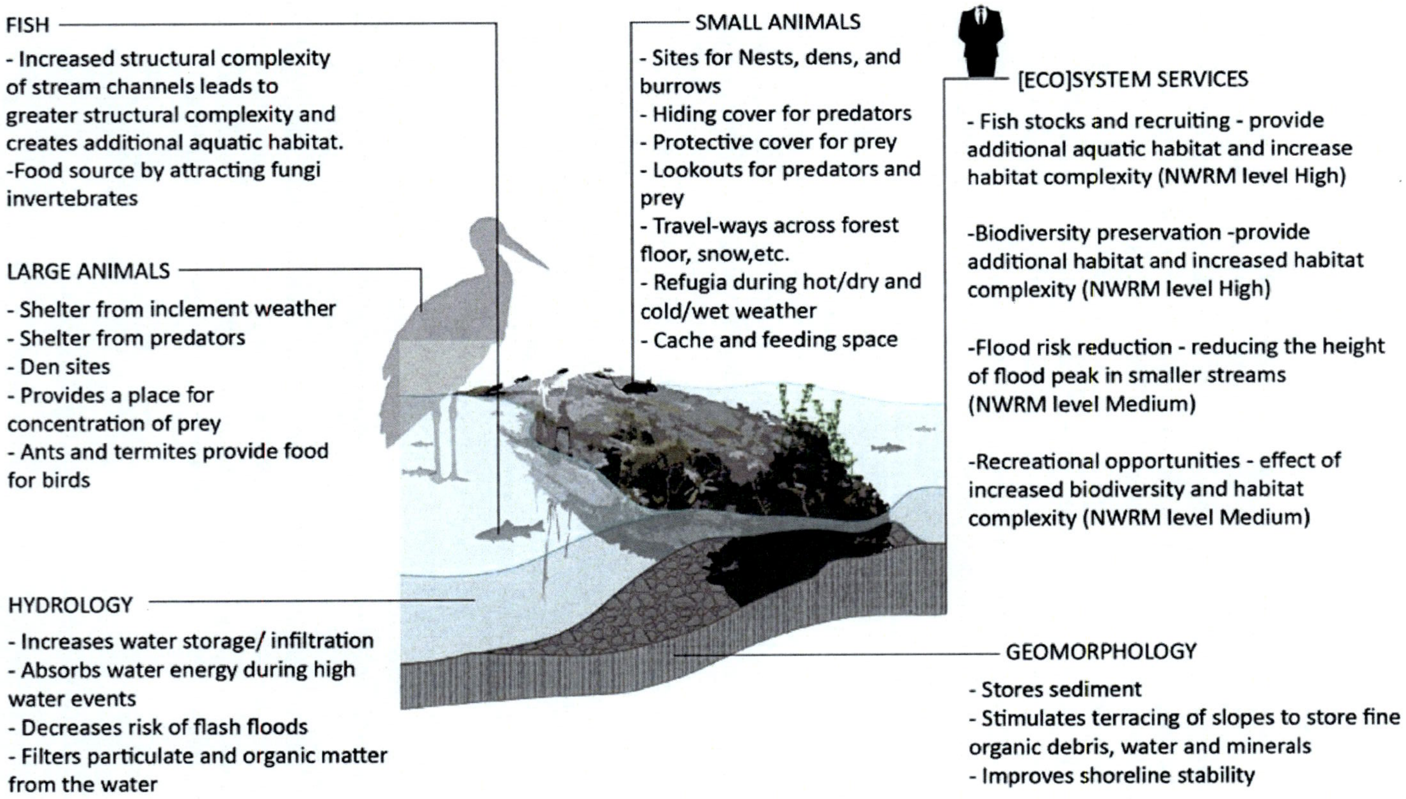

Fig. 7 Functional roles of coarse woody debris (Based on Maschhoff et al. 2001)

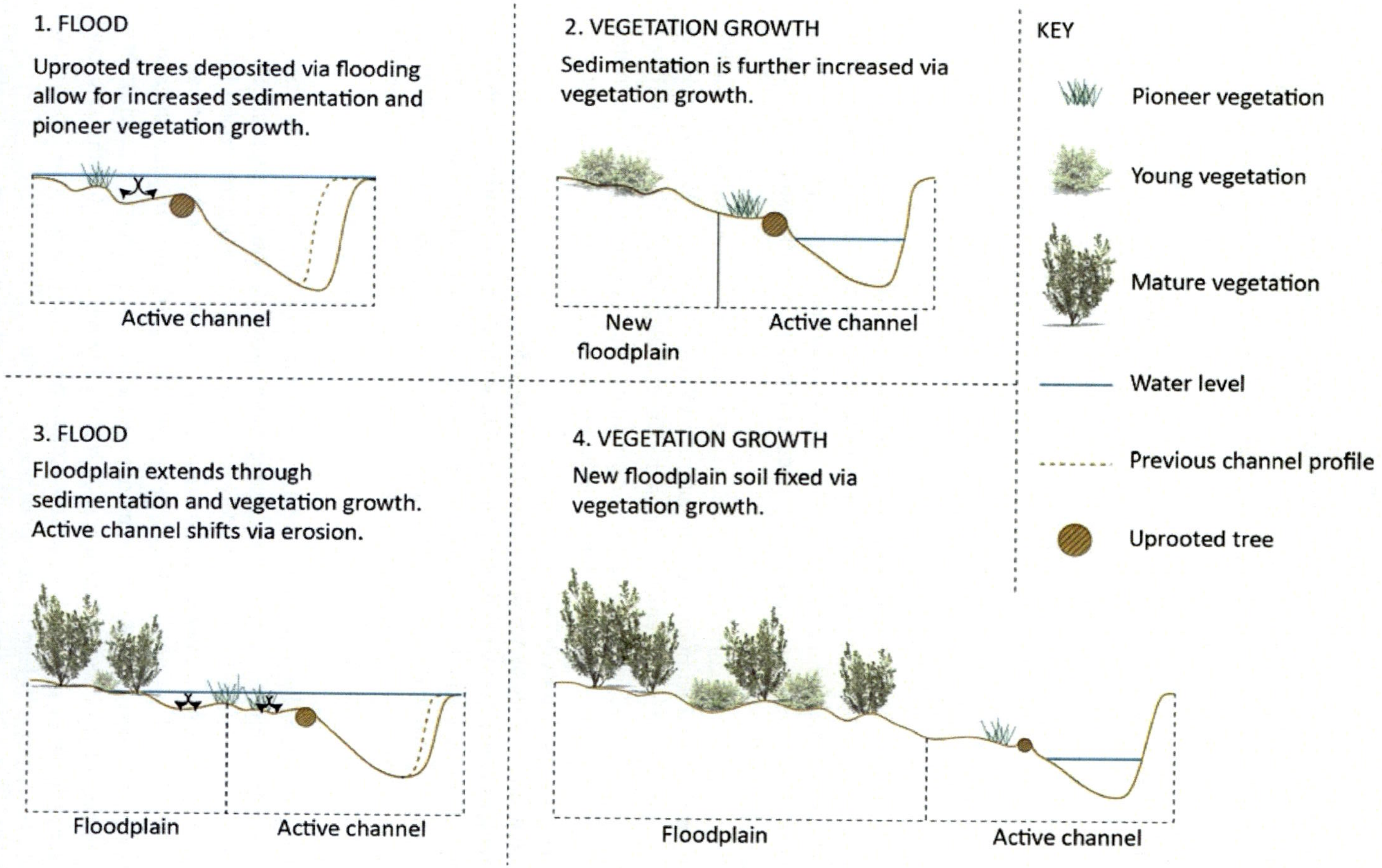

Fig. 8 Simplified conceptual representation showing stages of meander and floodplain development triggered by trees deposited via flooding (Based on Zen et al. 2017)

Dictionary defines ecological productivity as "the rate of production of new biomass by an individual, population or community, the fertility or capacity of a given habitat or area".

Productivity is an important factor in determining an ecosystems' level of flood resilience as, via its dependency on biodiversity, this parameter is a good indicator of the ability of an ecosystem to adapt and recover more easily from extreme climatic events such as flooding. Assessing flood resilience through such a factor may prove particularly relevant in terms of drawing parallels with the urban context, where, at a systemic level, the notion of productivity relates quantitatively to industrial/economic outputs, correlating with GDP and other economic factors as indicator variables.

Within natural ecosystems, in 2008, a Brown University study (Flombaum and Sala 2008) found significant positive correlations between ecosystem productivity and biodiversity. Furthermore, the study noted that the impact of biodiversity on productivity was more pronounced in undisturbed natural ecosystems than in those artificially constructed.

This relationship between species richness and productivity suggests that flood resilience at a systemic level is achieved via two distinct but interconnected factors. High species richness maximizes chances of species survival by diversifying adaptation strategies while the increased productivity, via a faster biomass production rate, allows for faster recovery of remaining species than in the case of systems with lower productivity levels.

The strong positive correlation between species richness and ecosystem resilience has been measured and observed in natural wetland ecosystems. For example, in 2012, experimental measurements within the Upper Paraná River floodplain in Brazil (Carvalho et al. 2012) revealed significant positive correlation between resilience and species richness. The study attributes the higher levels of resilience to the diversity of "resource use requirements".

In temporal terms, as we have seen in the previous ecosystem analysis sections, correlations between, for example, plant growth seasons and inundation events may in some cases lead to more pronounced negative effects. In a diverse system, not only will each species possess a different adaptation strategy and resource use requirement but also different reproductive and growth patterns. The impact of an extreme flooding event will not be equally detrimental to all species, allowing for adaptation to a wider range of climatic conditions via diversification.

However, it is also important to recognize that productivity as driver of species richness and implicitly flood resilience represents just one strategy through which ecosystems become more resilient. Coined by the American zoologist Robert Paine (Paine 1980), the term trophic cascade describes an indirect interaction, such as the removal or extinction of a predator species, which significantly impacts the entire ecosystem. The removal of a single predator species may lead to the extinction of a wide variety of plant and animal species, greatly reducing the system's diversity and implicitly its resilience level. Therefore, regardless of the diversity of the system, flood events may still lead to an overall decline of the system.

It is here that diversity, in terms of adaptation and resilience strategies, plays a vital role in expanding the range of scenarios in which the ecosystem can thrive.

As shown in the infrastructure and debris sections, within just a few examples discussed, several ways of nutrient delivery and new infrastructure emergence strategies were identified. The overall plant distribution pattern and infrastructure development are the results of several strategies which develop within separate timelines, their cumulative effect determining extent of infrastructure development and ratio between land and aquatic zones (Fig. 9).

The concomitant application of diverse strategies at multiple levels and across multiple timelines is, therefore, identified as a key overarching principle. The analysis of wetland infrastructure establishment, debris utilization and productivity (via diversity) yields a range of strategies and temporal correlations which enhance natural resilience and could be utilized in the context of urban design.

3 Speculation and Examples of Applied Temporal Correlations in Urban Contexts

The evolution of urban areas over time is a result of various social and political factors such as planning policies, economic potential and general public perceptions. These same factors also have an impact on prevailing approaches relating to flood design. Planning policies and economic investment in the recent past have largely been geared towards flood control measures (e.g. dams, levees, warning and forecasting systems, etc.) based on the general perception that floods are inherently hazardous.

Liao (2012) identifies the difference between the engineering and ecological definitions of resilience, surmising that the engineering approach (based on flood control) applied in past urban designs has led to cities becoming flood-resistant rather than flood resilient. As shown in Fig. 10, flood resilience implies adaptability to a range of flooding scenarios which the urban system is adapted to and makes use of. In contrast, the resistant city model is less flexible, being adapted to and making use only of scenarios in which flooding is averted almost completely.

In recent years, however, several cities around the world have adapted their approach towards a flooding resilient model, integrating and adapting ecological models in which seasonal flooding is interlinked within urban design strategies.

This section discusses how the temporal correlations and flood response strategies observed in the review of wetland ecosystems could be adapted and applied in urban design. Through a review of selected resilient projects, it is observed that successful existing flood resilient designs support the argument of using temporal correlation as a design constraint.

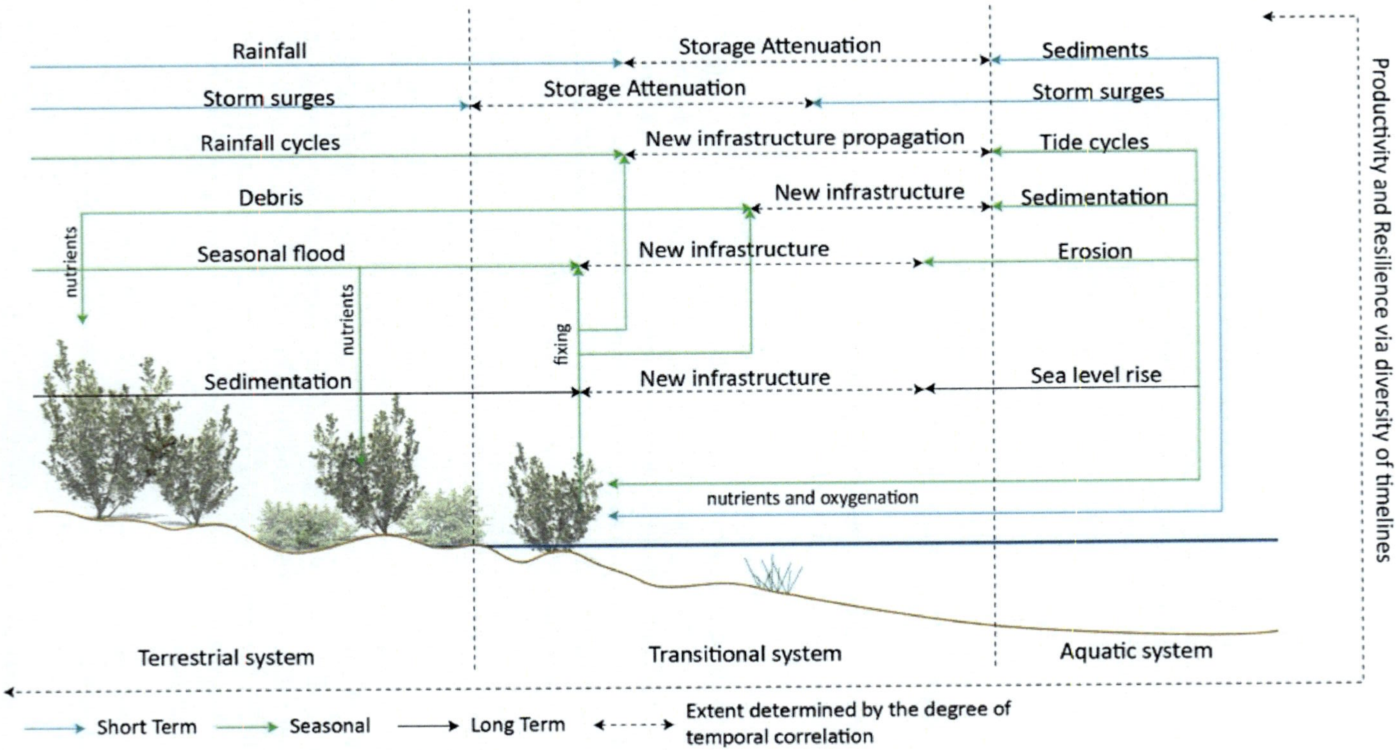

Fig. 9 Conceptual representation showing various timelines identified and the interaction between the degree of correlation and extent of aquatic or terrestrial systems

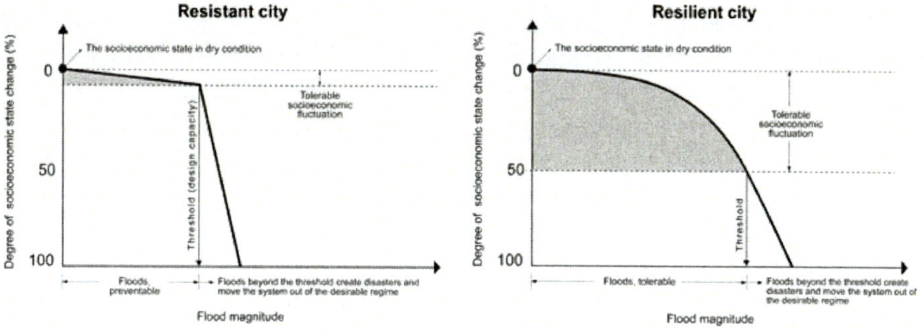

Fig. 10 (From Liao 2012) Comparison of degree of flexibility and flood impact in resistant and resilient city models

3.1 *Infrastructure—Growth and Symbiosis*

In flood-prone urban areas especially, the resistant infrastructure model presents two main drawbacks. As the urban landscape is ideally never flooded, its evolution in terms of design and strategy is rarely driven by flood patterns, and therefore, when flooding does occur, it has a greater negative impact (often disastrous). This correlates with the second aspect of flood control infrastructure often being centralized and of large scale, affecting natural patterns (hydrological regimes, ecosystem growth, etc.) and contributing to an increased flood vulnerability to extreme events.

While resistant designs can counter a range of flooding events, when considering the fluctuating climatic conditions predicted in the long term and economical design parameters (i.e. reasonable design flood-proofing levels and safety factors), it seems likely that flood control measures that do not adapt to changing inundation patterns will not provide sufficient future-proofing. By comparison, through temporal correlation between system growth patterns and short, seasonal and long-term flood patterns, wetland ecosystems adapt gradually to changes in the hydrological cycle showing localized adaptation strategies such as the localized adaptation of meander topography and tidal flat morphology discussed in the previous sections.

These adaptation strategies result in an optimized, gradual change of the ecosystem infrastructure where the degree of change in infrastructure is proportional to climatic/hydrological changes.

In short-term flood events such as storm or tidal surges, wetlands function as water storage basins, slowing down water flows and decreasing water levels in the main riverine or estuarine channels, thereby decreasing the impact of the flood. As shown in the ecological analysis section, the level of attenuation is dependent on the temporal correlation between peak wetland and peak channel water levels. This correlation is in turn determined by the elevation, location and size of the wetland. Furthermore, the wetlands water storage function can decrease the speed of saturation in adjacent land, implicitly decreasing the likelihood of a major flood event being triggered by average levels of rainfall.

There are three main approaches to integrating these strategies in urban design: restoration or creation of wetland areas within the city, the use of artificial systems which perform similar functions and a mix of the first two. The last version has thus far been more widely used as it aligns urban growth and ecological agendas boosting biodiversity and flood resilience at the same time. The integration of artificial systems also creates opportunities for the creation of flood-responsive urban typologies.

A good example is the "sponge city" concept which, since 2014, has gradually been implemented in pilot projects in flood-prone areas of China. The Sponge City Programme launched by the Chinese Government deals mainly with tackling pluvial flooding in urban areas by introducing planning policies which regulate the introduction of urban wetland areas, green roof and rain garden ratios, peak allowable run-off coefficients, capture ratios of annual rainfall, permeable paving percentages and other factors. The design principles of the policy draw inspiration from similar approaches applied in other countries (Li et al. 2018) such as low impact development strategies (USA), sustainable urban drainage systems (UK), water sensitive urban design (Australia) and others.

A 2019 study (Khadka et al. 2019), which analyses the effect on storm water storage capacity of different low impact management strategies on a site in southwestern Finland, exemplifies the impact of these types of designs in terms of flood reduction. The current scenario (0) and three design options with various degrees of low impact development (LID) strategies implementation were tested against four types of rainfall events.

The impact shown in Fig. 11 represents the worst-case scenario of a 1 in 100 years extreme event happening after an intensive rain event resulting in saturated soil conditions. Design option A introduces a vegetated swale. Design B includes the swale and some areas of green roof and permeable paving. Design C includes the swale, permeable paving in all areas, rain gardens and green roofing.

Although run-off and losses (from infiltration and evapotranspiration) show smaller variations, a dramatic reduction of flooding (from 50 to 3%) and an increase in storage capacity (from 0 to 42%) can be observed between the initial conditions and full LID implementation scenario.

While the above example deals with a relatively small catchment area, the local increase in water storage capacity has a strong impact on the large city scale. At the large city scale, the above strategies are usually complemented by the reintroduction or conservation of urban wetland areas, often designed as parks and inner-city leisure areas (Fig. 12). By storing and reutilizing water locally, the time lag between water beginning to flow into the wetland and the peak water levels (when maximum storage capacity is reached) is extended, thereby aiding the correlation between peak wetland water levels and peak channel water levels.

Understanding how this temporal correlation can be influenced (via wetland size, elevation and location) and enhanced (through localized small-scale measures) could prove a key design factor for future schemes. Especially in dense urban contexts, where it is often difficult to balance the space required for wetland versus urban development, optimizing temporal correlations could help to save vital space. For example, in coastal estuarine cities, a smaller area of intertidal wetland located further

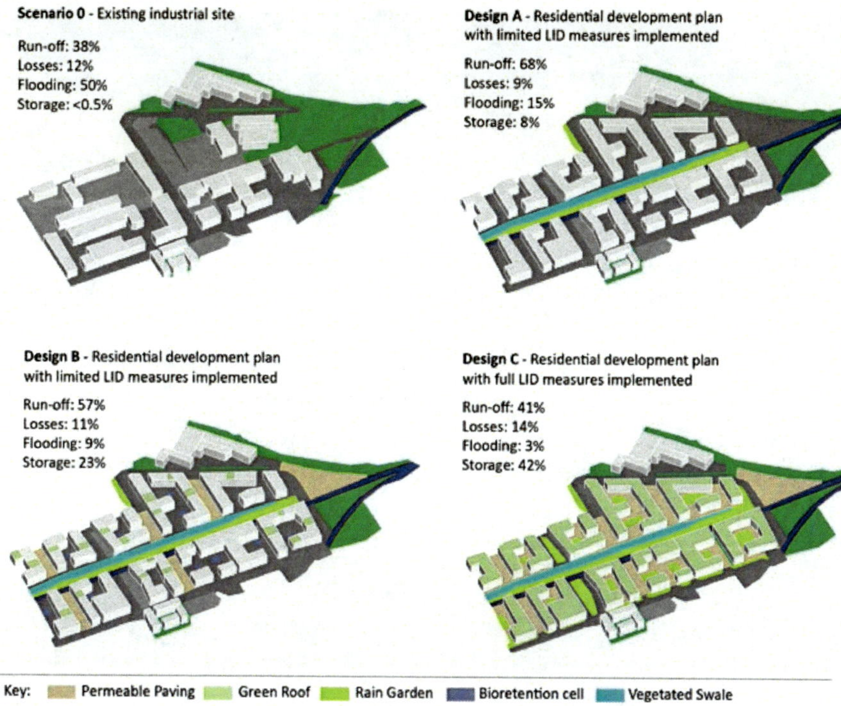

Fig. 11 Simulated design scenarios and worst-case scenario event impacts (Based on Khadka et al. 2019)

upstream along the estuary with a considered elevation could perform similarly or better than a larger-sized wetland located closer to the estuary mouth (Smolders et al. 2015). Coupled with localized measures, the size could further be reduced while obtaining the same level of flood mitigation.

Regarding seasonal temporal correlations, in the natural context, a correlation was identified between wetland infrastructure resilience/growth patterns and seasonal hydrological regime. This connection was established via an interdependency between infrastructure stability/emergence and plant lifecycles. This mechanism reveals a fundamental difference between the urban and natural systems.

The interdependency between plant growth and infrastructure implies that wetland growth at the large scale is achieved via localized adaptation. For example, if hydrological patterns change to the effect of more frequent flooding occurring over a specific area, plant numbers will dwindle, soil stability will diminish, and eventually, depending on the degree of change in hydrological patterns, the system area may shift to a different type of micro-ecology (from land to aquatic-based for instance). In contrast, urban design is not adaptable at a local scale with most designs being planned based on predicted future trends covering large regions. Given the fast, climatic changes and our current level of understanding underlying climatic mechanisms and cascade effects, accurately predicting future conditions is becoming less and less reliable.

Fig. 12 Skywalk above the storm water filtrating and cleansing ponds at Qunli stormwater park by Turenscape and Peking University, Beijing (Sponge city design awarded 2012 ASLA award of excellence) Photo: Kongjian Yu, Turenscape

The issue of localized adaptability in relation to flood response is highlighted by Liao (2012) and (2016) as a key principle to achieving a flood resilient urban model. The 2016 study investigates the flood response strategies used in two seasonally flooded Vietnamese hamlets in the Mekong Delta. It is noted that through localized adaptation to varying flood levels, the hamlets "remain functional at a prolonged flood as high as 2 m, which would likely devastate most cities around the world" (Liao et al. 2016).

One key adaptation in the hamlets is the dual nature of pedestrian pathways (dirt roads) which shift from land-based to aquatic waterways during flood season. A possible approach to integrating this type of adaptable infrastructure is explored by The Urban Design Program-Global Cities and Climate Change Studio at Columbia University.

The [Re]Flow Can Tho project included in the spring 2019 e-publication (GSAPP 2019) of the group proposes doubling principal road arteries with seasonally flooded waterways. Depending on the level of flooding, the waterways act as transport routes or ecological green corridors accommodating tourism and leisure activities.

These types of designs are inherently driven by adaptation to the seasonal timeline, creating correlation between the seasonal hydrological patterns and infrastructure functions. The resulting flood-dependant infrastructure provides the framework for

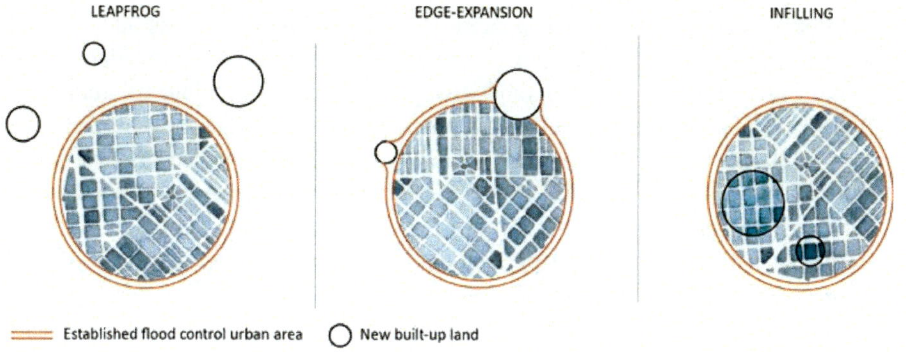

Fig. 13 Types of urban growth specific to the Yangtze river economic belt (Based on Han et al. 2019)

developing connected urban activities which increase overall flood resilience and encourage the growth of positive urban flood-related economy.

Aside from localized adaptation, in wetlands seasonal temporal correlations are also important for the establishment of new infrastructure. The lifecycle of pioneer species was shown to be optimally timed in relation to seasonal hydrological patterns to ensure that plant root establishment periods were correlated with milder dry season conditions.

Relating to the urban context, it is interesting to note that the same type of correlation shows significant impact on flood resilience of newly established urban areas. A 2019 study (Han et al. 2019) explores the level of flood vulnerability for different types of "pioneer" urban typologies based on a case study in the Yangtze River Economic Belt, China. Figure 13 shows the three types of urban growth observed in the area. It is noted that "in terms of expansion types, the edge-expansion type dominates 57.52% of the growth, followed by the leapfrogging and infilling expansions".

The study found that smaller emerging areas are more vulnerable to flooding than larger ones, and among expansion types, flood vulnerability was significant for edge and leapfrog expansion types, the latter being the most frequently flooded. Vulnerability was assessed relative to the protection levels of the originally built infrastructure which was protected via flood control measures and could withstand 1 in 50- to 100-year flood events, whereas edge-expansion and leapfrog types were found to be vulnerable to 1- in 10-year flood events.

The authors establish a link between the size of new-built expansion types, their economic development and the level of flood protection implemented. During the initial phases of new urban development, economic growth/stability may take several years to achieve. The fact that, due to proximity, edge-expansion types may partially rely on the economic systems of the original development could be a factor in determining why edge-expansion types were found less vulnerable compared to the leapfrog type.

Looking back to the growth pattern of wetland pioneer species and associated infrastructure, the urban economic growth timeline and associated flood vulnerability are conceptually comparable to the disturbance-free interval required by pioneer wetland species impacting the emergence and resilience of new infrastructure. While the economic growth timeline is recognized as critical to infrastructure growth and flood resilience, no attention is given to temporal correlations between critical establishment and seasonal hydrological patterns. The vulnerability of newly established urban areas could potentially be improved by temporally correlating the economic growth patterns of the area with seasonal hydrological regimes. In order to accelerate economic growth patterns, urban development could be phased to initially prioritize flood-related economic activities (fishing, aquaculture, etc.) and boost flood resilience by making use of and enhancing natural wetland functions as a cost-effective means of flood adaptation.

This type of seasonal correlation (achieved via the use of natural systems) would also have a marked impact on the long-term development and resilience of the site. There are numerous examples of restored habitats which are key to counteracting long-term coastal erosion and other harmful processes. The joint Danish and Indonesian mangrove re-planting initiative (Tonneijck et al. 2015) described below provides a pertinent example of how re-established mangrove forests can reverse coastal erosion processes and boost local economy. The project, which was prompted by rapid coastal erosion and its negative impact on local villages, is currently being implemented in various coastal areas across Indonesia's Demak region and follows the principles of "Building with Nature" standards. Between 2003 and 2009, due to mangrove deforestation to accommodate intensive aquaculture, the coastline retreated up to 1.5 km over relatively large inhabited areas of the coast.

The proposal aimed to reforest transition areas between land, aquaculture ponds and the sea via the use of strategically located "permeable dams". The dams perform a dual function in allowing the reestablishment of mangrove ecosystems via decreasing wave velocity and enhancing sediment accretion rates. At a physical level, the structure temporarily performs the role of a mature mangrove forest, allowing for the establishment of new specimens and the gradual development of new land infrastructure.

The project Design and Engineering Plan (Tonneijck et al. 2015) details not only the location and extent of permeable dam structures but also establishes temporal correlation between seasonal, design and long-term patterns, by proposing a design, implementation (data collection) and learning cycle which ensures the adaptive evolution of the design. Each year, monitoring data is analysed at the end of the monsoon season and integrated into an adjusted design proposal which is then implemented before the start of the monsoon.

As observed in the ecological analysis section, a number of recent studies suggest that some coastal wetland ecosystems may not survive the changing hydrological regimes forecasted over the next 80 to 100 years. Interventions such as the Demak mangrove restoration initiative may indeed prove essential in achieving new resilient urban development (Fig. 14).

Fig. 14 Local communities in Demak, Indonesia, supporting the construction of temporary semi-permeable barriers from poles and brushwood to dampen the waves and capture sediment, creating sheltered areas near the coast for accretion. Once the nearshore bed level rises enough, mangroves will regenerate naturally, developing a natural defence that will protect the hinterland from further erosion. Photo: by Nanang Sujana for the Building with Nature Indonesia project

The mangrove restoration initiative has successfully been running in Indonesia for some years now, demonstrating the potential positive impact of working with live ecosystems within a rural or agricultural setting.

Via a similar approach, APTUM Architecture's prototype "Isla Rhizolith" (Aptum 2019) provides some insight into how this type of system could be integrated into urban areas. The prototype consists of a floating concrete platform which is designed to support the establishment of mangrove propagules. The concrete structures function similar to the permeable dams above, in the sense that they enclose an initial area providing wave attenuation and, in time, allowing for the area to become populated by mangrove vegetation. The floating platforms, however, have a dual function, doubling up as a new type of floating urban space accessible from land (Fig. 15).

3.2 Debris—Lifespan and Functionality

In contrast to the functional nature of debris explained in the ecosystem analysis section, in the urban context, debris is not considered as having any functional use and is effectively classified as waste.

Fig. 15 Isla Rhizolith—floating concrete breakwater prototype Photo: Courtesy of APTUM Architecture and CEMEX Global Research and Development

Increasingly, flood-related design policies recognize the need for debris management and incorporate flood control strategies which limit the negative impacts of debris in flood scenarios. In the USA, the Federal Emergency Management Agency (FEMA) via the National Flood Insurance Program (NFIP) regulates the use of elevated ground floor areas and "breakaway walls" below the elevate ground floor for areas of increased flood risk. The breakaway walls are designed so that under specifically combined hydrostatic and wind pressures, the walls give way without affecting the main structural elements of the construction. Provisions are also made for designing unobstructed routes for floodwater to pass below the structure, allowing for slower water speeds but also for decreased hydrostatic pressure on adjacent supporting structures.

FEMA guidelines (FEMA 2008) specify that all materials used for breakaway walls should be "capable of withstanding direct and prolonged contact (i.e. at least 72 h) with floodwaters without suffering significant damage". It is interesting to note that the element's lifespan in terms of design and policy is considered only up to the "breakaway" moment, after which the wall, now debris, is no longer considered as a design element.

Including the lifespan as debris of elements such as breakaway walls as a design criterion could indeed prove a valuable resource in enabling faster recovery in the case of extreme flooding events. Potential strategies could range from assessing the decomposition speed of the specified materials, encouraging natural erosion and minimizing wastage, to design strategies which account for re and deconstruction via modular, re-usable floating breakaway wall types.

As we have seen in the ecosystem analysis section, in the natural world, the most important temporal correlation identified was between hydrological regime, rate of decomposition/lifespan of debris and ecological functions. In terms of applying this strategy in an urban context, a possible first step could therefore be the categorization of urban debris in terms of their potential function as debris and the length of their uses during and after a flood event. For example, in the natural world, large woody debris can play an important role in reducing water velocities especially in smaller streams.

In terms of urban flooding, controlling water speeds can prove a vital safety measure. The adjacent figure (Fig. 16) is a good example of the impact water speed can have on pedestrian and vehicular traffic as well as on lightweight structures. Even at very low water depths, floodwater may become unsafe to wade or drive through. This is additional to the risk posed by fast-moving debris that may be carried by the water. By designing likely hydrological flows at an urban level (e.g. inverse delta-shaped transport routes), debris accumulation points could be established. In this case, debris would accumulate at strategic points where their role in slowing water flows could be maximized.

A good example of how such a tactic could be applied is the SLIDE project (RCH Studios 2019) by Rios Clementi Hale Studios (RCH) which deals with flash flooding in the Los Angeles area (Fig. 17).

Via a series of strategically located steel micro-debris basins, the structures accumulate increasing amounts of debris with each flood, imitating the natural debris dam functions of slowing water speed and filtration. RCH further speculates that by tailoring each structure to filter a specific type of material, the basins could not

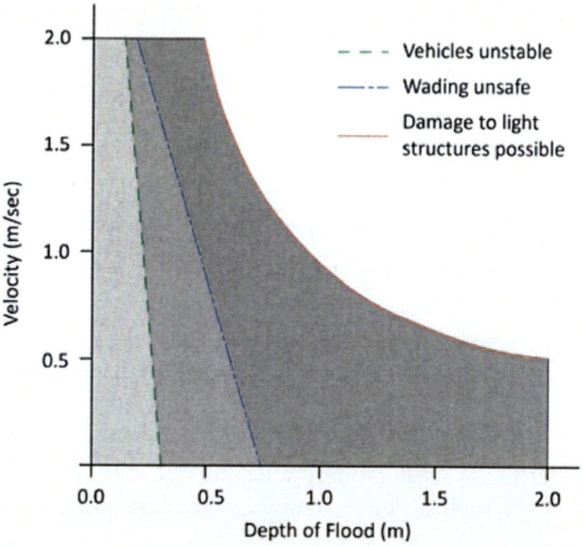

Fig. 16 Impact of water depth and velocity on the stability of vehicles and pedestrians under urban flood conditions. Based on NSW Government 2005

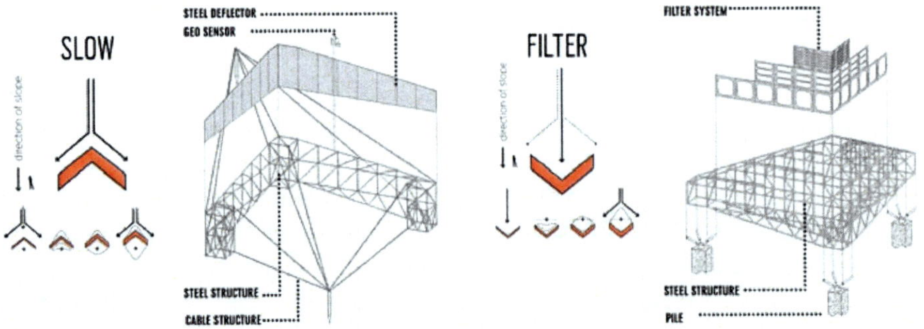

Fig. 17 SLIDE Step 1: Create micro-basin. Slow. Filter. The micro-basin serves two functions in mitigating extreme weather to create a networked, resilient and adaptive system. Image courtesy of RCH Studios

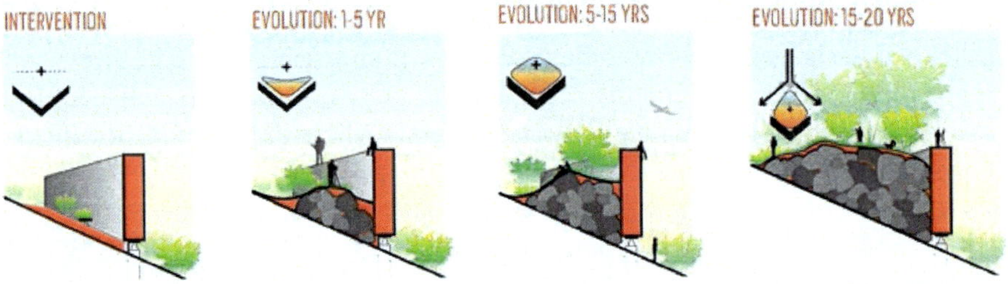

Fig. 18 SLIDE Step 2: Let the system grow. Accumulate. Through the repetition of extreme weather events, the micro-basins collect and grow from debris, allowing programme and habitat to emerge. Image courtesy of RCH Studios

only aid in mitigating the hazardous effects and forming new landscapes but that they could be used to generate revenue as mini quarries, loam farms or aqua basins (Fig. 18).

Much like the riverine floodplain example where meander bend topography was affected by the deposition of large woody debris, here the debris is integrated into strategic landscape design becoming the drivers for new infrastructure growth. In temporal terms, the design makes use of the long "lifespan" or slow decomposition (erosion) rate of rock and wood debris to establish a base for long-term infrastructure growth.

Although historically there are numerous examples of large urban populations who used seasonal flooding to increase crop yields, nowadays nutrient-rich silt deposited by floodwater has no productive role in the urban environment. It is here that some of the examples and strategies presented in the previous section could have a significant impact. Both the introduction of vegetated, seasonally flooded transport routes as well as the introduction of urban floodplains could provide opportunities for flood-fertilized seasonal crop areas to take root. In estuarine and tidal areas, halophyte crops such as chia or samphire could be utilized (Ventura et al. 2015).

Strategically located allotments could provide much-needed localized storage of flood-deposited, nutrient-rich silts. In riverine floodplains, submerged upstream structures could be limestone-based and designed for a certain level of erosion so as to take advantage of the natural property of calcareous debris, which aids in counteracting soil acidity.

One example for the implementation of such strategies within modern urban design is the Aranzadi Park Project (Aldayjover 2008) by landscape architects Aldayjover. Taking advantage of the seasonal flooding patterns around the Aranzadi meander of the Arga River in Pamplona, the project successfully implemented an urban flood-based orchard corridor design, which takes advantage of flood-deposited, nutrient-rich silts by directing water flows. In its initial state (Fig. 19), the meander bend floodplain would become entirely submerged, restricting access, with silt debris not performing any economically useful functions.

Following the implementation of the design strategy (Fig. 20), under similar meteorological conditions, the park floods in a controlled manner maintaining access and, via the strategic location of organic farming corridors, takes advantage of flood-deposited silts for crop fertilization.

Fig. 19 Aranzadi park initial state, flood February, 2012. 400 m^3/s, Q12. Photo: Gobierno de Navarra

Fig. 20 Aldayjover's Aranzadi park, flood April 19, 2018. 420 m^3/s, Q20. Photo: Gobierno de Navarra

3.3 Productivity—Diversity Types

In the urban setting, the term productivity generally refers to industrial productivity which is defined as "the effectiveness of productive effort, as measured in terms of the rate of output per unit of input" (Oxford Dictionary). In contrast, the ecological definition of "fertility or capacity of a given habitat or area" (Oxford Dictionary) seems to be geared towards qualitatively recognizing the features of the specific productive habitat.

The following examples draw upon the relationship between diversity, productivity and flood resilience, speculating on how the diversification strategies observed in the ecosystem analysis section could be applied in urban design. Productivity is, in this case, described as ecological rather than industrial productivity applied to an urban setting. In other words, productivity, here, refers to the capacity or economic potential of an urban development strategy rather than its effectiveness or rate of output.

As previously exemplified, a key aspect which allows wetlands to be more flood resilient is species diversity. The impact on flood resilience was found to derive from diversifying adaptation tactics, resource necessity and temporal growth patterns. In contrast, although urban flood adaptation strategies may apply diverse flood mitigation strategies, there is less diversity in terms of resource necessity or temporal growth patterns.

Biodiversity is surely an important design factor in urban development policy around the globe but applies only to natural or artificial ecosystems and farmlands. The connection between species richness and resilience to extreme events such as flooding is being increasingly recognized in the agriculture sector which plays an important role in overall urban flood resilience. Research from UN Food and Agriculture Organization (FAO) (Altieri et al. 2015) and other organizations such as Greenpeace (Tirado and Cotter 2010) recognizes the need for biodiverse crops in order to maintain food supply during flooding events and thus increase resilience.

The strategies for multiple functions, seasonally adapted transport routes and introduction of wetland as leisure and agriculture urban areas discussed in the infrastructure and debris sections, result in functionally and economically diverse urban approaches. These aspects are currently not regulated in a similar fashion to diversity in agriculture.

Looking back to productivity in the natural world, "temporal diversity" in which various strategies with the same output or goal were applied at different temporal scales (short, seasonal or long term) was identified as a key factor in increasing flood resilience. This suggests that the potential solutions discussed for infrastructure or debris should not be mutually exclusive but rather approached as complementary measures which increase the diversity of design urban adaptation timelines.

For long-term flood resilience, in the context of rising sea levels, economic and functional diversity of infrastructure could play vital roles. Infrastructure that is adapted to both land and aquatic uses could provide opportunities for faster adaptation to rising sea levels. At the same time, diversifying the economic landscape to include more flood reliant activities (transport by boat, aquatic tourism, aquatic farming and halophytic food crops) could have a bifold effect by increasing both local productivity and flood resilience. As observed via historical examples, maintaining a flood resilient economy can be crucial in determining population migration and minimizing displacement after flooding events.

In order to achieve productive flood resilient urban systems, it is therefore necessary to include diversity of function, economic growth, design criteria and design adaptation timelines as main driving forces of urban development.

4 Discussion

The examples presented above provide an initial insight into the potential utility of integrating "flood pulse design" as a temporal dimension of future coastal urban design. The term flood pulse design is used in a broad sense, to describe strategies for integrating periodic and non-periodic sequences of inundation as positive drivers of urban growth.

In the context of current predictions regarding climatic changes and high flood risks impacting coastal cities, identifying strategies that tackle the relationship between city and surrounding water bodies emerges as a crucial aspect for future designs.

By focusing on learning from natural wetland ecosystems, which have evolved to efficiently mediate the relationship between land and water, the strategies identified begin to define relevant timelines and correlations for the design of SeaCities as future urban interfaces between land and water.

Aside from their value as case studies, wetlands and coastal ecosystems represent the main natural environments or sites for the development of SeaCities. Understanding how these systems respond to climatic challenges may therefore prove vital in establishing a continuous urban development in the coastal context.

The ecosystem analysis section identifies a series of temporal correlations and strategies which boost flood resilience in natural wetlands and implicitly increase the ecosystem's potential for adaptation and long-term survival. Due to the interdependent relationship between urban fabric and wetlands as urban development sites, understanding and designing to enhance these temporal correlations could have a dual benefit.

Firstly, given the uncertainty of future seasonal periodic patterns and the increased responsiveness of natural ecosystems to such variations, the correlation of design, monitoring and re-design cycles (e.g. at policy assessment level) with seasonal cycles could prove beneficial in terms of increasing urban adaptation speed and effectiveness. The ecosystem would in this case become an indicator parameter for system evolution, with the associated design method providing a potentially viable alternative to current predictive design methods.

Secondly, the examples of applied temporal correlation discussed for urban contexts reveal the potential of combined design approaches where ecological and urban agendas can be met concomitantly via the integration of ecosystems as functional design elements that augment urban fabric. Temporal correlations could thus play a significant role in optimizing flood-adaptive designs in order to maximize space usage and capitalize on the added benefits of certain types of flooding.

Specifically, in terms of the three key factors discussed (infrastructure, debris and productivity), the examples from both natural and urban contexts highlight the potential for mitigating at least some of the negative flood-related effects by learning from and more closely correlating natural and urban cycles.

The following measures are generalizations based on the cases presented for each of the three factors which could help increase urban flood resilience and create new opportunities for urban growth:

- short-term storm water storage, using urban wetlands and permeable paving strategies (especially in riverine and estuary conditions) could be optimized by adjusting wetland location, elevation and size to synchronize main channel peak water flows with peak storage water flows,
- seasonal flooding timelines correlated with localized infrastructure and economic development can provide opportunities for the emergence of a new pioneer urban development typology,
- long-term resilience could be achieved by integrating adaptive planning and design processes that are seasonally cyclical and based on measured data feedback,

- flood-related debris could become functional design elements through debris collection point design and correlation between debris decomposition rates and functional lifespan,
- diversity of function, economic growth, design criteria and design adaptation timelines could become main driving forces of flood resilient urban development.

Although the timelines, associated temporal correlations and potential areas of application identified in this review, highlight the potential value of time-base design approaches, further research into more precise definition of what and how to correlate in each specific case as well as further comparison with specifically established design methodologies would be needed in order to establish an approach beyond the proof of concept presented here.

Given that diversity proves to be a parameter which has wide impacts on various trophic chain types of dependencies, the measures described above should be considered as complementary approaches which aid to diversify urban adaptation strategies and concomitantly tackle short, seasonal and long-term timescales.

References

Adnitt C, Brew D, Cottle R, Hardwick M, John S, Leggett D, McNulty S, Meakins N, Staniland R (2007) Saltmarsh management manual

Aldayjover (2008) Aranzadi Park. http://www.aldayjover.com/en/component/articulo/?idcategoria=17&idarticulo=251

Altieri MA, Nicholls CI, Henao A, Lana MA (2015) Agroecology and the design of climate change-resilient farming systems. Agron Sustain Dev 35, 869–890. https://doi.org/10.1007/s13593-015-0285-2

APTUM (2019) Isla Rhizolith. http://www.aptumarchitecture.com/content/isla-rhizolith

Barbosa R, Castilho C, Perdiz R, Damasco G, Rodrigues R, Fearnside P (2017) Decomposition rates of coarse woody debris in undisturbed Amazonian seasonally flooded and unflooded forests in the Rio Negro-Rio Branco Basin in Roraima, Brazil-NC-ND license (http://creativecommons.org/licenses/by-nc-nd/4.0/). Forest Ecol Manage 397:1–9

Carvalho P, Thomaz SM, Kobayashi JT, Bini LM (2012) Species richness increases the resilience of wetland plant communities in a tropical floodplain. Austral Ecol 38. https://doi.org/10.1111/aec.12003

Columbia GSAPP Urban Design (2019) Water urbanism: can tho. Global Cities Clim Change Studio. https://www.arch.columbia.edu/books/reader/419-water-urbanism-can-tho#reader-anchor-3

Fekete A (2019) Critical infrastructure and flood resilience: cascading effects beyond water. WIREs Water 6:e1370. https://doi.org/10.1002/wat2.1370

FEMA (2008) Technical bulletin 9, design and construction guidance for breakaway walls below elevated coastal buildings. https://www.fema.gov/sites/default/files/2020-07/fema_tb9_design_construction_guidance_breakway_walls.pdf

Fivash G, van Belzen J, Temmink R, Didderen K, Lengkeek W, van der Heide T, Bouma T (2019) Elevated micro-topography boosts growth rates in Salicornia procumbens by amplifying a tidally-driven oxygen pump: implications for natural recruitment and restoration. Annals Botany. https://doi.org/10.1093/aob/mcz137

Flombaum P, Sala O (2008) Higher effect of plant species diversity on productivity in natural than artificial systems. Proc Natl Acad Sci USA 105:6087–6090. https://doi.org/10.1073/pnas.0704801105

Gingerich R, Panaccione D, Anderson J (2015) The role of fungi and invertebrates in litter decomposition in mitigated and reference wetlands. Limnologica–Ecol Manage Inland Waters 54. https://doi.org/10.1016/j.limno.2015.07.004

Han Y, Huang Q, He C, Fang Y, Wen J, Gao J, Du S (2019) The growth mode of built-up land in floodplains and its impacts on flood vulnerability. Sci Total Environ 700:134462. https://doi.org/10.1016/j.scitotenv.2019.134462

IPCC (2019: Summary for Policymakers. In: Pörtner H-O, Roberts DC, Masson-Delmotte V, Zhai P, Tignor M, Poloczanska E, Mintenbeck K, Nicolai M, Okem A, Petzold J, Rama B, Weyer N (eds) IPCC special report on the Ocean and cryosphere in a changing climate, (in press). https://report.ipcc.ch/srocc/pdf/SROCC_SPM_Approved.pdf

Jardine T, Bond N, Burford M, Kennard M, Ward D, Bayliss P, Davies P, Douglas M, Hamilton S, Melack J, Naiman R, Pettit N, Pusey B, Warfe D, Bunn S (2015) Does flood rhythm drive ecosystem responses in tropical riverscapes? Ecology 96:684–692. https://doi.org/10.1890/14-0991.1

Khadka A, Kokkonen T, Niemi TJ, Lähde E, Sillanpää N, Koivusalo H. (2019) Towards natural water cycle in urban areas: Modelling stormwater management designs. Urban Water Journal, 1–11 https://doi.org/10.1080/1573062X.2019.1700285

Kim J, Johnson L, Cifelli R, Thorstensen A. Chandrasekar V (2019) Assessment of antecedent moisture condition on flood frequency: an experimental study in Napa River Basin, CA. J Hydrol Reg Stud 26. https://doi.org/10.1016/j.ejrh.2019.100629

Lawrence J, Blackett P, Cradock-Henry NA (2020) Cascading climate change impacts and implications. Clim Risk Manage 29:100234. ISSN 2212-0963. https://doi.org/10.1016/j.crm.2020.100234

Li Z, Dong M, Wong T, Wang J, Kumar A, Singh R (2018) Objectives and indexes for implementation of sponge cities-a case study of Changzhou City, China. Water 10(5):623. https://doi.org/10.3390/w10050623

Liao K-H, Le TA, Nguyen KV (2016) Urban design principles for flood resilience: learning from the ecological wisdom of living with floods in the Vietnamese Mekong Delta. Landscape Urban Plan 155:69–78. https://doi.org/10.1016/j.landurbplan.2016.01.014

Liao K-H (2012) A theory on urban resilience to floods—a basis for alternative planning practices, Ecology and Society 17(4):48. http://dx.doi.org/10.5751/ES-05231-170448

Maschhoff JT, Dooley JH (2001) Functional requirements and design parameters for restocking coarse woody features in restored wetlands, ASAE Meeting Presentation, Paper No: 012059

Mehlig U (2006) Phenology of the red mangrove, *Rhizophora mangle* L, in the Caeté Estuary, Pará, equatorial Brazil. Aquatic Botany—AQUAT BOT 84:158–164. https://doi.org/10.1016/j.aquabot.2005.09.007

Monteiro M (2017) Natural and anthropogenic control of water quality of an Amazon estuary: the Caeté estuary (Brazil). PhD Thesis

Natural Water Retention Measures (NWRM) (2013) Individual NWRM coarse woody debris http://nwrm.eu/sites/default/files/nwrm_ressources/f10_-_coarse_woody_debris.pdf

Paine RT (1980) Food webs: linkage, interaction strength and community infrastructure. J Animal Ecol 49:667–685

Plyer A (2016) Facts for features: katrina impact https://www.datacenterresearch.org/data-resources/katrina/facts-for-impact/

Ramsar Convention on Wetlands (2018) Global wetland outlook: state of the world's wetlands and their services to people. Ramsar Convention Secretariat, Gland, Switzerland

RCH Studios (2019) SLIDE: a resilient strategy for extreme climate in Los Angeles and beyond. https://www.rchstudios.com/news/slide-a-resilient-strategy-for-extreme-climate-in-los-angeles-and-beyond/

Rodríguez J, Saco P, Sandi S et al (2017) Potential increase in coastal wetland vulnerability to sea-level rise suggested by considering hydrodynamic attenuation effects. Nat Commun 8:16094. https://doi.org/10.1038/ncomms16094

Rothero E, Lake S, Gowing D (2016) In: Floodplain meadows: beauty and utility—a technical handbook. Milton Keynes, Open University. http://www.floodplainmeadows.org.uk/floodplain-mea

Smolders S, Plancke Y, Ides S, Meire P, Temmerman S (2015) Role of intertidal wetlands for tidal and storm tide attenuation along a confined estuary: a model study—discussion paper. Nat. Hazards Earth Syst. Sci. 15:1659–1675. https://doi.org/10.5194/nhess-15-1659-2015

State of Queensland Government (2011) Understanding floods: questions and answers

Nadia TdL, Morellato P, Machado I (2012) Reproductive phenology of a northeast Brazilian mangrove community: environmental and biotic constraints. Flora—Morphology Distrib Funct Ecol of Plants 207:682–692. https://doi.org/10.1016/j.flora.2012.06.020

Tiner RW (2016) Wetland indicators: a guide to wetland formation, identification, delineation, classification, and mapping, 2 edn. CRC Press. ISBN-10: 9781439853696

Tirado R, Cotter J (2010) Ecological farming: drought-resistant agriculture, Greenpeace Research Laboratories, University of Exeter, UK GRL-TN 02/2010

Tonneijck FH, Winterwerp H, Weesenbeeck VB, Bosma R, Debrot AO, Noor YR, Wilms T (2015) Building with nature Indonesia: securing eroding delta coastlines

Trenberth K (2008) The impact of climate change and variability on heavy precipitation, floods, and droughts. In: Anderson MG, McDonnell JJ (eds) Encyclopedia of Hydrological Sciences. https://doi.org/10.1002/0470848944.hsa211

van Eck WHJM, Lenssen JPM, van de Steeg HM et al (2006) seasonal dependent effects of flooding on plant species survival and zonation: a comparative study of 10 terrestrial grassland species. Hydrobiologia 565:59–69. https://doi.org/10.1007/s10750-005-1905-7

Ventura Y, Eshel A, Pasternak D, Sagi M (2015) The development of halophyte-based agriculture: past and present. Annals Botany 115(3):529–540. https://doi.org/10.1093/aob/mcu173

WHO Emergency and Humanitarian Action SEA Regional Office (2007) Flood fury A recurring hazard, emergency and humanitarian action focus, vol 1

Woetzel J, Pinner D, Samandari H, Engel H, Krishnan M, Boland B, Cooper P (2020) Can coastal cities turn the tide on rising flood risk? (Case Study), McKinsey Global Institute

Zen S, Gurnell A, Zolezzi G, Surian N (2017) Exploring the role of trees in the evolution of meander bends: The Tagliamento River, Italy. Water Resour Rese 53. https://doi.org/10.1002/2017WR020561

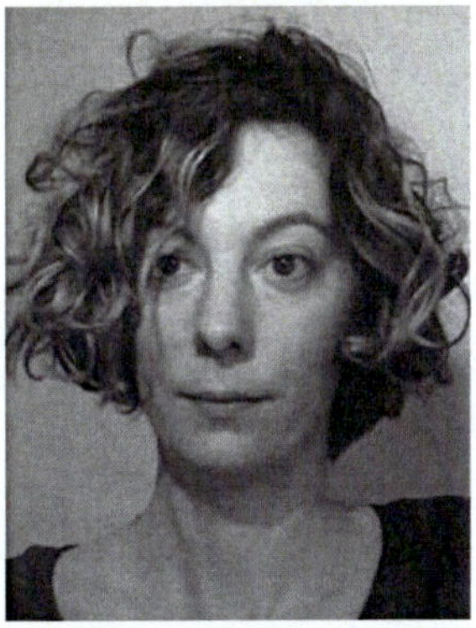

Ioana Corina Giurgiu Ioana is a SeaCities PhD researcher, currently developing a thesis on hybrid water-based urban system designs. She graduated with Honours from the Architectural Association School of Architecture in London and, as a Certified Passivhaus Designer and registered architect, has worked on a number of Passivhaus and energy-efficient projects in UK and Romania. Her research focus is on decreasing the environmental footprint of the built environment via hybrid designs which create symbiotic relationships between natural and built environments.

A Comparison Between Alternative Relocation Options for the Pacific Islands Based on a Human-Centred Approach

Despina Linaraki

Abstract Extreme weather conditions, rising sea levels, and floods will make many low-lying Pacific Islands uninhabitable by 2050. Frequent floods affect food production and freshwater availability, enhance shoreline erosion, and cause the destruction of the coastal infrastructure. Current adaptation strategies require a considerable amount of funds, materials, technological, and human resources. As a result, researchers argue that local adaptation strategies will continue for the next years, but relocation will be proved to be the most effective solution. Based on the argument above, this research paper is analysing the criteria for efficient relocation based on a human-centred approach and specifically on the notion of islandness. Following, it is comparing alternative relocation options for the Pacific Islands, as per the relocation criteria and the land availability. The analyses showed that the distance from the land of origin is critical for the efficiency of the relocation. The further the distance, the less effective the relocation will be.

Keywords Relocation · Pacific Islands · Human-centred · Islandness · Expansion

1 Introduction

Researchers argue that around 200 million people who currently live on low-lying islands and coastal areas will have to relocate inland or in higher elevation countries by 2050 due to the climate change impacts (McLeman 2011; Nunn 2013; Perumal 2018; Wyett 2014). The consequences of climate change include extreme weather events and disasters, storms, cyclones and droughts, sea level rise, and floods (Ourbak and Magnan 2018). The frequency and intensity of these events will soon make some low-lying islands and coastal areas uninhabitable (Gemenne 2011; Perumal 2018). The latest predictions from the Intergovernmental Panel on Climate Change (IPCC) on sea level rise project an average rise between 0.29 and 1.10 m by 2100 (Oppenheimer et al. 2019). However, researchers argue that this prediction can rise to

D. Linaraki (✉)
SeaCities, Cities Research Institute, Griffith University, Gold Coast, Queensland, Australia
e-mail: d.linaraki@gmail.com

J. Baumeister et al. (eds.), *SeaCities*, Cities Research Series,
https://doi.org/10.1007/978-981-15-8748-1_10

2 m or even more, due to environmental and anthropogenic stressors (Bamber et al. 2019; NOAA 2012). If sea level rises above 2 m globally, many of the low-lying islands and coastal metropolis will be underwater (Jamero et al. 2017; Yamamoto and Esteban 2017). As a result, most of the Pacific Islands, Indian Ocean Islands, Caribbean Sea Islands, and coastal cities around the world such as Amsterdam, Dubai, Abu Dhabi, Mumbai, New Orleans, Miami, Jakarta, and Bangladesh will be affected (Connell 2013; FloodMap 2020).

2 Pacific Islands' Response to Climate Change

This research is focusing on the low-lying Pacific Islands due to its high rate of vulnerability. Specifically, among the 20 countries with the highest risk in the World Risk Index on 2019, 11 are islands: Vanuatu, Antigua and Barbuda, Tonga, Solomon Islands, Papua New Guinea, Brunei Darussalam, Philippines, Cape Verde, Fiji, Timor-Leste, and Kiribati, while nine out of the 11 are specifically Pacific Islands (Day et al. 2019). Research evidence has repeatedly indicated that these islands are expected to be severely impacted by climate change while some of them have already disappeared (Connell 2015; Kueffer and Kinney 2017; Nunn 2013; Walshe and Stancioff 2018).

2.1 Geomorphological Characteristics and the Anthropogenic Pressures

The reason for the low-lying islands' vulnerability is mainly their natural geomorphological characteristics, along with the pressure from anthropogenic activities (Duvat and Magnan 2019; Royle 2014; Wong 2011; Woodroffe et al. 1999) (Fig. 10.1). According to research evidence, 27% of the Pacific Islands have a maximum elevation distance of 5 m from the current sea level (Nunn et al. 2016). Consequently,

Vulnerability of Low-lying Pacific Islands to climate change impacts		
Unconsolidate materials	Land availability	Low elevation heights
- Morphodynamic changes	- Overpopulation and density	- Frequent coastal flooding
- Unstable structures	- Unmanagable city planning	- Affected by sea level changes
- Vulnerable to erosion	- Limit the adaptation options	- Lack of mountains to protect

Fig. 10.1 This figure summarizes the vulnerability of low-lying Pacific Islands to climate change impacts, which is mainly resulted from the natural geomorphological characteristics and the anthropogenic pressures (*Source* Nunn 2009a, b and 2016, Webb and Kench 2010, Royle 2014, Wong 2011; Woodroffe et al. 1999)

even a small increase in sea level is immediately reducing the area above the surface of the water. At the same time, the shortage of land is reducing the defence to floods and storms because there is not efficient space to protect the inland areas. Moreover, limited available land leads to density and overpopulation. For example, Tarawa at Kiribati is over-dense and overpopulated with not efficient setbacks from the coastline, which increases its vulnerability to extreme events.

Moreover, one-third of the Pacific Islands are classified as coral reef islands, such as the Federated States of Micronesia, French Polynesia, Kiribati, Marshall Islands, and Tuvalu (Nunn et al. 2016). The coral reef islands have been creating as part of biogenic processes (Ramalho et al. 2013). Specifically, the low-lying coral islands' have been created from sediments produced by corals, coralline algae, foraminifera, molluscs, and Halimeda (Kench et al. 2014). These islands are extra vulnerable to coastal erosion, land loss, and morphological changes, due to these unconsolidated biogenic structural materials (Jamero et al. 2019; Nunn 2009a; Nunn et al. 2016; Webb and Kench 2010). Another 39% of the Pacific Islands have been created by volcanic activities (Nunn et al. 2016). Volcanic islands are not as vulnerable as reef islands. First, because they can extend higher than 5 m, consequently, they do not have an immediate impact on land reduction due to sea level rise. Furthermore, constant lava production armours their perimeter. However, their geological formation is also easily affected by coastal erosion (Ramalho et al. 2013). As soon as volcanic activity ceases, the islands begin losing their durability, and they start sinking (Nunn and Kumar 2018; Weigelt et al. 2014). To conclude, more than two-thirds of the Pacific Islands will be affected by climate change. However, for the reasons analysed above, low-lying coral reef islands are found to be at the forefront of climate change impacts (Fig. 10.1).

2.2 *The Impact of Sea Level Rise and Flood on Food Production and Freshwater Availability*

Except for the land erosion and vulnerability, climate change is also affecting food production and eliminates the freshwater, which is causing extra stress to the islanders. Specifically, the increase in carbon dioxide emission is lowering the ocean's PH and is causing acidification (Neill 2015 p. 69). The decrease of the ocean's PH and the deoxygenation affect the coastal biodiversity, the species growth, and reproduction and impact the food chain (IPCC, 2014; Oppenheimer et al. 2019). Also, coral bleaching, saltwater intrusion due to sea level rise and floods is affecting the food production in many low-lying islands (Nunn 2013). Specifically, coastal flooding from waves extends far inland, and saltwater intrusion can impact food production for months. Due to that reason, some communities have been forced to relocate, such as in parts of Papua New Guinea and Vanuatu (Piggott-McKellar et al. 2019). Moreover, frequent floods are affecting freshwater supply. For example, on the Island of Pele, 11 out of 23 groundwater wells were affected by salinization since 1990

(Perumal 2018). In the Pacific Ocean, about 2.1 million people are expected to be affected by 2050 (Ash and Campbell 2016; Nunn 2013).

2.3 Current Adaptation Strategies

Current adaptation strategies that defend the coastal areas focus on protection, either by using seawalls, breakwaters, or dikes. Also, local adaptation methods include accommodation to the floods either by raising the infrastructures on stilts or by converting from agriculture to aquaculture (Jamero et al. 2019). At the same time, locals are exploring soft protection strategies, such as the replanting of mangroves along the coast to reduce storm impacts (Paeniu et al. 2015).

The protection strategy each country uses is based on available funds, materials, technological, and human resources. Insufficient availability on any of them can lead to unstable constructions that damage both the environment and human lives (Nunn 2013). For example, research has proven that hard infrastructures are not as protective at low-lying Pacific Islands as they are on the Western countries (Nunn 2009b, 2013; Oppenheimer 2019). First, due to their design and material, hard protection strategies have been accused many times that instead of effectively protect the coastline, they reflect the wave energy and cause erosion in adjacent areas. Moreover, there is not enough available data at these areas for effective modelling of the protective infrastructure (Nunn 2009b).

Also, the protection strategies against sea level rise and floods require an enormous amount of funds, not only because of the materials and construction but also because of the maintenance (Oppenheimer 2019). People are spending millions of dollars every year towards the defence of the urban infrastructure. For example, it has been estimated that by 2040, the annual cost for adaptation strategies will be approximately USD 3 million in Palau, USD 17 million in Kiribati, and USD 329 million in Fiji (WorldBank 2016). Gathering this amount of funds is not easy, especially for the low-lying Pacific Islands, where the average income is very low. For example, in Kiribati, according to the World Bank, the annual income per capita is USD 1,625 (WorldBank 2019).

At the same time, resources of materials for construction, such as sand, are being eliminated, causing extra stress to the impact of sea level rise and floods on islands (Oppenheimer 2019). Consequently, at some point shortly, the funds, the human resources, and the sediment supplies will not be enough to protect coastal areas and especially the low-lying islands. Based on the argument above, researchers argue that relocation will be the most effective option for many Pacific Islands by the end of this century. In situ adaptation strategies will continue for the next 20 to 40 years, but locals will soon realize that relocation is their only effective solution (Nunn 2013).

2.4 The Cases of Kiribati, Fiji, and Vanuatu

Kiribati is a significant example of a country in the middle of the Pacific Ocean that is already facing multiple environmental challenges such as occasional floods and severe coastal erosion. It is composed of 33 reef islands (32 atolls and a raised coral island, known as Banaba) (Thomas 2009). The unique geomorphological characteristics of Kiribati's reef islands make them extremely vulnerable to the climate change impacts and especially sea level rise and floods. First, the highest elevation point of most of Kiribati's islands is only 3 m above sea level (Wyett 2014). Consequently, the projection of 1 or 2 metres of sea level rise will significantly reduce the available area of the low-lying islands. At the same time, most of Kiribati's atolls form a long and narrow strip (approximately a few metres to less than one kilometre) of land (Fig. 10.2). Therefore, in case of extreme events such as floods, there is not an available inland area that people could move to protect themselves. Also, the lack of mountains makes these areas completely unprotected from all the climatic events. Moreover, as referred above, these reef islands are composed of unconsolidated materials. As a result, they are vulnerable to morphological changes caused by environmental conditions (Biribo and Woodroffe 2013). Consequently, coastal erosion and floods are severely impacting this region.

According to Kiribati's government, current in situ adaptation strategies have not been proved to be successful (Wyett 2014). At the same time, Kiribati's population

Fig. 10.2 Tarawa atoll, the capital of Kiribati, is formed by a long and narrow reef strip. (*Source* D. Linaraki, adapted from Google Earth © 2020 Maxar Technologies)

Fig. 10.3 South Tarawa atoll, Betio at the left and Bairiki at the right side as seen from satellite images. (*Source* D. Linaraki, adapted from Google Earth © 2020 *Maxar Technologies*)

is expected to increase to 208,000 people by 2050 (Bedford 2016). Most of these people will be distributed to South Tarawa, the capital of the Republic of Kiribati, in search of more opportunities. While future overpopulation is expected at Tarawa, researchers argue that Kiribati will need to relocate soon due to climate change (Wyett 2014). However, there is not enough space for locals to move inland due to the narrow natural geomorphology of Tarawa (Fig. 10.3). The other atolls of Kiribati have similar geomorphological characteristics such as low elevation heights, comprised of narrow and long stripes of land, which immediately makes them vulnerable to sea level rise and floods.

Consequently, the Government of Kiribati is preparing for future relocation in other countries. They have already purchased 6,000 acres of land in Naviavia, Fiji, for USD 7 million to prepare for relocation (Jamero et al. 2017; Yamamoto and Esteban 2017). At the same time, Kiribati's government has started a programme called "Migration with Dignity". This programme aims to educate local citizens in order to be able to move and work in other countries (Yamamoto and Esteban 2017). However, even though the Government of Kiribati is preparing a future relocation, Kiribati's residents do not wish to relocate from their islands, and they prefer in situ adaptation strategies (Wyett 2014).

Similarly, to Tarawa, Viti Levu Island in Fiji is home to 28,000 people. According to predictions, at least half of the population will move by the end of this century (Piggott-McKellar et al. 2019). The town of Nadi has experienced an increase in floods due to sea level rise, river infilling, deforestation, unmanaged coastline construction, and poor drainage. Researchers argue that the town will be uninhabitable in the next 15-20 years (Nunn 2013). Adaptation strategies have been explored, such as river dredging, reforestation, or even redirection of the river delta, but still, based on the evidence that sea level is accelerating and extreme events will be more frequent, researchers argue that relocation will be their best option (Nunn 2013).

At the same time, four indigenous communities at Fiji have already been relocated. The village of Vunidogoloa has been relocated 1 km inland due to erosion, floods, and saltwater intrusion (Fig. 10.4). Similarly, Denimanu has relocated inland (500 m), to search for a better quality of life and food production. The relocation of both villages was effective. Locals got newly constructed houses and infrastructure, including solar power, rainwater tanks, fishponds, pineapple plants, and household facilities (Jamero et al. 2019). Moreover, their new locations are within walking distance from their land of origin, and consequently, locals were able to keep their everyday traditions such as food production and fishing (Piggott-McKellar et al. 2019). However, Denimanu new location is suffering from erosion and landslide issues that occurred after the construction of the new village (Piggott-McKellar et al. 2019).

After experiencing extreme events, such as cyclones, floods, and severe erosions, the Government of Fiji released on 2018 relocation guidelines for the communities in need (Government 2018). These guidelines were released to ensure that in the extreme scenario of relocation, the cultures and livelihoods of the Fijians will remain intact. Furthermore, the Fijian government recognized over 80 communities in need of future relocation. Similarly, to Fiji, Vanuatu has developed guidelines on climate change and disaster-induced displacement strategies (Piggott-McKellar et al. 2019). Kiribati, Vanuatu, and Fiji are at the forefront of climate change impacts due to their

Fig. 10.4 Vunidogoloa village relocation to Kenani as seen from Google Earth on 2018. The original village location has been covered with vegetation. (*Source* D. Linaraki, adapted from Google Earth © 2020 *Maxar Technologies*)

location, geomorphological characteristics, and anthropogenic pressures. Frequent extreme events, sea level rise and floods are forcing the low-lying islands and coastal areas of Pacific to mandatory relocation.

3 Human-Centred Approach and the Notion of Islandness

There is a need to determine whether relocation is an efficient option for islanders. As mentioned, due to climate change, many people will be forced to leave their homes, as low-lying islands and coastal areas will be marked uninhabitable. Many criteria need to be taken into consideration for efficient relocation. This research is identifying and analysing these criteria based on a human-centred approach and specifically on the notion of *islandness*. Islandness is a terminology that is used by many researchers worldwide to describe the state someone is while having lived on an island (Baldacchino 2018). Research evidence suggests that islanders have developed a strong sense of community and a strong connection to their environment due to the physical isolation of the island (Conkling 2007). The smaller and the more isolated the island, the bigger this connection exists. This feeling is accelerating because people depend on each other to survive. Consequently, based on the identity of islandness, the relocation of small villages or even whole islands can prove to be challenging for many reasons that are analysed below.

3.1 Psychological, Sociological, and Physiological Impacts

Relocating on different environments and intergrading into other cultures and societies can have significant psychological, physiological, and sociological impacts. Research evidence has shown that immigrants experience unemployment, homelessness, landlessness, and food insecurity (Barnett and O'Neill 2012; Tabucanon 2012). Specifically, the relocation of Pacific Islanders is particularly difficult due to the fear of losing their identity and because of the cultural, social, political, and economic costs that will be involved (Bukvic 2018; Perumal 2018). For example, people from the island of Tubigan were forced to relocate, after the 2013 earthquake that caused land subsidence and floods. However, they were opposed to relocation due to the fear that they will lose their livelihoods (Jamero et al. 2017). Another example comes from Maat people who were forced to relocate after a volcanic eruption. Most of them are still wishing to return to their ancestors' lands and are still considering themselves Ambrymese, even 15 years after the resettlement (Perumal 2018). These psychological impacts are the results of islandness and particularly the strong connection that islanders have with their ancestors' land.

Another case of resettlement comes from Banabans, whom they relocated to Rabi Island. Banabans started saving funds after they realize that their island would sooner or later run out of food due to extreme phosphate mining. After 30 years of savings, they started the relocation process in 1945. Banabans resettled on Rabu Island. Rabu Island provided more land, food, water, and material resources than the island of Banaban. Still, most of the refugees were reported unhappy. Especially, the elderly was feeling depressed and hopeless. At the same time, people had to quickly adapt to new methods of food production, such as reef fishing and coconut plantation, a situation that caused them extra stress (Kumar et al. 2006; Tabucanon 2012). Except for the psychological issues, they experienced physiological challenges as well. The new island was wetter and colder than their island of origin. Immigrants were not prepared for this harsh winter. As a result, many people died from pneumonia (Tabucanon 2012). Consequently, the relocation of Banabans was challenging not only because of psychological, but also because of physiological issues.

Similarly, researchers interviewed people from Vanuatu to find out if they would agree to relocate due to severe weather events. The results showed that in case they had to relocate, they would instead relocate inland so they could maintain their culture and livelihoods (Perumal 2018). For islanders, moving away from their land means that they will lose their islandness and everything that this notion accompanies, such as their culture and livelihoods. The strong connection that they have with their ancestors' land and the water is severely affecting their emotions and causing them psychological, sociological, and physiological unbalances. In some extreme cases where relocation is unavoidable, islanders would instead relocate close to their land of origin, in environments like their current locations. Similar environments to their environments of origin could provide the security that their culture, social tights, and networks could be maintained the same (Perumal 2018).

3.2 Cultural Disagreements

Furthermore, many concerns have been raised both from the refugees and the hosts. One recent example comes from Fijians regarding the resettlement of people from Kiribati at Fiji. Locals are concerned that overpopulation will increase pollution in their area (Yamamoto and Esteban 2017). Similarly, the relocation from the island of Gaua to the island of Vanua Lava was unsuccessful due to cultural disagreements between the hosts and the guests (Perumal 2018).

Research evidence has shown that the hosts concern when it comes to migration issues. This feeling, which is known as xenophobia (derived from the Greek words ξένος = foreigh and φόβος = fear), is being developed between living organisms, and it is resulting from biological uncertainties and the fear of the predator (Delanty et al. 2011). The feeling of uncertainty is accelerating, especially when a large amount of people needs displacement, which is currently the case for most low-lying Pacific Islands.

Moreover, the accommodation of a large group of people results in the loss of otherwise available land. This land was available for use from the host country either for agriculture or for accommodation. Furthermore, more people will have to use the same resources for freshwater, food, or material supply, which in the case of the small Pacific Islands are limited due to their small size and their extremely isolated locations. Integrating two societies into one will be of a great challenge for the governments and the residents of Pacific Islands. When resettling and integrating into an existing society, new uncertainties will occur due to different environments, societies, and cultures between the residents. Consequently, for Pacific Islanders, relocating into environments where the integration of two different societies will not be necessary would be more efficient.

3.3 Global Agreements for Pacific Islands

In addition, currently, there is no agreement between counties to accommodate climate change refugees. Consequently, there is a big concern about where and how Pacific Islanders will relocate (Yamamoto and Esteban 2017). The USA is granting access to some of the Pacific Islands. This agreement is known as the Compact of Free Association (COFA), and it involves the Federal States of Micronesia, Marshall Islands, and Palau. Similarly, New Zealand grants access to the Cook Islands, Tokelau, and Niue (Ash and Campbell 2016). Besides, to the USA, many experts argue that Pacific Islanders should relocate to Australia and New Zealand (Tabucanon 2012).

Australia has been seen many times in history as a place to resettle people in need because there is plenty of available habitable space both inland and on the coastal zones. However, most of this available space is unproductive, since more than half of the surface is composed of a desert. For example, Nauruans were offered relocation to Australia, New Zealand, or the UK, but they declined the offer due to the fear of losing their identity (Tabucanon 2012; Tabucanon and Opeskin 2011). Relocating people from a small island to a vast continent is not a convenient solution. Topological differences are resulting in livelihood changes. Specifically, islanders have learned to live close to the water. As a result, their culture and livelihood have been developed based on water. Consequently, if the new area is not close to the water, it will result in losing their livelihood.

3.4 Funds

Furthermore, relocation requires many funds to be successful. The process of relocation in a new land requires cost for preparation, construction of infrastructure, and moving expenses of the families. Also, the further the distance from the land of origin, the more the cost for the relocation will be (Campbell 2014). Gathering a huge amount of funds is not an easy task, especially for Pacific Islanders, where the average income is very low. It has been estimated that the cost of relocation in Fiji could be as high as USD 10,000 per household, which will include land acquisition and building of infrastructure and moving (Oppenheimer 2019). However, according to the World Bank, the annual income per capita at Fiji is USD 6,267, a lot less than the relocation cost (World Bank 2019). Consequently, for relocation to take place, locals should use external funds and donors. However, that raises another issue, as many times, donors make their own decisions without considering the needs of the people. To sum up, relocation cost can create extra pressure on the islanders. Also, it is important to notice that relocating as close as possible to the land of origin reduces the relocation expenses.

3.5 *Efficient Relocation Criteria*

Consequently, deciding the place for relocation is equally crucial to the process itself. Based on the analyses described above, relocating as close as possible, in environments similar to the land of origin, could be proven to be beneficial for Pacific Islanders, whereas relocating people due to climate change in other countries could be challenging, mainly because of the strong notion of islandness. As described above, the strong connection that Islanders have with their land, the water, and their culture could derive psychological, sociological, and issues in case of relocation. In addition, the impact of relocation cost, the cultural differences, and the legal agreements between the places are equally crucial. Moreover, relocation criteria should cover the basic needs for human survival and development, such as food production, freshwater availability, security, shelter construction, health, and employment. All these criteria should be taken into consideration when decisions are taken for relocation (Fig. 10.5).

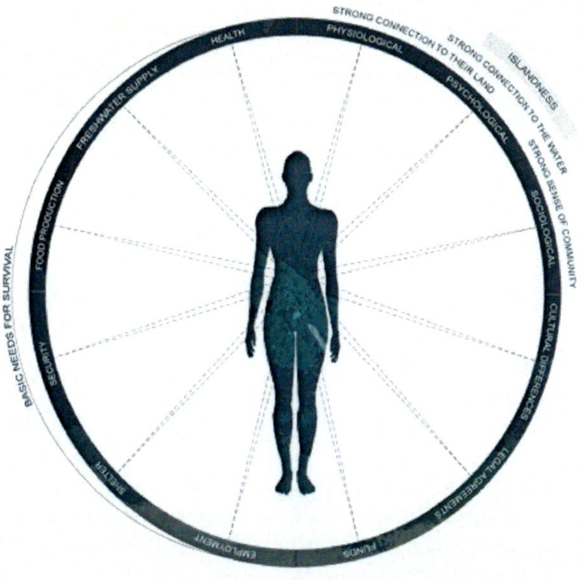

Fig. 10.5 This diagram represents the criteria that should be taken into consideration in case of relocation. These criteria are based on a human-centred approach and specifically on the notion of islandness

4 Relocation Options

Taking into consideration the urgency of the low-lying islands and coastal areas of Pacific to relocate, this research is exploring relocation scenarios that respond to the criteria analysed above. The relocation options that exist vary according to the available land resources of the affected country (Fig. 10.6). For example, for coastal areas that are part of larger islands, it is relatively easy to relocate inland, because there is available space like it happened with the relocation of Vunidogoloa. Similarly, in some islands, there is the possibility to relocate in higher elevations, such as Apia in Samoa, Honiara in the Solomon Islands, or Sura in Fiji (Nunn 2013). Sikaianans and Ontong Java are also considering relocating to higher elevations (Connell 2013). However, even if there is available land for relocation, it may not be habitable. For example, Avarua at the Cook Islands or Pago Pago in American Samoa, even if there is available space in higher altitudes, the landscape is so steep that it is not convenient for humans to live there (Nunn 2013). A significant example that was mentioned earlier is the relocation of Denimanu. The new relocation area has been proved to be unsuitable for habitation due to the landslide issues that occurred afterwards.

Particularly for low-lying reef islands, relocation inland is often inconvenient since there is not enough available land. Also, most of the coral reef islands, as discussed above, have an approximate elevation of 3 m above sea level, which immediately makes them vulnerable to sea level rise and floods. As a result, they are unsuitable for relocation. Furthermore, limited land resources for agriculture and limited freshwater have opposed a scenario of internal relocation. Consequently, in case there is no available land, people have to relocate to other cities or other countries, as it happened with the case of Banabans.

However, in case of relocation, people have opposed a scenario of moving in other cities or countries, integrating with other cultures, or moving in environments different from their current environments, where they would have to readapt their livelihoods and cultures. Consequently, this research argues that if there is no available land for relocation, adaptation strategies should focus on expansion, to enhance

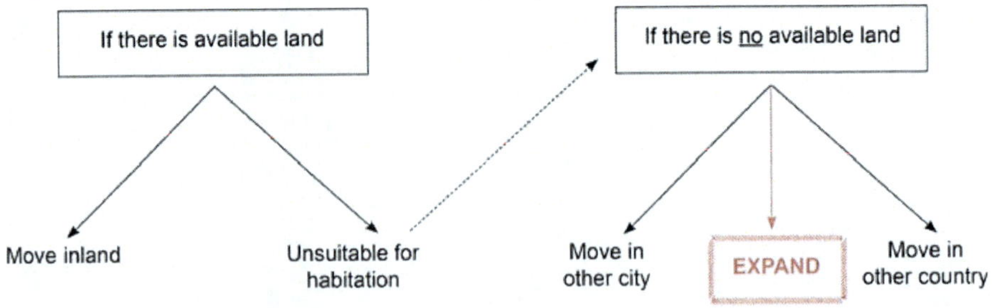

Fig. 10.6 Relocation scenarios. If there is available land, people can move inland in higher elevations. However, even if there is available land, this land is not always suitable for relocation. If there is no available land, people will have to move to other cities or other countries. However, this research suggests expanding, to create available land for habitation close to the land of origin

the opportunity for relocation as close as possible to their current land and the water (Fig. 10.6).

4.1 Expansion

Three strategies are examined in this research and are associated with expansion into the water. First is about horizontal expansion in the water, to create protection barriers and to increase the land resources. The second strategy is about vertical expansion by raising the inland area in order to relocate people in higher elevations. The third option is to create new islands as close as possible to the islands of origin that could accommodate the immigrants (Fig. 10.7). The selection should be based on available land resources and suitable environmental conditions.

First, expanding horizontally in the water could create a protective barrier to eliminate the wave action before reaching the coastal infrastructure. In this way, it can protect the coastal areas from extreme events and frequent flooding, while at the same time, it increases the available land. The expansion could be done by using soft infrastructure, for example, by the planting of vegetation such as mangroves or by expanding the reef zones. Currently, many Pacific Islands are following the strategy

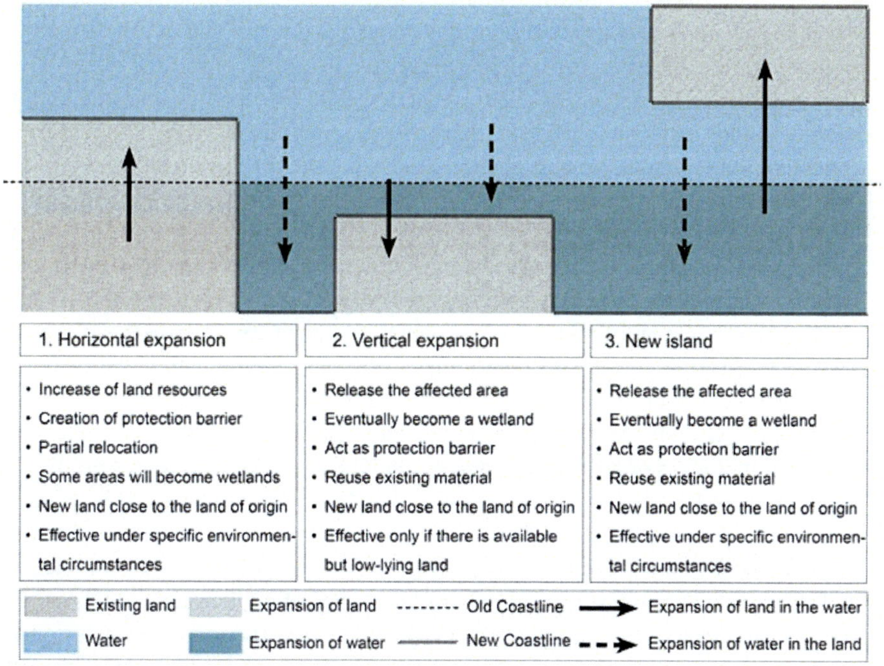

Fig. 10.7 Expansion scenarios in plan view. The first option shows the scenario of horizontal expansion in the water. The second is about the vertical expansion of inland areas, and the third concept is about the creation of a new island and the release of the existing land to nature

of planting vegetation as a defence system (Paeniu 2015). However, the areas that are located in low elevation heights will eventually need to relocate in higher elevations due to sea level rise. Also, for the horizontal expansion to act as a protective barrier, the new land should extend far inshore to dissipate the wave energy successfully.

The second scenario is to expand vertically. By raising and levelling the inland areas, people could move in higher elevation. As a result, they could be protected from rising sea levels and floods. At the same time, the affected coastal area could be released to the water, creating in this way a big wetland, which would protect the inland areas. The existing infrastructure materials and land resources could be reused to elevate the inland areas. This method could be proven efficient, for low-lying islands with available land resources but not enough elevation height.

The third concept is to create a new island close to the island of origin. This option could be effective when there is no available land to move inland or in higher elevations, for example in the case of Tarawa, Kiribati. In the case of Tarawa atoll, a new island in the middle of the lagoon would be protected from waves by the long and narrow strips of the perimeter zone (Fig. 10.2). In case of full relocation, the affected area would become a wetland and be used for aquaculture, while like the second scenario, the existing materials and land resources could be reused to create the new land.

There are many advantages, as well as limitations following this strategy. As argued above, by developing the strategy of expansion, people could relocate closer to their land of origin. However, to create a new land that could accommodate the whole village or the island would require a considerable amount of funds, materials, and technological resources, which are not always available to the Pacific Islands. The challenge of these concepts is to create new land by using the minimum requirements of funds and resources. Moreover, the design of the new land should take into consideration the environment. Pacific Islands are unique, biodiverse environments that host significant aquatic and terrestrial living organisms (Berry and Gillespie 2018) and need to be protected in case of new land creation. Consequently, the new land should not extend on top of existing ecosystems and should not overexploit the natural resources for the construction.

Also, there is many times the case where unmanageable expansion in the water creates more problems for the community instead of eliminating them. For that reason, the construction method and morphology of the land expansion have a significant role as well. For example, in Tarawa atoll, the capital of Kiribati, people used land reclamation in order to connect or extend the reef islands. These solid fill lands limit lagoonal flushing and cause erosion (Biribo and Woodroffe 2013). Similarly, in Mumbai, people connected the original seven islands with debris, stopping in this way, the natural flow of water and causing erosion and pollution (Zope et al. 2017). In a similar vein, Dubai's palm islands were accused of pollution, due to the stagnate waters that resulted from the enclosed shape of the construction (Martín-Antón et al. 2016). Consequently, materials, construction methods, and morphology should be equally considered for the concept of land expansion.

5 Conclusion—Further Discussion

This research analyses the relocation options for Pacific Islanders based on a human-centred approach. As argued, efficient relocation should take into consideration the notion of islandness and provide the basic needs for survival. The analyses showed that the distance from the land of origin is critical for the efficiency of the relocation. The further the distance, the less effective the relocation would be. Moreover, the environment of the new land has an essential part in maintaining the islandness. Relocating people as close as possible to their land of origin in environments similar to their current environments could keep their culture and livelihood intact. However, based on the available land resources, relocation close to their land is not always achievable. That is the case on many low-lying coral reef islands. Consequently, this research argues that if there is no available land, adaptation strategies should focus on expansion.

References

Ash J, Campbell J (2016) Climate change and migration: the case of the Pacific Islands and Australia. J Pacific Stud (ERA 2012, 2015, 2018 J (s) Listed) 36(1):53-72

Baldacchino G (2018) In: the routledge international handbook of island studies. Routledge

Bamber JL, Oppenheimer M, Kopp RE, Aspinall WP, Cooke RM (2019) Ice sheet contributions to future sea-level rise from structured expert judgment. Proc Natl Acad Sci USA 166(23):11195–11200. https://doi.org/10.1073/pnas.1817205116

Barnett J, O'Neill SJ (2012) Islands, resettlement and adaptation. Nature Climate Change 2(1):8–10. https://doi.org/10.1038/nclimate1334

Bedford R (2016) Population change and migration in Kiribati and Tuvalu, 2015–2050: hypothetical scenarios in a context of change. New Zealand Population Rev 42:103–134

Berry AJ, Gillespie RG (2018) Evolution. In: The routledge international handbook of Island studies, Routledge

Biribo N, Woodroffe CD (2013) Historical area and shoreline change of reef islands around Tarawa Atoll Kiribati. Sustain Sci 8(3):345–362. https://doi.org/10.1007/s11625-013-0210-z

Bukvic A (2018) Towards the sustainable climate change population movement: the relocation suitability index. Climate Dev 10(4):307–320. https://doi.org/10.1080/17565529.2017.1291407

Campbell JR (2014) Climate-Change Migration in the Pacific. Contemp Pacific 26(1):1–28. https://doi.org/10.1353/cp.2014.0023

Conkling P (2007) On Islanders and Islandness. Geogr Rev 97(2):191–201. https://doi.org/10.1111/j.1931-0846.2007.tb00398.x

Connell J (2013) Soothing Breezes? Island perspectives on climate change and migration. Aust Geogr 44(4):465–480. https://doi.org/10.1080/00049182.2013.852497

Connell J (2015) Vulnerable Islands: climate change, tectonic change, and changing livelihoods in the western Pacific. Contemp Pacific 27(1):1–36. https://doi.org/10.1353/cp.2015.0014

Day SJ, Forster T, Himmelsbach J, Korte L, Mucke P, Dr. Radtke K, Weller D et al (2019) WorldRiskReport 2019, Focus: water supply. Retrieved from https://reliefweb.int/sites/reliefweb.int/files/resources/WorldRiskReport-2019_Online_english.pdf

Delanty G, Wodak R, Jones P (2011) Identity, belonging and migration. Liverpool University Press, Liverpool

Duvat VKE, Magnan AK (2019) Rapid human-driven undermining of atoll island capacity to adjust to ocean climate-related pressures. Sci Rep 9(1):15129–15116. https://doi.org/10.1038/s41598-019-51468-3

FloodMap (2020) Flood Map. Retrieved from www.floodmap.net
Gemenne F (2011) Climate-induced population displacements in a 4°C + world. Philos Trans Royal Soc A: Math Phys Eng Sci 369(1934):182–195. https://doi.org/10.1098/rsta.2010.0287
Government F (2018) Fiji: planned relocation guidelines - a framework to undertake climate change related relocation. Retrieved from https://cop23.com.fj/wp-content/uploads/2018/12/CC-PRG-BOOKLET-22-1.pdf
IPCC (2014) Climate change 2014 synthesis report summary for policymakers. Retrieved from https://www.ipcc.ch/site/assets/uploads/2018/02/AR5_SYR_FINAL_SPM.pdf
Jamero ML, Onuki M, Esteban M, Billones-Sensano XK, Tan N, Nellas A, Valenzuela VP (2017) Small-island communities in the Philippines prefer local measures to relocation in response to sea-level rise. Nature Climate Change 7(8):581–586. https://doi.org/10.1038/NCLIMATE3344
Jamero ML, Onuki M, Esteban M, Chadwick C, Tan N, Valenzuela VP, Avelino JE (2019) In-situ adaptation against climate change can enable relocation of impoverished small Islands. Marine Policy 108:103614. https://doi.org/10.1016/j.marpol.2019.103614
Kench PS, Owen SD, Ford MR (2014) Evidence for coral island formation during rising sea level in the central Pacific Ocean. Geophys Res Lett 41(3):820–827. https://doi.org/10.1002/2013GL059000
Kueffer C, Kinney K (2017) What is the importance of islands to environmental conservation? Environ Conserv 44(4):311–322. https://doi.org/10.1017/S0376892917000479
Kumar S, Terubea T, Nomae VD, Manepora'a A (2006) Poverty and Deprivation amongst Ethnic Minorities in Fiji: The Case of Ni Solomoni and Rabi Islanders. Fijian Stud A J Contemp Fiji 4(1):125–142
Martín-Antón M, Negro V, María José, José María de C, López-Gutiérrez JS, Esteban MD (2016) Review of coastal Land Reclamation situation in the World. J Coastal Res 1(75):667–671. https://doi.org/10.2112/SI75-133.1
McLeman RA (2011) Settlement abandonment in the context of global environmental change. Glob Environ Change 21(1):S108–S120. https://doi.org/10.1016/j.gloenvcha.2011.08.004
Neill P (2015) The once and future Ocean, notes towards a new hydraulic society. Leete's Islands Books, Maine
NOAA. (2012). Global sea level rise scenarios for the United States national climate assessment (OAR CPO-1)
Nunn PD (2009a) In: Island formation. 1 edn. University of California Press, pp 490
Nunn PD (2009b) Responding to the challenges of climate change in the Pacific Islands: management and technological imperatives. Climate Res 40(2/3):211–231. https://doi.org/10.3354/cr00806
Nunn PD (2013) The end of the Pacific? effects of sea level rise on Pacific Island livelihoods. Singap J Trop Geogr 34(2):143–171. https://doi.org/10.1111/sjtg.12021
Nunn PD, Kumar L, Eliot I, McLean RF (2016) Classifying Pacific Islands. Geosci Lett 3(1):1–19. https://doi.org/10.1186/s40562-016-0041-8
Nunn PD, Kumar R (2018) Origins and environments. In: The routledge international handbook of island studies, Routledge
Oppenheimer M, Glavovic BC, Hinkel J, van de Wal R, Magnan AK, Abd-Elgawad A, Cai R, Cifuentes-Jara M, DeConto RM, Ghosh T, Hay J, Isla F, Marzeion B, Meyssignac B, Sebesvari Z (2019) Sea level rise and implications for low-lying islands, coasts and communities. In: IPCC special report on the ocean and cryosphere in a changing climate. Retrieved from Monaco
Ourbak T, Magnan AK (2018) The Paris Agreement and climate change negotiations: small Islands, big players. Reg Environ Change 18(8):2201–2207. https://doi.org/10.1007/s10113-017-1247-9
Paeniu LIV, Jacot Des Combes H, De Ramon N'Yeurt A, Korovulavula I, Koroi A, Sharma P, Hobgood N, Chung K and Devi A (2015). Coastal protection: best practices from the Pacific
Perumal N (2018) "The place where I live is where I belong": Community perspectives on climate change and climate-related migration in the Pacific island nation of Vanuatu. Island Stud J 13(1):45–64. https://doi.org/10.24043/isj.50
Piggott-McKellar A, McNamara K, Nunn P, Sekinini S (2019) Moving people in a changing climate: lessons from two case studies in Fiji. Soc Sci 8(5):133. https://doi.org/10.3390/socsci8050133

Ramalho RS, Quartau R, Trenhaile AS, Mitchell NC, Woodroffe CD, Ávila SP (2013) Coastal evolution on volcanic oceanic islands: A complex interplay between volcanism, erosion, sedimentation, sea-level change and biogenic production. Earth Sci Rev 127:140–170. https://doi.org/10.1016/j.earscirev.2013.10.007

Royle SA (2014) Islands: nature and culture. Reaktion Books, London

Tabucanon GM (2012) The Banaban resettlement: implications for pacific environmental migration. Pacific Stud 35

Tabucanon GM, Opeskin B (2011) The resettlement of nauruans in Australia: an early case of failed environmental migration. J Pacific History 46(3):337–356. https://doi.org/10.1080/00223344.2011.632992

Thomas FR (2009) Historical ecology in Kiribati: linking past with present. Pacific Sci 63(4):567–600. https://doi.org/10.2984/049.063.0406

Walshe RA, Stancioff CE (2018) Small Island perspectives on climate change. Island Stud J 13(1):13–24. https://doi.org/10.24043/isj.56

Webb AP, Kench PS (2010) The dynamic response of reef islands to sea-level rise: evidence from multi-decadal analysis of island change in the central Pacific. Global Planet Change 72(3):234–246. https://doi.org/10.1016/j.gloplacha.2010.05.003

Weigelt P, Jetz W, Kreft H (2014) Bioclimatic and physical characterization of the world's islands vol 110, p 15307 (2013); In: Proceedings of the national academy of sciences of the United States of America, vol 111(51), pp 18400-18400. https//doi.org/10.1073/pnas.1422548112

Wong PP (2011) Small island developing states. Wiley Interdisciplinary Rev Climate Change 2(1):1–6. https://doi.org/10.1002/wcc.84

Woodroffe CD, McLean RF, Smithers SG, Lawson EM (1999) Atoll reef-island formation and response to sea-level change: West Island, Cocos (Keeling) Islands. Marine Geol 160(1):85–104. https://doi.org/10.1016/S0025-3227(99)00009-2

WorldBank (2016) Climate and disaster resilience. Retrieved from http://pubdocs.worldbank.org/en/720371469614841726/PACIFIC-POSSIBLE-Climate.pdf

WorldBank (2019) GDP per capita (current US$)-Pacific island small states. Retrieved from https://data.worldbank.org/indicator/NY.GDP.PCAP.CD?amp%3Bview=map&locations=S2

Wyett K (2014) Escaping a rising tide: sea level rise and migration in Kiribati. Asia Pacific Policy Stud 1(1):171–185. https://doi.org/10.1002/app5.7

Yamamoto L, Esteban M (2017) Migration as an adaptation strategy for atoll Island states. Int Migr 55(2):144–158. https://doi.org/10.1111/imig.12318

Zope PE, Eldho TI, Jothiprakash V (2017) Hydrological impacts of land use–land cover change and detention basins on urban flood hazard: a case study of Poisar River basin, Mumbai, India. Nat Hazards 87(3):1267–1283. https://doi.org/10.1007/s11069-017-2816-4

Despina Linaraki Architect engineer, Despina Linaraki is a PhD researcher at SeaCities, Cities Research Institute, in Queensland. She has completed a Master of Science, 2015 from Columbia University in New York, at Advanced Architectural Design and the Development of Global Cities. Furthermore, she holds a Master of Architectural Engineering, 2013 from Technical University of Crete, in Greece. Since 2014 she is a registered Architect in Greece. Having lived experience in different coastal cities around the world, she is researching nature-based solutions for the adaptation of low-lying islands and coastal cities to sea level changes and floods through the interdisciplinary of architecture with the fields of biology and geology.

Modelling of SeaCities: Why, What and How to Model

Edoardo Bertone and Oz Sahin

Abstract In order to realise SeaCities as depicted in this book, several individual projects, at a building (e.g. floating houses) and community (e.g. energy–water–food supply, transportation) levels, need to be designed to solve the inherent issues of such relatively unexplored urban concept. In this chapter, the focus is on the importance of modelling in providing a critical set of support tools to better inform SeaCities design and decision-making. The chapter is intended for the reader interested in SeaCities and relatively unfamiliar with the modelling world. The chapter is structured as follows: First, a brief justification of the need for modelling in SeaCities is provided ("WHY modelling"); this is followed by examples of systems, elements or parameters that will require modelling ("WHAT modelling"). Next, different modelling categories are introduced, based on different criteria, such as the type of inputs available or the type of output desired ("HOW modelling"). The chapter concludes with two case studies, selected as examples of an integrated, participatory systems approach which is suggested as being highly applicable to many of the future SeaCities modelling needs.

1 WHY Modelling of SeaCities

The ideal "SeaCity" would, eventually, fully take advantage of its interface between land and sea, and benefit of the sea for its incredible amount of resources and opportunities that it provides, rather than merely adapting to its relentless rise due to climate change.

E. Bertone (✉) · O. Sahin
School of Engineering and Built Environment, Griffith University, Brisbane, QLD, Australia
e-mail: e.bertone@griffith.edu.au

Cities Research Institute, Griffith University, Brisbane, QLD, Australia

E. Bertone
Australian Rivers Institute, Griffith University, Brisbane, QLD, Australia

O. Sahin
Griffith Climate Change Response Program, Brisbane, QLD, Australia

J. Baumeister et al. (eds.), *SeaCities*, Cities Research Series,
https://doi.org/10.1007/978-981-15-8748-1_11

Although the ultimate SeaCity might be an urban concept which is yet to be realized, examples from past to present cultures and civilisations, as well as from a new influx of research and innovation around the world, provide evidence of beneficial, feasible individual design solutions that a current coastal city can deploy to slowly move towards the SeaCity dream. A SeaCity cannot be created overnight; however, individual solutions can be tested and progressively implemented.

The type of key questions that need to be addressed to ensure that such design solutions are successfully implemented can be:

- What are the benefits and for whom?
- What are the risks?

New projects typically require a clear understanding and quantification of the benefits. Would the government have a return of investment, if floating houses are allowed to be built? What are economic costs and benefits of an offshore desalination plant or of a floating solar farm? What about less tangible cost and benefits, such as to society and to the environment?

Moreover, any project, especially unconventional designs, will inherently carry a number of risks. What are the water quality impacts of a floating city? What about a large, modular aquaculture area? What would happen to a floating development during high seas events? What are the social and large-scale economic implications of bringing people away from land to the water? How would the real estate market be affected?

Any individual SeaCity-related project will have to properly address all these types of questions, as well as several other issues such as regulatory frameworks and governance structures. Although seemingly daunting, government, academic, and industry experts do have a unique opportunity at this time in history: We have unprecedented knowledge, facilities, infrastructures and computational power, to enable us to accurately quantify and account for all the pros and cons, benefits and risks, of any innovative, disruptive research or design project, before final approval and implementation. In other words, we can *model*, and predict, how effective and feasible SeaCities-related projects will be.

As a consequence, this chapter will explore this essential step required to ultimately realise the SeaCity objective. The chapter will first look more specifically at why modelling is crucial and what is needed to be modelled, such as engineering, environmental and socio-economic aspects of a SeaCity project. Subsequently, the focus will shift to how it is possible to model and predict such complex behaviours or systems with a certain level of rigour or accuracy. Specifically, an overview of available modelling types will be presented and discussed. The chapter will conclude with a set of recommendations and case studies, especially advocating the need of a participatory, holistic, "system" thinking and modelling approach.

2 WHAT Modelling in SeaCities?

Whether a proposed SeaCities project is about floating structures, floating modular aquaculture systems or amphibious developments for instance, there are a number of common aspects that need to be modelled to comprehensively quantify risks and opportunities. These are described in the sections below.

First of all, it is important to identify and describe the problem, or risk, that SeaCities aim to provide solutions for. The main one is sea level rise and consequently, exacerbated flooding risks for several low-lying coastal cities both in the short, medium and long term. Climate change studies have recently updated their sea level rise estimations (Oppenheimer et al. 2019), dramatically increasing the predicted number of cities and people that in a few decades will be temporarily or permanently under water; such predictions include, for instance, 190 million people occupying land that will be below the projected high tide lines in 2100, under the more conservative low carbon emission scenarios (Kulp and Strauss 2019). Although this should be enough to trigger global attention towards the gravity of this problem and the benefits of potential SeaCities-related projects, modelling is needed, at a local level, to downscale such predictions and provide a more detailed estimation of costs and benefits for a local council or government. As discussed in Lin et al. (2019), with some of the larger cities in the world such as London, Miami, Tokyo and Jakarta expanding their territory by up to 10% in urban waters in the next few decades; from an urban planning perspective, it is important to evaluate which areas of the city must be kept dry, which areas can be changed from dry to wet, and which waters are most suitable for city expansion. Relevant modelling outputs include the specific areas of a city which are likely to be underwater regularly or during storm surges, as well as an economic assessment of related costs for the council and government. The economic assessment should also quantify traditionally more qualitative measurements such as loss of health and well-being of locals impacted. Clear, numerical assessment of the expected costs of the do-nothing option is crucial to convince political institutions, which are usually motivated for achieving short-term goals in line with their political mandates, rather than investing heavily upfront in exchange for a long-term benefit.

Once the problem is clear and quantified, it becomes priority for governments and urban planners to assess the different possible solutions. Studies have already shown (Haasnoot et al. 2012) how traditional "protective" projects, such as raising dikes, are orders of magnitude more expensive than solutions such as floating houses, which look at working together with the sea rather than against it. Land reclamation as well becomes proportionally more expensive with depth, unlike floating developments (Ko 2015). However, modern floating houses/villages are not that widespread as one would expect, and only in the last decade, the world has witnessed an increase in floating architectural projects (the curious reader can refer to Lin et al. (2019) for a few examples). A reason for such slow uptake is that being a relatively unexplored field of research and investment, there is a feeling that potential negative effects have not been comprehensively identified and quantified. As a consequence, such knowledge gaps have not allowed water authorities and councils to develop policy frameworks clearly defining regulations for new similar projects (de Lima et al. 2015a, b), with often unfounded concerns hindering the implementation of new floating development concepts (de Lima et al. 2015b). As a result, focusing specifically on floating structures, there is an urgent need to model and quantify all the potential risks, so that the project can be either approved, or optimised, or even rejected in favour of an alternative, better one.

Specifically, for floating development projects, the first, key element to analyse would be the structural engineering feasibility. The Archimede's principle is a starting point to help with the design of the structure, with tilting and rotation also critical elements to control (Mohamad et al. 2013); strong external loading affecting the structure includes wind and waves, rain, ice and solar radiation (Stopp and Strangfeld 2010) and needs to be quantified. Other critical elements to quantify include the centre of gravity, centre of buoyancy and moment of inertia (Ishaque et al. 2014). Floatability, floodability and stability need to be guaranteed and verified (Shekhorkina 2015) for normal and worst case scenarios for the location and the project in consideration. In general, the height of buildings is quite limited due to stability issues, based on current engineering knowledge, although research and knowledge applied to, for instance, floating oil rigs could be relevant in expanding current height constraints. Structural calculations need to be performed to determine the dimensions and the amount of materials (e.g. reinforced steel) required. An issue with materials is that they could get corroded based on salt content and pH of surrounding water (Stopp and Strangfeld 2010; Strangfeld and Stopp 2014), and this could lead to structural issues in the long term. At the same time, floating structures are protected from earthquakes (Ko 2015), so there would be no seismic requirements on the structure, regardless of the location/seismic area. Potentially, it could also deal better with post-earthquakes events such as tsunamis due to their ability to float and follow the sea level variations. If the floating structure is very large, hydrodynamic modelling challenges arise, because the structure cannot be modelled as a rigid body, and because the computational power and time required for the model would increase dramatically (Lamas-Pardo et al. 2015; Yeh et al. 2015).

In addition to engineering feasibility, environmental feasibility is also critical and still relatively understudied (de Lima et al. 2015b). Floating structures (especially when very large, such as houses or buildings) will block sunlight, providing different types of disruption to the marine ecosystem (Ko 2015). This will indirectly cause a change in temperature of the upper layer of water, causing a reduction of dissolved oxygen too, usually produced by photosynthesis (Burdick and Short 1999) which will be also reduced due to the lower amount of incident solar radiation. This can result in large amounts of dead organic material, which decreases dissolved oxygen further through mineralisation and increases turbidity, and in turn, further decreases sunlight penetration, through a positive feedback loop. Wind, which has a significant impact on circulation of water bodies (Ji 2017), will also change its pattern due to compression of air between the houses. If not recycled, wastewater should also be treated before being released to the environment; if this cannot be done on-site, pipelines to the mainland would have to be adequately designed to withstand external forces and moments (Ko 2015).

Finally, social/health impacts need to be considered. It will be more demanding to design for a comfortable indoor climate, especially in warmer months, due to for instance a lack of vegetation, an increase in reflected radiation and limited use of blinds due to stronger winds (Stopp and Strangfeld 2010). Hence, residents' well-being would be affected, unless more energy is used, or new design solutions are applied to guarantee the same indoor comfort that they now enjoy on land. On the other hand, there is potential to improve the outdoor environment with a better connection to water, and with the potential for new urban design solutions which, unlike the "urban sprawl" concept, can improve cohesion, interactions and physical activities.

All the described potential engineering, environmental and social issues need to be modelled and quantified as best as possible, to provide a holistic assessment of the project and in turn inform the economic assessment to be presented to the stakeholders. Importantly, feasibility must be verified not only for the actual floating structure/buildings, but also for all the critical services which must be provided to the residents.

On the bright side, despite the potential risks, there are also considerable opportunities that, through research and innovation, can make floating developments more feasible. Surrounding water could be used for heating and cooling (Lin et al. 2019), using techniques such as evaporation and heat pipes or running water through the building envelope (Stopp and Strangfeld 2010). Rainwater harvesting would help meet the freshwater demand (Bhattacharjee and Mukherjee 2017), with more potential for storage (e.g. underwater), and floating and/or mobile desalination plants have been already designed (Lin et al. 2019). For food production, it is possible to produce hydroponically grown crops and rely on fishing, especially cage fishing and aquaculture in general (Bhattacharjee and Mukherjee 2017), though the water quality issues mentioned before (e.g. drop in dissolved oxygen and water temperature, potential contamination from discharged treated wastewater) need to be accounted for. More research into fish aggregating devices could also be beneficial. Wastewater coming from the aquaculture sites can be naturally treated to improve its quality through the use of floating plant bed systems (Nduwimana et al. 2007). Modelling of high-frequency water quality sensor data also showed, unexpectedly, a limited negative impact on dissolved oxygen levels under floating houses in Delft (Foka et al. 2015), with abundant marine life still achievable in proximity of the floating structures (de Lima et al. 2015b), thus showing promising signs despite the theoretical water quality concerns mentioned earlier.

Vertical vegetation could help reduce the external wall temperatures during warmer months, thus helping to achieve comfortable indoor environment with a lower energy demand. For waste reduction, concepts exist that aim at recycling waste nutrients and carbon dioxide to produce food and biofuels through the use of floating algae systems (Roeffen et al. 2013). For energy production, solar panels, wind turbines and biodigesters have been suggested for floating houses (Bhattacharjee and Mukherjee 2017); solar panels are known to work better if the temperature is not too high, hence placing them in the water would improve their efficiency and thus, provided the perfect orientation and tilt angle can be still guaranteed, it would lower the number of panels (and space) required; floating solar panels would also benefit water quality by reducing algal growth (Sahu et al. 2016) although a comprehensive assessment of the impact on the water environment underneath should be conducted. Floating projects examples around the world summarised by Lin et al. (2019) achieved (1) energy self-sufficiency with wind turbines and solar panels; or (2) no disruption by wind and waves through the use of dampers, cables and springs.

Structurally, as summarised in Lin et al. (2019), floating foundations are usually heavy and stable, also allowing the structure to adapt to sea level moving up and down. As a consequence and as previously mentioned, the effects of extreme events such as tsunamis (Lin et al. 2019) or earthquakes (Lamas-Pardo et al. 2015) are usually much less severe in open waters. From a socio-economic perspective, the direct experience of the natural environment provided by the surrounding water would lead to better lifestyle and well-being, in turn making properties more attractive and valuable (Stopp and Strangfeld 2010).

Most of the proposed concepts are still in their infancy and need to be modelled and refined. A number of modelling approaches will be described in the next sections, to help the reader choose the most appropriate technique based on the modelling goal and scope, and to eventually achieve a holistic, comprehensive assessment of a floating development project.

3 HOW Modelling? Modelling Categories

There are endless distinctions, in the literature, between models based on their unique features. However, what all the models have in common is the need for some type of inputs and the delivery of some type of outputs. In SeaCities, the type of inputs available for modelling can vary considerably, as well as the desired type of output. Hence, in this book chapter, we categorise different models in simple terms based on the inputs and based on the outputs they produce (Fig. 1).

3.1 Model Choice Based on Inputs

We first analyse the left-hand side of Fig. 1 and introduce different modelling categories based on their inputs.

3.1.1 Physical Models

A physical model is essentially a copy of an object or a system, typically of a different size. For systems and objects investigated in a SeaCity, these are typically large, and thus, the physical model would have a smaller scale. The main input to a physical model, following its preliminary design, is *material*.

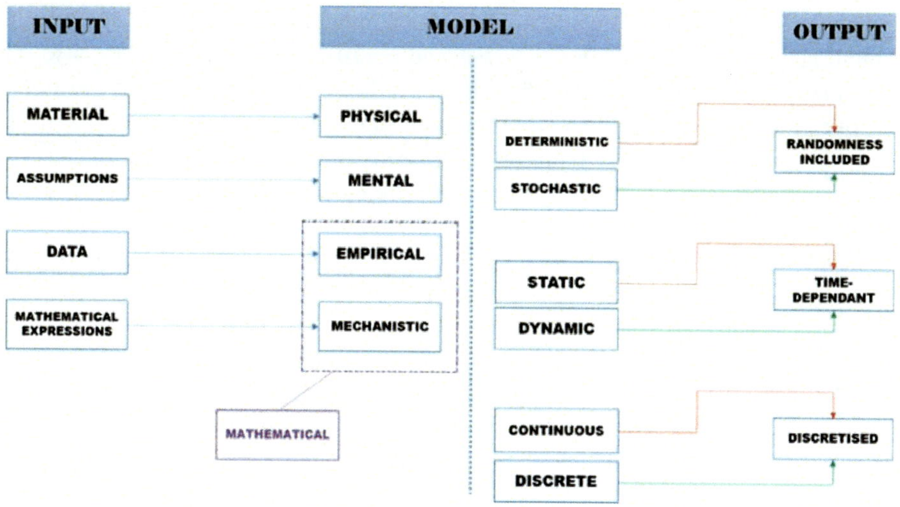

Fig. 1 Types of models based on inputs and outputs. Red arrows imply a negative relationship, green arrows a positive relationship. Dashed blue line separates model categories based on inputs or outputs

Physical models can be used in conjunction and in support of numerical models. For SeaCities, an important combination is between physical models and computational fluid dynamics (CFD) models, i.e. a type of process-based models. Physical models can be placed in wind tunnels or wave tanks to validate the CFD simulation outputs in relation to how flows of fluids might affect the model, and thus, in turn, they can help in optimising its own design (e.g. shape). Scaling is a critical issue for physical models, in relation to the dimensions of the instrumentation available for experiments (e.g. wave tank). Boundary conditions become very delicate and important to consider. Froude number and Reynolds number scaling approaches are typically used.

Relevant examples include from authors who have proposed 3D numerical models of floating structures to estimate wave drift forces and hydro-elastic responses and compared the results with experiments conducted with physical models; creating opportunities for comparison of the influence of difference factors (Ma et al. 2001).

3.1.2 Mental Models

Although there is no clear, agreed definition of mental models across different disciplines (Rook 2013), a robust definition has been provided by Rook (2013): *"a concentrated, personally constructed, internal conception, of external phenomena (historical, existing or projected), or experience that affects how a person acts"*. The curious reader is directed to Rook (2013) for an interesting analysis of different mental model definitions and concepts. The main input for this type of models is *assumptions*, i.e. conjectures of how and why a certain object, process or system behaves in a certain way in the real world. A mental model has a number of limitations: firstly, a human mind has limited computational power and abilities; mental models are also only qualitative thus not predictive; they do not indicate which parts of the system are active (and which are not) and are reliant on subjectiveness. But, a well-elicited (to avoid cognitive biases) mental model is typically the first step in modelling any new system/process, after problem formulation; hypotheses must be made and verified in the real world, before the mental model can be validated and a more complex and accurate numerical model developed. There are different types of mental models, and some of them, such as casual loop diagrams, are the foundations of complex system thinking and system dynamics models, described later in this chapter. In the context of SeaCities, mental models as the foundation for more complex quantitative modelling were applied in Australian coastal locations to assess adaptation options (Smith et al. 2015).

3.1.3 Mathematical Models: Empirical and Mechanistic Models

Mathematical modelling in general has been described as "*the art of translating problems from an application area into tractable mathematical formulations whose theoretical and numerical analysis provides insight, answers and guidance useful for the originating application*" (Neumaier 2004). The vast majority of models typically developed in contexts relevant for SeaCities are mathematical, and several different types exist. The main distinction, however, can be made between empirical and mechanistic models. Such subdivision resembles what occurred in the field of philosophy when defining the foundational principle of knowledge; in the seventeenth and eighteenth centuries, two philosophical schools emerged in Europe: one (the *empiricists*, led by Locke, Berkeley and Hume) asserting that knowledge can be only gained by experience of reality (Priest 2007), and the second one (the *rationalists*, led by Descartes, Leibniz and Spinoza) claiming that meaningful facts about reality can be inferred by use of reason rather than experience (Garrett 1999).

Empirical, or inductive, modelling in general terms relies on observations to develop, and to validate, a model. In most cases, data-driven models can be considered part of this category of models. Data-driven models essentially rely on data (usually quantitative, but also qualitative in some cases) to seek relationships between different variables and processes and to eventually extrapolate such relationships for different purposes, such as predictions, optimisation, or scenario testing. Clearly, the main input for this type of model is *data*. Models under this category can be parametric or non-parametric. Parametric models have a fixed number of predictors based on certain assumptions. and only coefficients need to be defined. An example is regression models. Non-parametric models do not rely on any assumption, and the most suitable "form" of the model, as well as the number and form of each predictor, is entirely learnt from the data. Under this umbrella sit most of artificial intelligence and specifically machine learning models. Several examples relevant to SeaCities exist, and here, we will mention a couple of interesting ones. Multi-objective evolutionary algorithms were developed and deployed by Ugurlu et al. (2015) in order to optimise a floating settlement (e.g. shape, functions) in a way to minimise the need of roads and to maximise accessibility, wind protection and visibility based on a number of constraints. A very similar multi-objective optimisation model for urban planning was developed in Levi et al. (2019) for a very large floating structure.

Mechanistic, or deductive, modelling instead relies on the systematic application of mathematical formulations of causal mechanisms to explain a process or a system. The main input here is therefore *mathematical expressions,* and the assumption that a system can be decomposed and analysed by examining its individual components. Mechanistic modelling is quite commonly applied in different modelling aspects of SeaCities, such as to understand the water-energy nexus as reviewed in Kenway et al. (2011), to predict urban flooding as reviewed in Mignot et al. (2019) or to predict the performance of marine renewable energy devices as reviewed by Day et al. (2015).

Advantages and disadvantages of empirical and mechanistic models are quite distinct, and similarly to the philosophical debates occurring 2–3 centuries ago, with available data increasing dramatically every day, modellers are now more than ever asked to take an informed decision on what type of model would suit best a particular application. In general, the suitability will depend on the amount of relevant data available as well as the goal and features of the model (e.g. scenario analysis, classification, prediction). A number of dedicated scientific resources are available for the interested reader, to be trained to select the best model type, including the specific one within the individual category (i.e. which particular data-driven model). However, due to the increasing development and application of smart technology and sensors able to collect very large amount of data (see, e.g. Fig. 2), coupled with the increasing capabilities to handle them through machine learning and increased computational power, empirical models seem to have more opportunities for large-scale deployment in future years.

3.2 Model Choice Based on Outputs

We now analyse the right-hand side of Fig. 1 and discuss different modelling categories based on their outputs.

Fig. 2 Vertical profiling systems (VPS) rely on high-frequency sensors to monitor in real-time the water quality of lakes and reservoirs, at multiple sites and depths

3.2.1 Deterministic/Stochastic Models

A deterministic model is a model that does not include any randomness, and thus, the model outputs are fully defined by the initial conditions and applied equations/algorithms. Conversely, a stochastic model inherently includes some degree of randomness, so that even with the same sets of parameters and initial conditions, different results would be obtained in different simulations. Of relevance for SeaCities, a recent review of flood loss models (Gerl et al. 2016) noticed that 96% of existing models were deterministic, seemingly over simplifying the uncertainties encapsulated in the real systems.

Our world is filled with stochasticity, and also our computers, despite being able to handle large computational loads and calculations, will inherently add an element of randomness to any deterministic model when, for instance, rounding decimals. Some of these minor changes in initial conditions might not affect certain systems, but for complex, chaotic systems, these variations might lead to totally different predictions even in a relative short-time horizon. This concept is at the foundation of chaos theory and was discovered by Edward Lorenz in the field of meteorology, who called this phenomenon the (famous) "butterfly effect" (Lorenz 2000). This also explains why it is so difficult to generate accurate weather forecasts. In conclusion, whether the model explicitly accounts for some degrees of uncertainty or not, it is important to be able to interpret the modelling outputs correctly, and to fully understand potential assumptions and sources of errors leading to the results. Trust the model, but do not trust it first.

3.2.2 Static/Dynamic Models

Another way to classify a model is based on whether time is considered in the way modelling outputs are calculated. In the case of static models, they describe a system at a specific point in time, and thus, there is no accounting of previous states of the system when calculating the next. Dynamic systems are instead, to some degree, time dependant, and thus, they maintain a "memory" of inputs and outputs of the system at earlier, relevant time steps. In SeaCities, for instance, assessing the evolution of a heavy rainfall event and potential related dam spilling and river flooding, inevitably lends itself to a dynamic modelling approach due to inherent lags between rainfall, runoff and flows at different locations. Of relevance for SeaCities research, in a study comparing a static and a dynamic model to evaluate building (and people) vulnerability to impulsive floods, the dynamic model was found to perform better when assessing them against historical data (Pilotti et al. 2016).

3.2.3 Discrete/Continuous Models

Although a discrete/continuous distinction can also refer to different methods to run temporal simulations in certain models, here we refer to what form the final output of the model takes. If discrete, this means that only a limited number of states can be assigned to the output parameter. An example of a type of discrete model which has seen widespread application in last couple of decades is Bayesian networks (Fenton and Neil 2013). With Bayesian networks, even when numerical, continuous variables are used for input; these have to be discretised in relevant intervals. For instance, in case of SeaCities and specifically coastal flooding, the "mean sea level" in metres could be discretised in "low", "medium" and "high" based on meaningful thresholds. This can be seen as a limitation of such models for most applications; however, often a simpler, categorical representation of a model output can facilitate interpretation by a wide interdisciplinary range of stakeholders; in addition, research is rapidly improving the discretisation capabilities of Bayesian networks (e.g. Mayfield et al. 2019). Continuous model outputs would instead not need to be discretised and the precise numerical output would be calculated. The users should be wary, for fully deterministic models, of such precise estimation and, as discussed previously, make sure to interpret the results correctly. For stochastic models, a number (usually millions) of different simulations are run, and the results can be represented in the form of probability distributions. Although they can be less user-friendly for some, they allow to account for the uncertainty present in the system and in the model. Probabilistic outputs can be summarised in helpful ways; for SeaCities, for instance, "risk maps" of, e.g. coastal flooding could be generated by elaborating a coastal model's outputs. An example is given by Mojaddadi et al. (2017) where a complex spatial model's outputs were effectively summarised by generating flood risk maps.

3.3 System Thinking and Integrated Approaches

There exists a range of modelling techniques, and the most suitable choice will depend on a range of factors. The advantages and disadvantages of several common techniques were discussed above. The shortcomings of the techniques as discussed above are evidence that individually, most of the time none of them is suitable for meeting all the modelling requirements of any complex system. For example, no single technique can easily incorporate the expert opinions that could be required for predictions, as well as accounting for complex feedback loops and relationships required to analyse the systems' behaviours. To achieve the required balance of complexity and system transparency, an integrated modelling is needed to eliminate the weaknesses and utilise the strengths of each technique (Voinov and Gaddis 2008). It is easy to envisage the synergy achieved by integrating multiple techniques in order to provide extra dimensional modelling power, as well as enhanced abilities (Sahin 2011). As such, an "integrated modelling approach" is proposed herein, which can effectively help model complex systems embedded within SeaCities.

Behind this modelling approach, *system thinking* forms the underlying foundations. Unlike mechanistic analysis, where the goal is to separate the individual pieces of what is being studied (in fact, the word "analysis" actually comes from the root meaning "to break into constituent parts"), in system thinking the focus is how what is being studied interacts with the other constituents of the system of which it is a part. A human body, or a car, provides good examples of such systems. The purpose of a car can be generally defined as to bring its driver from point A to point B faster than what the driver could achieve on foot. A car is made of an engine, four wheels, a transmission system and several other individual components. However, if one puts all these components inside a box at random, the car will not fulfil its purpose; even if, individually, it has the most powerful engine or the tires with the best grip. What is critical is not only the performance of its individual components, but how such components are connected and interact with each other. The key idea is that a system is not the sum of its components, but the product of its interactions.

Under the system thinking framework, a number of modelling steps are usually completed; as a consequence, such framework cannot be included in some of the specific modelling categories described above. A system thinking-based approach will have a variety of features, dependent on the project-specific directions and approaches used.

In system thinking, the first modelling step is to articulate the problem and conceptualise the system. Such conceptual models (described earlier in this chapter) can be developed with an integrated, participatory approach where stakeholders' workshops are designed to achieve an agreed common understanding of the system being modelled. The first step in developing such a conceptual model with a participatory approach is to agree upon the project goals and thus key variables among the stakeholders. A common shortcoming of this conceptual modelling step is the tendency to focus on parameters and connections which are within the domain of expertise of the specific stakeholder; thus, selecting the appropriate "expert" is crucial. In addition, tools and methods previously developed (e.g. Godet 2001) can facilitate a more rigorous approach to the model conceptualisation process. Of key importance is also a clear definition of the system's boundaries (i.e. scope of the model), based on a fine balance between need for model complexity and demand for data and computational power.

The following step, once the conceptual model has been finalised, is its conversion to a mathematical model. Based on the different modelling needs and thus modelling features required, different types of models can be used. Commonly used ones are either Bayesian networks or system dynamics models -or a combination of them (e.g. Bertone et al. 2016a, 2017).

Bayesian networks (BN), previously mentioned due to being discrete, are probabilistic, graphical models of the system of interest and use the Bayes theorem as the mathematical framework for updating probabilities based on evidence. A BN consists of variables (i.e. "nodes") interconnected (parent to child node) through arcs. For nodes having "parent" nodes, the relationships between connected nodes are quantified through Conditional Probability Tables (CPTs). Populating the CPTs is one of the most challenging parts of developing a BN (Druzdel and Van Der Gaag 2000), and at the same time one of the most important. This is because the input data used to populate the CPTs can be empirical, quantitative data, but also qualitative (e.g. based on experts' opinion), based on other models' outputs, or spatial (e.g.

geographic information system, or GIS, data), to name a few. Missing data can be appropriately handled with BNs, as this can either be overcome with expert input or by reflecting this in the level of uncertainty behind specific CPTs, which is clearly represented. Additional advantages of BNs are (Uusitalo 2007), as a consequence, the ability of incorporating inputs of experts from different disciplines (critical for modelling complex systems), fast simulations facilitating rapid feedback to participants when testing different scenarios, thus supporting effective decision-making; and the ability to detect and explore inconsistencies and different perceptions among stakeholders. Moreover, BNs benefit from the key feature of the Bayes theory, i.e. the probability of a hypothesis being true given updated evidence can be inferred, and also that uncertainty can be quantified. The main limitations, instead, include the difficulty to conduct temporal analyses. Furthermore, feedback pathways are not supported in the structure of BNs due to its "directed *acyclic* graph structure" necessary for the Bayes's theorem to be applied, thus restricting them to static analysis or very basic dynamic ones (e.g. Bertone et al. 2019). Finally, variables are discretised, with discretisation often limited to only a few states per variable (e.g. Richards et al. 2013; Bertone et al. 2020). This results in a coarsely parameterised model, thus potentially reducing its utility or range of applications (Uusitalo 2007); though recent improvements and potential beneficial applications exist and were mentioned in previous sections.

System dynamics (SD) modelling, in contrast, represents an approach to numerically quantify the dynamics of complex systems, by facilitating the development of computer simulations of such systems (Sterman 2000). SD, unlike BN, is dynamic (e.g. they include time as a variable, as well as feedback loops). However, they are less flexible when dealing with model inputs and assumptions.

Table 1 summarises pros and cons and features of each approach, showing how BN and SD complement each other and a combination of the two would allow to meet different data availability and output needs. In addition, it must be said

Table 1 Features of Bayesian networks and system dynamics

Feature	Bayesian networks	System dynamics
Numerical data input	Yes	Yes
Qualitative data input	Yes	No
Handling missing data	Yes	No
Output type	Probabilistic Discrete Static	Deterministic Continuous Dynamic
Feedback loops	No	Yes
Computation time	Short	Short
Conceptual model needed	Yes	Yes

that each modelling approach has intrinsic ways to overcome their limitations. For instance, though SD provides deterministic outputs, sensitivity analysis and Monte Carlo modelling are typically used to quantify uncertainty.

Recently, attempts have been made to combine BN and SD, in order to merge the benefits of the two approaches. The potential for integrating "fuzzy logic" in SD, thus creating "fuzzy system dynamics" to enhance input flexibility, was already highlighted by Tessem and Davidsen (1994). Such "fuzzy logic"-based approach can be used to quantify relationships and values for a number of nodes wherever empirical evidence is not available. Fuzzy logic was deployed to account (to some extent) for the uncertainty in systems characteristics, as summarised in a review paper by Labadie (2004) in relation to water resource optimisation research, and by Khanzadi et al. (2012) with a coupled fuzzy logic-systems dynamics approach. Bertone et al. (2016b) proposed a fuzzy logic approach with BN node decomposition as an effective way to complete and deploy a SD model whenever several variables do not have empirical data available, and they successfully applied such approach to study future health-related water quality risks for a large Australian reservoir (Bertone et al. 2016a).

4 Case Studies

Below, we present two case studies demonstrating successful examples of an integrated, system modelling approach relevant for SeaCities.

4.1 A Hybrid GIS-System Dynamics Approach to Assess Vulnerability to Sea Level Rise

Environmental problems are neither static nor do they exist in isolation; indeed, they are in a series of continuous dynamic interactions, driven by multi-causality. The complexity and interdependencies of their elements define their characteristics. Consequently, as the environment changes in time and space, so do its problems and the solutions to these problems. As a result, the dynamic characteristics of environmental systems must be fully considered to enable them to be addressed effectively. When addressing these problems, it is essential to use an approach that elicits a variety of relevant information. When designing a suitable purpose-specific model, it is necessary to select the most suitable assessment and decision-making techniques. With this in mind, the aforementioned system approach provides a suitable platform in the development of a dynamic model that assesses whole environmental systems and the many interacting components (Sahin and Mohamed 2014).

The world's coastal areas have been a focus when considering adaptation strategies for sea level rise (SLR). To assess coastal vulnerability and the adaptation alternatives

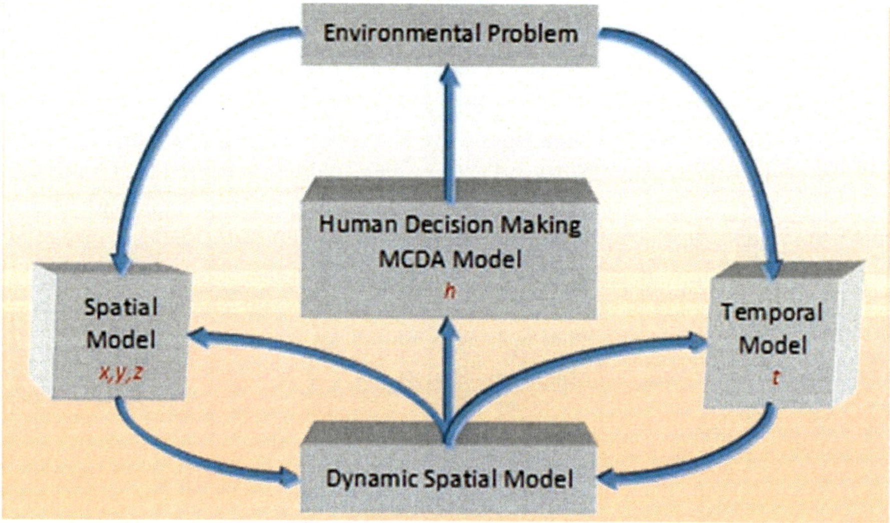

Fig. 3 Five dimensional (x, y, z, t, h) spatio–temporal decision modelling framework

to SLR, a spatial–temporal decision modelling framework as illustrated in Fig. 3 was developed (Sahin and Mohamed 2013). The framework is based on five dimensions of the decision process for environmental dynamics. Space (x, y, z) and time (t) constitute the first four dimensions and provide a common base where all natural and human processes occur. This was important since, in order to model and simulate changes in coastal zones it is necessary to consider the spatial and temporal dimensions of rising sea level, as any single approach is deficient in one feature or the other. Therefore, an integrated approach is the most appropriate way of dealing with various complex problems. Combining SD and GIS approaches provides the power to meet these requirements as previously mentioned and also emphasised by several authors (Grossmann and Eberhardt 1992; Ruth and Pieper 1994; Ahmad and Simonovic 2004; Gharib 2008).

This approach is crucial in generating adequate information from which decision makers can devise realistic adaptation strategies. For this reason, it is essential to incorporate the first four dimensions into a fifth dimension, i.e. the element of human decision-making (h).

It must be noted that, in this study, the term vulnerability refers to people at risk and the loss of property due to exposure to SLR and related storm surges. Using the spatio–temporal framework, we analysed the current conditions first; then, these systems were examined further, under various scenarios, so as to identify how SLR will affect the already distressed systems, over time. We used a range of SLR scenarios and sensitivity analyses to address the uncertainty issues associated with the projections of sea level rise.

This approach facilitates the use of creativity and experience to clearly structure the complex problems and pursue their solution within a systematic framework. In developing the framework and subsequent model, the following modelling steps were adopted:

1. Problem definition: Identifying and classifying the objectives provide a basis upon which to build a model and solve a problem using mathematical solutions.
2. Model development: The model is formulated, and a conceptualised model of the system under question is developed. All essential system variables are identified and processed. The relationships among the variables are also established, with some assumptions being made for the sake of simplification. Subsequently, the required data about the system is collected. Finally, the equations and functions needed to explain the relationships among these variables are determined.
3. Model deployment: Through the implementation of the model, the solutions are achieved.
4. Model validation and optimisation: After obtaining a solution, the results are examined to determine whether the model works effectively and satisfy the problem objectives.
5. Final model results interpretation and communication: The results are interpreted to provide a conclusion and recommendations to the stakeholders.

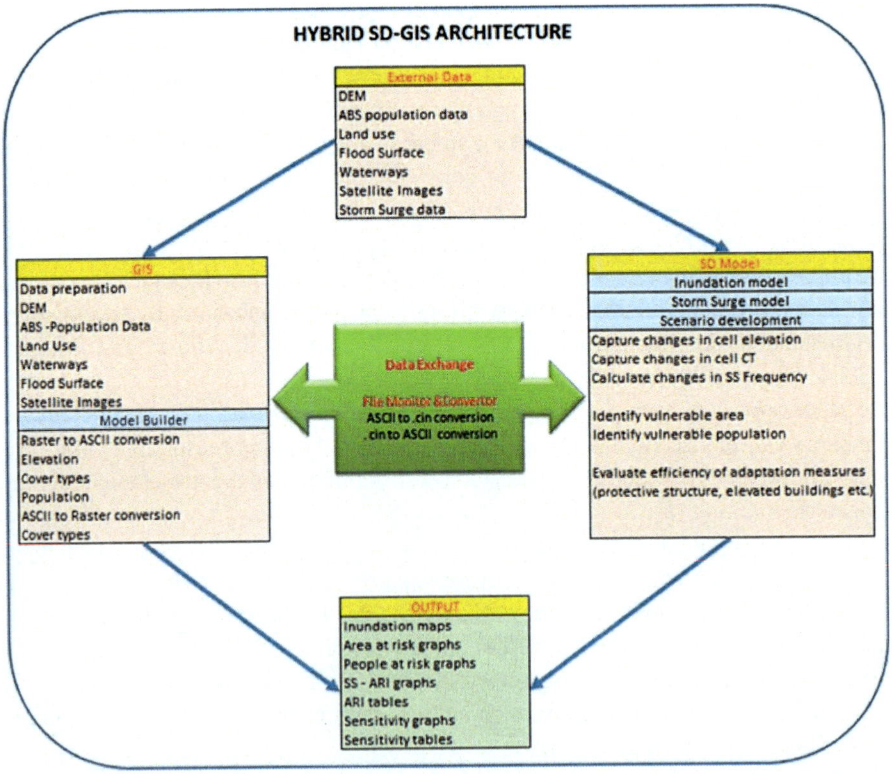

Fig. 4 Hybrid GIS-SD architecture

In this approach, GIS and SD assessed different aspects of the complex problem. While GIS handles spatial data, SD processes the dynamics of the complex system, revealing its causal structure and the relations of the system components. The use of these two methods is appropriate for tackling different environmental problems (both spatial and temporal). This versatile method, firstly, allows the processing of details separately; and, secondly, it then combines the results of the different methods. Therefore, the coupling of these two approaches provides immeasurable potential to solve problems. Importantly, the coupling of SD and GIS provides a hybrid model which can address temporal and spatial aspects of dynamic processes as illustrated in Fig. 4.

For brevity, a detailed discussion of this method is not given here; the interested reader is referred to Sahin (2011), Sahin and Mohamed (2013) for detailed description of the methods and results of the case study.

4.2 A Spatial BN Approach to Evaluate Future Coastal Erosion of a Pacific Island

The focus of this second case study was on predicting, and planning for, rare or unprecedented coastal erosion events related to longer term climate change effects on coastal regions. As discussed in the previous section, SLR on coastal regions is expected to exacerbate the impact of coastal hazards such as erosion, inundation, land loss, saline intrusion and flooding from storm surge with increasing frequency and intensity. Accordingly, the resulting negative impacts on coastal settlements and infrastructure will be exacerbated. Therefore, identification and assessment of coastal hazards are critical for planning and developing effective management strategies to protect lives and properties exposed to coastal hazards.

Coastal erosion is the movement of sediments in response to waves and currents. Erosion is a naturally occurring coastal process without which many coastal landforms, in particular beaches would not exist. Generally, coastal adaptation planning requires long-term strategies addressing various aspects such as infrastructure, land use planning and flood protection (Wong 2014). Therefore, suitable approaches must encapsulate uncertainty effectively through a range of scenarios considered for long-term adaptation planning. In previous studies, a range of qualitative and quantitative methods was used to assess the impacts of SLR on shoreline change, which include: a bathtub approach that provides simple vulnerability prediction (McGranahan et al. 2007) but not addressing uncertainty (Gesch 2009) and a quantitative coastal erosion assessment (Gornitz et al. 2001) based on Bruun rules (Rosen 1978). However, many of these methods fail to specifically address the considerable uncertainty involved in quantifying how factors influence the shoreline (Gutierrez et al. 2011).

Gutierrez et al. (2011) developed a BN model to demonstrate the probability of shoreline change rate on the US Atlantic coast, while Dlamini (2011) developed a

Fig. 5 Tanna Island's reliance on the sea is affected by sea level rise and coastal erosion

spatially explicit BN model to assess climate change impact on vegetation in Swaziland. Following the suggested research directions in these two previous studies, we coupled GIS and BN to generate probabilistic risk maps for predictions of coastal erosion in a small Pacific Island, namely Tanna Island, Vanuatu (Fig. 5).

Key distinctive characteristics of the novel approach developed and demonstrated in this study are:

(a) use of high-resolution raster datasets (10 m) for more accurate prediction;
(b) development of a BN (structure and parameters) from data rather than using expert knowledge; and
(c) rapid and interactive demonstration of probabilistic predictions through BN and GIS coupling

As illustrated in Fig. 6, the first step of BN and GIS integration is to define a BN structure to understand the coastal system qualitatively. The structure was created using Netica software package (Norsys 2018). The input data used in the BN was based on GIS layers. Thus, as a second step, all datasets required to populate the BN input variables were converted to spatially explicit raster grids with a resolution of 10 m using ArcGIS (ESRI 2014). The next step was to generate ASCII prediction maps using Netica for shoreline mobility.

The prediction of future shoreline change is a challenging issue due to a range of complex factors contributing to inherent uncertainties and the absence of high-resolution data required for sophisticated physical modelling. The complexity requires integrated robust approaches to understand the causal relationships between the system components. The BN approach provides a robust platform for modelling the complex interactions of these factors influencing the shoreline changes and to quantify the prediction uncertainty. The proposed spatial BN, through the integration of BN with GIS, offered a novel integrated modelling approach to generate probabilistic prediction maps of likely climate change related to erosion impacts.

For further details, the interested reader is referred to Sahin et al. (2019).

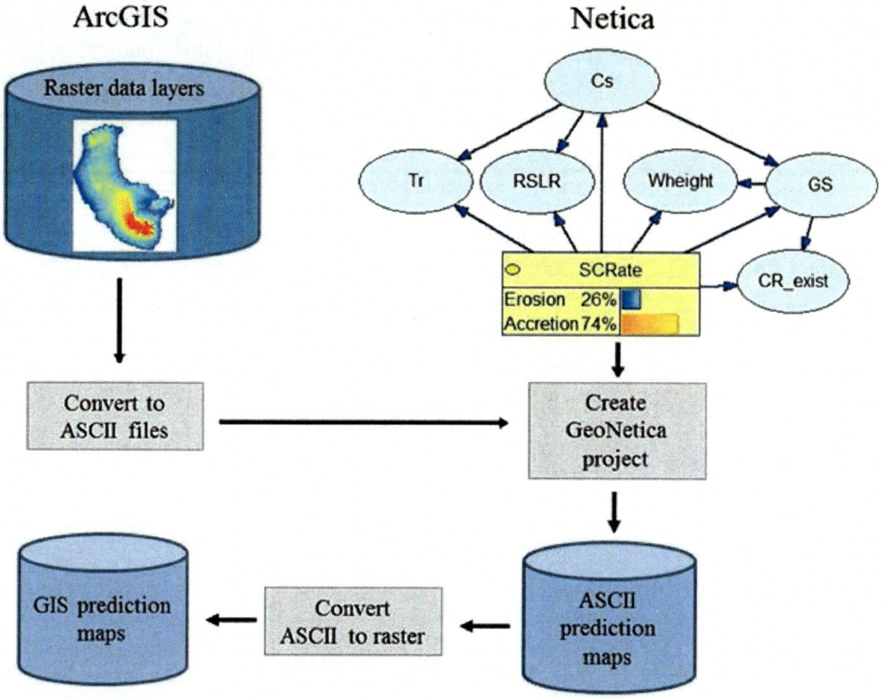

Fig. 6 Hybrid BN and GIS architecture

5 Conclusions

Modelling is a spontaneous human process to fulfil our need to understand, predict and control the world around us. SeaCities, and in general coastal cities of the future, will disrupt our current way of living. As such, our ability to predict and control such new living environment will be lower than what we are capable of in traditional coastal, land-based environments. Therefore, the need for modelling will increase even further.

Though other existing books and resources can provide a more in-depth illustration of the modelling techniques described herein, this chapter wanted to introduce the interested reader with limited modelling knowledge, to a range of modelling categories, highlighting pros and cons and their suitability based on available inputs and desired outputs. The two most important conclusions are:

(a) We are now in the world of smart technology and artificial intelligence. As such, empirical data-driven models will see a surge in their application opportunities. Widespread deployment of high-frequency sensors and enhanced computational capabilities of our machines will allow the understanding and modelling of complex environmental or human patterns and behaviours, which will in turn facilitate a better design of SeaCities.

(b) Although, individually, data-driven models are expected to become by far the leading modelling type to be deployed, an integrated, "participatory" approach is also recommended when modelling complex systems. A step-by-step, hybrid modelling process, where sensor data is merged with experts' input, and different modelling types are linked together, will likely lead to a more comprehensive, and accurate, understanding and modelling of the system analysed.

Whist the full concept of SeaCities might not be realised for a while, the increased dangers to coastal cities coming from sea level rise and overall climate change, with the most negative scenarios becoming more and more likely, are increasingly making such concept worth of discussion, especially for those coastal cities where adaptation is not even an option anymore and retreating appears the only way forward. Modelling approaches described herein will increasingly become the right-hand man to SeaCities and will assist in understanding, predicting and improving the economic, environmental and social viability of SeaCities-related projects.

References

Ahmad S, Simonovic SP (2004) Spatial system dynamics: new approach for simulation of water resources systems. J Comput Civil Eng 18(4):331–340

Bertone E, Sahin O, Richards R, Roiko A (2016a) Extreme events, water quality and health: a participatory Bayesian risk assessment tool for managers of reservoirs. J Clean Prod 135:657–667

Bertone E, Sahin O, Richards R, Roiko A (2016b) Modelling with stakeholders: a systems approach for improved environmental decision making under great uncertainty, 8th International Congress on Environmental Modelling & Software (iEMSs). Toulouse, France

Bertone E, Sahin O, Stewart R, Zou P, Alam M, Hampson K, Blair E (2017) Role of financial mechanisms for accelerating the rate of water and energy efficiency retrofits in Australian public buildings: hybrid Bayesian network and system dynamics modelling approach. Appl Energy. https://doi.org/10.1016/j.apenergy.2017.08.054

Bertone E, Purandare J, Durand B (2019) Spatiotemporal prediction of *Escherichia coli* and Enterococci for the Commonwealth Games triathlon event using Bayesian Networks. Mar Pollut Bull 146:11–21

Bertone E, Luna Juncal MJ, Prado Umeno RK, Peixoto DA, Nguyen K, Sahin O (2020) Effectiveness of the early response to COVID-19: data analysis and modelling. Systems 8(2):21

Bhattacharjee S, Mukherjee N (2017) Floating houses: a design for flood resilience innovations in Bangladesh, Edinburgh

Burdick DM, Short FT (1999) The effects of boat docks on eelgrass beds in coastal waters of Massachusetts. Environ Manage 23(2):231–240

Day AH, Babarit A, Fontaine A, He YP, Kraskowski M, Murai M, Penesis I, Salvatore F, Shin HK (2015) Hydrodynamic modelling of marine renewable energy devices: a state of the art review. Ocean Eng 108:46–69

de Lima R, Boogaard F, de Graaf R, Pires MD, Sazonov V (2015a) Monitoring the impacts of floating structures on the water quality and ecology using an underwater drone, The Hague, Netherlands

de Lima R, Boogaard F, Graaf Rd (2015b) Innovative dynamic water quality and ecology monitoring to assess about floating urbanization environmental impacts and opportunities. International Waterweek

Dlamini W (2011) Probabilistic spatio-temporal assessment of vegetation vulnerability to climate change in Swaziland. Glob Change Biol 17(3):1425–1441
Druzdel MJ, Van Der Gaag LC (2000) Building probabilistic networks: "Where do the numbers come from?". IEEE Trans Knowl Data Eng 12(4):481–486
ESRI (2014) ArcGIS Desktop: Release 10.3 Environmental Systems Research Institute, Redlands, CA
Fenton N, Neil M (2013) Risk assessment and decision analysis with Bayesian networks. CRC Press, New York
Foka E, Rutten M, Boogaard F, de Graaf R, Lima R, Giessen N (2015) The effect of floating houses on water quality
Garrett D. (1999) Teleology in Spinoza and early modern rationalism. New essays on the rationalists, pp 310–336
Gerl T, Kreibich H, Franco G, Marechal D, Schröter K (2016) A review of flood loss models as basis for harmonization and benchmarking. PLoS ONE 11(7):e0159791
Gesch DB (2009) Analysis of lidar elevation data for improved identification and delineation of lands vulnerable to sea-level rise. J Coast Res, pp 49–58
Gharib S (2008) Synthesizing system dynamics and geographic information systems in a new method to model and simulate environmental systems. Doctoral thesis. The University of Bergen. http://hdl.handle.net/1956/3296
Godet M (2001) Creating futures. Economica
Gornitz V, Couch S, Hartig EK (2001) Impacts of sea level rise in the New York City metropolitan area. Global Planet Change 32(1):61–88
Grossman WD, Eberhardt S (1992) Geographical information systems and dynamic modelling. Ann Reg Sci 26:53–66
Gutierrez BT, Plant NG, Thieler ER (2011) A Bayesian network to predict coastal vulnerability to sea level rise. J Geophys Res-Earth Surface 116
Haasnoot M, Middelkoop H, Offermans A, Van Beek E, Van Deursen WP (2012) Exploring pathways for sustainable water management in river deltas in a changing environment. Clim Change 115(3–4):795–819
Ishaque F, Ahamed M, Hoque M (2014) Design and estimation of low cost floating house. Int J Innov Appl Stud 7(1):49
Ji Z-G (2017) Hydrodynamics and water quality: modeling rivers, lakes, and estuaries. Wiley
Kenway S, Lant P, Priestley A, Daniels P (2011) The connection between water and energy in cities: a review. Water Sci Technol 63(9):1983–1990
Khanzadi M, Nasirzadeh F, Alipour M (2012) Integrating system dynamics and fuzzy logic modeling to determine concession period in BOT projects. Autom Constr 22:368–376
Ko KKM (2015) Releasing a floating city a feasibility study of the construction of a floating city, Delft University of Technology, Civil engineering and Geosciences, Specialisation Hydraulic Structures
Kulp SA, Strauss BH (2019) New elevation data triple estimates of global vulnerability to sea-level rise and coastal flooding. Nat Commun 10(1):4844
Labadie JW (2004) Optimal operation of multireservoir systems: state-of-the-art review. J Water Resources Plann Manage 130(2):93–111
Lamas-Pardo M, Iglesias G, Carral L (2015) A review of Very Large Floating Structures (VLFS) for coastal and offshore uses. Ocean Eng 109:677–690
Levi Y, Bekhor S, Rosenfeld Y (2019) A multi-objective optimization model for urban planning: The case of a very large floating structure. Transp Res Part C: Emerging Technol 98:85–100
Lin Y-H, Chih Lin Y, Tan H-S (2019) Design and functions of floating architecture–a review. Mar Georesour Geotechnol 37(7):880–889
Lorenz E (2000) The butterfly effect. World Sci Ser Nonlinear Sci Ser A 39:91–94

Ma N, Hirayama T, Ishikawa K (2001) Practical design of ships and other floating structures. Elsevier, pp 187–195

Mayfield HJ, Bertone E, Smith C, Sahin O (2019) Use of a structure aware discretisation algorithm for Bayesian networks applied to water quality predictions. Math Comput Simul

McGranahan G, Balk D, Anderson B (2007) The rising tide: assessing the risks of climate change and human settlements in low elevation coastal zones. Environ Urbanization 19(1):17–37

Mignot E, Li X, Dewals B (2019) Experimental modelling of urban flooding: A review. J Hydrol 568:334–342

Mohamad MI, Nekooie MA, Ismail ZB, Taherkhani R (2013) Amphibious urbanization as a sustainable flood mitigation strategy in South-East Asia. Trans Tech Publ, pp 1696–1700

Mojaddadi H, Pradhan B, Nampak H, Ahmad N, Ghazali. AHb (2017) Ensemble machine-learning-based geospatial approach for flood risk assessment using multi-sensor remote-sensing data and GIS, Geomatics. Nat Hazards Risk 8(2):1080–1102

Nduwimana A, Yang X-L, Wang L-R (2007) Evaluation of a cost effective technique for treating aquaculture water discharge using Lolium perenne Lam as a biofilter. J Environ Sci 19(9):1079–1085

Neumaier A (2004) Modeling languages in mathematical optimization. In: Kallrath J (ed). Kluwer, Boston

Norsys (2018) Netica—Bayesian network software, Norsys Software Corp. Available: http://www.norsys.com/. Vancouver, Canada

Oppenheimer M, Glavovic B, Hinkel J, van de Wal R, Magnan AK, Abd-Elgawad A, Cai R, Cifuentes-Jara M, Deconto RM, Ghosh T (2019) Sea level rise and implications for low lying Islands, coasts and communities

Pilotti M, Milanesi L, Ranzi R (2016) People and buildings vulnerability to floods in mountain areas

Priest S (2007) The British empiricists. Routledge

Richards R, Sanó M, Roiko A, Carter R, Bussey M, Matthews J, Smith TF (2013) Bayesian belief modeling of climate change impacts for informing regional adaptation options. Environ Model Softw 44:113–121

Roeffen B, Zanon BDB, Czapiewska K, De Graaf R (2013) Reducing global land scarcity with floating urban development and food production

Rook L (2013) Mental models: a robust definition. The learning organization

Rosen PS (1978) A regional test of the Bruun Rule on shoreline erosion. Mar Geol 26(1):M7–M16

Ruth M, Pieper F (1994) Modeling spatial dynamics of sea-level rise in a coastal area. Sys Dynam Rev 10(4):375–389

Sahin O (2011) Dynamic assessment of coastal vulnerability and adaptation to sea level rise: an integrated spatial-temporal decision making approach. Griffith University, Gold Coast, Australia

Sahin O, Mohamed S (2013) A spatial temporal decision framework for adaptation to sea level rise. Environ Model Softw 46(2013):129–141

Sahin O, Mohamed S (2014) Coastal vulnerability to sea-level rise: a spatial-temporal assessment framework. Nat Hazards 70(1):395–414

Sahin O, Stewart RA, Faivre G, Ware D, Tomlinson R, Mackey B (2019) Spatial Bayesian Network for predicting sea level rise induced coastal erosion in a small Pacific Island. J Environ Manage 238:341–351

Sahu A, Yadav N, Sudhakar K (2016) Floating photovoltaic power plant: a review. Renew Sustain Energy Rev 66:815–824

Shekhorkina S (2015) Basic principles of low-rise residential floating buildings design. Lviv Polytechnic Publishing House

Smith T, Low Choy D, Thomsen D, Serrao-Neumann S, Crick F, Sano M, Richards R, Harman B, Baum S, Myers S (2015) CRC Press

Sterman JD (2000) Business dynamics: systems thinking and modeling for a complex world. The McGraw-Hill Companies, Unied States of America
Stopp H, Strangfeld P (2010) Floating houses–chances and problems. WIT Trans Ecol Environ 128:221–233
Strangfeld P, Stopp H (2014) Floating houses: an adaptation strategy for flood preparedness in times of global change. WIT Trans Ecol Environ 184:277–286
Tessem B, Davidsen PI (1994) Fuzzy system dynamics: an approach to vague and qualitative variables in simulation. Syst Dyn Rev 10(1):49–62
Ugurlu C, Chatzikonstantinou I, Sariyildiz S, Tasgetiren MF (2015) Identification of sustainable designs for floating settlements using computational design techniques. IEEE, pp 2303–2310
Uusitalo L (2007) Advantages and challenges of Bayesian networks in environmental modelling. Ecol Model 203:312–318
Voinov A, Gaddis EJB (2008) Lessons for successful participatory watershed modeling: a perspective from modeling practitioners. Ecol Model 216(2):197–207
Wong PP, Losada IJ, Gattuso J-P, Hinkel J, Khattabi A, McInnes KL, Saito Y, Sallenger A (2014) Coastal systems and low-lying areas. Field CB, Barros VR, Dokken DJ, Mach KJ, Mastrandrea MD, Bilir TE, Chatterjee M, Ebi KL, Estrada YO, Genova RC, Girma B, Kissel ES, Levy AN, MacCracken S, Mastrandrea PR, White LL (ed) Cambridge University Press, Cambridge, United Kingdom and New York, NY, USA, pp 361–409
Yeh N, Yeh P, Chang Y-H (2015) Artificial floating islands for environmental improvement. Renew Sustain Energy Rev 47:616–622

Dr. Edoardo Bertone is a researcher with a civil engineering background, and a current focus on data-driven modelling, Bayesian network and system dynamics modelling applied to the water resources management, water treatment optimisation, water–energy nexus, sustainable cities, environmental health and climate change adaptation fields. He completed his Ph.D. in water resources engineering in 2015, and he is currently a lecturer at Griffith University. He collaborates with several large Australian water utilities, Government departments and city councils, and he is a member of the SeaCities group under Griffith's Cities Research Institute.

Dr. Oz Sahin is a Senior Research Fellow at Griffith Climate Change Response Programme. He is a systems modeller with specialised expertise in a range of modelling approaches to problems of natural resource management, risk assessment and climate change adaptation. Prior to his current role, after completing his Ph.D. in the field of climate change adaptation and vulnerability assessment, he worked on numerous research projects with a range of large interdisciplinary teams including: Urban and Environmental Planning, Environmental Health and Engineering. His research interests include: climate change adaptation research; integrated water, energy and climate modelling, integrated decision support systems using coupled system dynamics and GIS modelling, Bayesian network modelling, multiple criteria decision analysis, operational research methods and models. Dr. Sahin has been an active member of a number of international modelling and simulation societies, including: the International Environmental Modelling and Software Society (iEMSs); the Modelling and Simulation Society of Australia and New Zealand (MSSANZ); the System Dynamics Society and the International Centre for Sustainable Development of Energy, Water and Environment Systems (SDEWES).

Overlapping Worlds: An Indigenous Jurisprudential Approach to SeaCities

C. F. Black

1 The Winding Path of Commentary

This chapter is more of a commentary than straight discourse. It is a winding path of storytelling and discourse of considerations on the influences that are shaping the public discourse around sea rise which in turn influences risks and opportunities when planning the urban tactics for flooding in cities such as the Gold Coast, the Gold Coast being the traditional lands of my mother's ancestors.

As an indigenous jurisprudence scholar, I endeavour to understand major cultural shifts through an analysis of influential 'popular culture' narratives of the past and the present, to ascertain a society's principles and value that guide practice, decision making and in the case of the topic of this book.

But first, I will give an example of indigenous legal thinking with a quote from the book—*The Land is the Source of the Law.*[1]

> A further realization of land as the law comes from the understanding that law is not dependent on human action for its continuity, for the Law is already posited in the land. If anything, man would appear to be redundant species when it comes to maintaining the law, as shown in the following dialogue between Malnic and Senior Law Man Mowaljarla:
>
> What about the area where there are no Aboriginal people still surviving or at least not living traditionally there any longer?
>
> You're wrong there thinking like that. The land remained, you can't get away from that. It acts for the people and the imprint is still there. If the land sinks into the ocean, the symbols will still be there. Only if the whole continent is blown to pieces and nothing is left of it, then it will be finished. (Mowaljarlai and Malnic 1993: 192)

This quote is an example of how sea rise does not mean the end of relatedness to a landscape, but a continuing relationship which many coastal engineers are learning

C. F. Black (✉)
Griffith Centre for Coastal Management, Cities Research Institute, Griffith University, Brisbane, QLD, Australia
e-mail: c.black@griffith.edu.au

[1] Black (2011a).

J. Baumeister et al. (eds.), *SeaCities*, Cities Research Series,
https://doi.org/10.1007/978-981-15-8748-1_12

when dealing with indigenous peoples. I will therefore turn to an indigenous legal fiction or story to bring context and meaning to the commentary.

2 Turning to a Story—Measurement and Speed

Indigenous people can be forgiven for thinking that an influential group of non-indigenous have a strong relationship with measurement and speed. I will therefore tell a story to make my point. This story is about two creator beings. They were the creator beings who gave birth to the profession called engineering. These creator beings were called Measurement and Speed.

Measurement and Speed came to the Earth from 'a galaxy far, far away' to measure every aspect of existence on Earth, right down to the nanoscale. They did this with the greatest of speed. The intention being to ensure that they knew everything about the planet Earth and could return to their home planet and tell them all about Earth and discuss its potential for exploitation and invasion, but more importantly, the reshaping of the landscape to look like their home planet.

At the same time this was happening, a young Aboriginal man was learning the law.

3 Straight Lines and Winding Paths

There was this young man standing on a high cliff. He looked out across the landscape for a waterhole as he had a mighty thirst. There he saw a waterhole off in the distance, so he worked out in his mind the fastest way to get there. This invariably led him along a straight line. Little did he know he was being watched by a group of elders. The young man jumped down from the cliff and happily made his way in quick time to the waterhole. Upon arriving at the waterhole, he was met by a group of angry elders. They proceeded to tell him he was about to be punished for walking in a straight line instead of walking along the natural paths made by the Earth which were winding and, in some cases, difficult to get across. They further explained that he had not let the land teach him how to approach with an integrated sense of how the water and land interact, nor of the various species of food he might find on the winding path, that he would miss on the straight path. Instead, all he had thought about was 'the risks and opportunities' to quench his thirst.

As Yungaporta in his doctoral thesis[2] originally argued:

> There were stories of ancestral heroes who had broken the law by travelling in a straight line. They were punished for this, as the right way to travel is taking a winding path. The idea of a winding path as a map for learning was explored in local cultural contexts like painting,

[2] Yungaporta (2009).

> and it was found across many activities relating to native knowledge and learning. It came out in shapes in the landscape as well, in winding rivers and old pathways.

Therefore, the health of the landscape or seascape depends on a jurisprudence which is built on the *Law of Relationships*.[3] The relationship is built on the knowing through thousands of years of engagement that the human is one of the many species of the Earth and is subject to the whims of the Earth. I know this is hard to swallow and blasphemy in some religious quarters. But God did not hand over the real estate rights to the newcomers just because they planted a flag and thought up something called "squatters right" and then engaged in a war, that is still going on 200 years later.

This long-term relationship then builds up knowledge that is empirical and relational, rather than instrumental and exploitative, as evident in the management of Australian rivers. The Murray Darling river being one of the most obvious debacles.

The recent arrivals, however, have used instruments to come here and have been using them ever since. However, with one major blind spot. They still seem to think they live in the Northern hemisphere. Just take the fact they refer to the seasons as being four seasons. The cyclical six seasons of some areas of Australia, or the seemingly annual cycle of flood, drought and fire from a casual observation might seem more likely to be the norm, rather than something like the arrival of spring should be on or near the first of September. This rigidity and boxing in of the weather is frightening, as it is so illogical.

When a knowledge of a subject comes via instruments rather than engagement or relational, a disconnection and objectification come about. This includes the objectification of other human beings.

I know that many readers are thoughtful and environmentally conscious participants. The goal of the Dutch sea rise management concept of 'Room for the River' is to give the river more room to be able to manage higher water levels.[4]

These types of concepts overlap well with indigenous thinking. However, what concerns me is that this wonderful thinking is based on the following thinking:

> If focus is turned back towards this asset, this natural resource, we could start treating it better and improving the condition of the waterways.[5]

This type of thinking is still one in which humans are saving the day and are in partnership with the Earth. A fundamental flaw by the instrumentalist is that they believe that they can be (equal) partners. This is false thinking as humans are not partners. The Earth does not need them for its survival. Take Mars for example; for that planet to exist, it is not dependent on any species at all. But rather, species are dependent on the planet for the development of an ecosystem in which to survive. We are in the fortunate position that the Earth allowed the enhancement of the ecosystem to allow the development of humans. But there has developed over recent times a

---

[3]Black (2011b).

[4]https://www.ruimtevoorderivier.nl/english/.

[5]https://www.dezeen.com/2015/12/09/floating-architecture-buildings-will-offer-improved-way-of-living-amphibious-housing-houseboats/.

culture of garbage generating. This cultural behaviour is now rampant on the Earth's surface, with small pockets of the original civilised humans living in the Amazon, Africa and other parts of the world where Coca Cola and coal are not the staple diet. Many are trying hard to change this cultural tendency, but it is a losing battle, especially if they live in cities.

My suggestion is that if they want to survive, they will need to step down from the false notion that they are the apex species, and humbly stand to one side, and work out how to fit in with all the species of the Earth. After all most of the innovation in the world is based on mimicking nature, i.e., stealing the IP of the species of the Earth rather than humbly acknowledging the source of the basic concept.

This brings me back to the book's theme to recognise the risks and opportunities of the Earth's turnings. How do we adjust and mitigate the impact on our cities? Can I suggest, as argued early, we begin by shifting the apex of design and planning from the desires of the humans to that of one of the many 'stakeholders'. I do believe the Native American use of the terms such as Stone Nations, Bear Nations, Spider Nations which is a way of constantly reminding the humans they are not the only stakeholders in the need for survival, but rather treats the non-humans, as political entities with their own rights and responsibilities to the Earth.

To extrapolate the reasons for the need for an appreciation of the relation to the sea and land, let me turn to the issue of why I believe the Earth is making it very uncomfortable for consumer societies.

Last November, I carried out research in Copenhagen, Denmark, in relation to the works of Hans Christian Andersen and Norse Myths. One cannot but help when in such a city to notice the sea and the canals. When travelling the canals, one must duck his/her head to go under the bridges due to sea rise. And I must say it is quite a tight fit going under some of those bridges. But of course, they have built dykes, so they live in a self-satisfied environment that is safe due to a technological fix. In other words, their relationship with the sea is one of defence, rather than following a traditional 'law of relationship'.

And more importantly the popular myth is that the sea will rise at a rate, and they have projected it on their models. And yet Nature will have her way:

> Even with a 25-year data record, detecting acceleration is challenging. Episodes like volcanic eruptions can create variability: the eruption of Mount Pinatubo in 1991 decreased global mean sea level just before the Topex/Poseidon satellite launch, for example. In addition, global sea level can fluctuate due to climate patterns such as El Ninos and La Ninos (the opposing phases of the El Nino-Southern Oscillation), which influence ocean temperature and global precipitation patterns.[6]

Yet the scientist's rhetoric is often showing data that assure us.... *All the really bad stuff is way off in the future.* No wonder the politicians do not listen! Why do we not talk of abrupt climate change, e.g., the abrupt arrival of the bushfires,[7], the

[6] https://climate.nasa.gov/news/2680/new-study-finds-sea-level-rise-accelerating/. Accessed 10 April 2020.

[7] https://www.theguardian.com/australia-news/2020/mar/05/bushfire-crisis-conditions-eight-times-more-likely-under-2c-warming-analysis-shows. Accessed 10 April 2020.

floods brought on by sudden inundation of a month's rain in a matter of hours[8] and most current of all the arrival of Coronavirus, in downtown Southport on the Gold Coast, Australia, in the local shopping mall.[9] Just like an earthquake and its ability to destabilise solid earth, it is in fact destabilising our global economy. The virus is down the road, not way off in a foreign country or all snug across the road in the hospital.

There 'far, far away in 2100' rhetoric which is devoid of the both the immediacy of location, let alone the possibility of events that allow for the continued political passivity and apathy towards listening to the risks and opportunities of sea rise. Even when they do see the flooding, there is this 'deer staring at the head lights' syndrome. Is this a healthy approach and does it colour how sea rise will be dealt with in the future?

4 Wilful Ethics

The Earth is a wilful being. Close attention to the dramatics of a volcanoes, bushfires, and polar vortexes soon dispels the notion of the benevolent mother Earth who deems humans as special. If anything, it is probably only a certain group of humans who think they are so special. Indigenous shamanic and medicinal knowledge soon alerts us to the knowing that other species of the Earth see us as either prey, predator or just plain dumb, as there is no other species on the planet that is constantly filling their backyards and oceans with garbage.[10]

If I turn to a podcast by NOAA, the National Oceanic and Atom Administration in the US I learn:

> The words 'garbage patch' accurately describes what it is, because these are patches of ocean that contain our garbage. But they're not areas where you can easily go through and skim trash off the surface. First of all, because they are tiny micro plastics that aren't easily removable from the ocean. But also just because of the size of this area. We did some quick calculations that if you tried to clean up less than one percent of the North Pacific Ocean it would take 67 ships one year to clean up that portion. And the bottom line is that until we prevent debris from entering the ocean at the source, it's just going to keep congregating in these areas. We could go out and clean it all up and then still have the same problem on our hands as long as there's debris entering the ocean.[11]

These garbage patches across the oceans are a prime example of why there is a need for an Indigenous jurisprudential approach, which asks questions to future designers

---

[8]https://www.abc.net.au/news/2019-12-12/brisbane-storms-damage-six-months-rain/11791512. Accessed 10 April 2020.

[9]https://www.goldcoastbulletin.com.au/news/new-coronavirus-cases-found-on-gold-coast/news-story/0b70ddc5844bd70df7c356c0de161b39. Accessed 10 April 2020.

[10]Luna (1984).

[11]*Garbage Patches: How Gyres Take Our Trash Out to Sea* (NOAA Ocean Podcast: Episode 14) https://oceanservice.noaa.gov/podcast/mar18/nop14-ocean-garbage-patches.html. 26 February, 2020.

and planners that give them a pathway of thought, yes a pathway of thought…which takes into account other priorities, which in turn as Mary Graham who comments on governance, postulates is far more economically efficient in the long term.[12]

5 Major Cultural Shifts

An example of that pathway of thought would include questions such as *is design built on fear?* For example, has there come about a culture of fear of oceans and their relentless encroachment on what has always been considered prime real estate? Even presidential real estate is threatened such as Mar a Lago in Florida, let alone the capital cities such as Djakarta, which are heeding the call and moving north. Well maybe the ruling elite are moving north, but how will it be possible for the poor to move? When it hits the Corporate Kings does that change the attitudes and designs in response to sea level rise? Or will the response to climate change generate another form of inequality? These are the questions many are asking.

Or on a totally different pathway, if I return to my comments on my recent research trip to Denmark: what voices evoke a response? I therefore ask:

Is it when we read in the media, that six-year-old children now worry about climate change?[13]

Or is it when a Scandinavian youth, suffering from Asperger's, has to miss school to bring to the attention of the politicians and global leaders of industry, that the future generations want to be able to breath clean air and fish in the sea, without bringing up tones of plastic rubbish?[14]

Whose concerns change design, those of the large or those of the small? I could further point out that planning and design now have an extra imperative: how to avoid becoming a target for a future generation's anger? As we may well return to the 1960s and the global social upheaval of that era? Especially, considering the global economic fallout from the coronavirus.

And rather curiously, why have Viking sagas become so popular with the viewing public?

Is there something the Norse people know that we do not know?

Are their descendants about to hop on their ships and go *Viking* again? Is the genetic memory of their descendant, Greta Thunberg telling her Ragnarok is about to come upon her people? Ragnarok being the equivalent of the End Times when the giants were on the move. The giants being the natural forces. And the gods were facing 3 years of continuous winter which would mean the end of the good life. Will

[12]Graham (1999).

[13]Kuebler and Russell (2019).

[14]Cwienk, J. Greta Thunberg the teenage eco-activist who took the world by storm. https://www.dw.com/en/greta-thunberg-the-teenage-eco-activist-who-took-the-world-by-storm/a-50018055. Accessed 10 April 2020.

the north see three years of continuous winter or is it that winter will no longer return to the frozen north?

Furthermore, why is the Disney animation Frozen 1 & 11[15] sweeping the world and influencing the young, like some modern-day Pied Piper. The story is loosely based on the Hans Christian Andersen story of the Snow Queen. The original story by Hans Christian Andersen[16] being that of a young girl named Gerda, who travels across Scandinavia to save her friend Kai, who has been captured by the Snow Queen. The Snow Queen being the epitome of the analytical mind, which freezes out any sense of relatedness to the natural world. The Snow Queen allures Kai by offering instruments which distort both human relations and Nature. He is caught up in her icy palace trying to put back together a shattered ice.

> Kai was cleverly arranging his pieces in the game of ice-cold reason. To him the patterns were highly remarkable end of up most important[17]

In other words, obsessed with scientific discovery. Whilst young Gerda must rely on Nature and good relations with others, to get her across the Scandinavian landscape to the icy palace of the Snow Queen. During the time of this venture, the young girl matures into a young woman. She eventually reaches Kai with the help from a reindeer. Her tears finally melt his icy heart and bring him back into alignment with the world around him.

Is that what we are watching in young Greta Thunberg, who tries to work with Nature and calls on the help of other to bring her 'mature' message to those bent on the instrumentalization and exploitation of the Earth and its peoples?

Furthermore, why is Frozen 11 incorporating the Norse mythological concept of Ragnarok by referencing the need to revisit the past to find out why climate change is occurring in the little kingdom that sits on a fjord. We find in the story that by revisiting the past, through the help of Princess Elsa's magical power, they find that the King had cheated the indigenous peoples by offering to build a dam, which in turn destroyed their lands. The King did this because he feared their close relationship to nature, which he saw as magic. This story alludes to the real-life controversy relating to the dam building and Saami rights. The Saami tried to stop the building of the Alta Dam on the Alta–Kautokeino River in the far reaches of Arctic Northern Norway in the early part of the 1980s.[18]

This story could also be a timely reminder in modern times when the latest imaginings of the great oceanographers is to build giant dams across the North Sea. In this case, two oceanographers have proposed that dramatic action needs to be taken to continue 'business as usual' for 25 million people to protect the 'important economical regions' in northern Europe from sea level rise, rather than considering how

---

[15]Lee, J. & Buck, C. (Dir) *Frozen 11*, Walt Disney Studios, USA.

[16]Andersen (1844).

[17]Ibid.

[18]Andersen et al. (1984).

many lives of the species who live in the sea will cease, because humans need to continue polluting the seas, so they can live a 'civilised life'.[19]

6 The Vikings and Ragnarok

The north would appear to be calling us to look to revisit the ancient past. What was the meaning in the Norse myths of Ragnarok, but more importantly what were the principles and values that guided the Norse people as they faced the time of Ragnarok? I turn to the independent scholar Daniel McCoy and his website[20] which goes into detail to teach the general public about the finer attributes of the Norse culture. He references the major Norse scholars such as Australia's Professor Clunies Ross.

The Norse peoples of northern Europe and Scandinavia lived in a world full of sacred meaning, much akin to the Australian Aboriginal world. There was an acceptance of 'the way things are, where nature and culture were one seamless whole rather than a pair of opposites'.[21]

This did not turn them into nature worshipers as modernity likes to portray any culture which appears to understand they are part of nature and not the bosses of nature. These people accepted that they were not there to 'save the world' or make it over in their own image but 'to encounter the deep enchantment in the world went hand in hand with their hard-headed pragmatism'.[22]

After all, the pagan Norse peoples believed that 'the unfolding of events was directed by a blind, implacable fate'. In their worldview, you ultimately could not decide what happened to you; all you could do was react to your lot with 'honour and greatness or dishonour and smallness'. In such a perspective, there is not really any room for "saving the world"—only for doing the best you can during all the changes.[23]

Such an attitude to live was not about 'the inevitability of death and misfortune but instead spurred one onto noble attitude and to do noble deeds—the kind worthy of being recounted by bards many generations after we ourselves are gone'.[24]

---

[19]Groeskamp and Kjellsson (2020).

[20]McCoy, D. Norse Mythology for Smart People. https://norse-mythology.org/the-enchanted-world/. accessed 20 March 2020.

[21]Ibid.

[22]Ibid.

[23]Mc Coy, Ibid.

[24]Ibid.

7 Conclusion

Therefore, when approaching the issue of *flooding risks and opportunities*, perhaps a more mature attitude is needed when dealing with the public. Rather than pandering to their fears and trying to constantly 'plug the dyke', like the little Dutch Boy, and perhaps we should learn from the Coronavirus pandemic, just how quickly people conform if the threat is real,

and how quickly the Earth then cleans its air. Therefore, do we find new pathways to start telling people—the ocean is more powerful than human desire, and just like the virus, it does not care for your economic worries and you must move now! And furthermore, other possible catastrophes are a certainty, not a perhaps.

This would require a re-thinking of just how we tell the stories, write the reports and develop PR material for the public and the business community.

We need to put forward a pathway of stories to show people that they must adjust, just as we have seen with the government's response to the pandemic. As for foreign real estate developers, they also need to be more accountable. New development must move inland, as the beach-side real estate is no longer feasible. It is a matter of common sense and long-term investment gain.

And finally, I would ask the individual: Do we go *Viking* and align with the Earth and all her disasters and beauty and be thankful that we can feel that almighty power under our feet?

References

Andersen HC (1844) The snow queen. https://andersen.sdu.dk/vaerk/hersholt/TheSnowQueen_e.html. Accessed 10 Apr 2020

Andersen SS, Midttun A, Andersen S (1984) Conflict and local mobiliaztion: the alta hydropower project. Acta Socilogica 28(4):317–335

Black CF (2011a) The land is the source of the law: a dialogic encounter with an indigenous jurisprudence. Routledge, UK, p 52

Black CF (2011b) The land is the source of the law: a dialogic encounter with an indigenous jurisprudence, p 15. Routledge, UK

Graham M (1999) Some thoughts about the philosophical underpinnings of aboriginal worldviews. Worldviews Environ, Culture, Religion, 3

Groeskamp C, Kjellsson J (2020) The Northern European Enclosure Dam for if climate change mitigation fails. Bull Am Meteorol Soc. https://doi.org/10.1175/bams-d-19-0145.1. https://journals.ametsoc.org/doi/pdf/10.1175/BAMS-D-19-0145.1. Accessed 24 Mar 2020.

Kuebler M, Russell R (2019) Five answers for kids concerned about climate change. https://www.dw.com/en/five-answers-for-kids-concerned-about-climate-change/a-51561183. Accessed 10 Apr 2020

Luna LE (1984) The healing practices of a Peruvian Shaman. J Ethnopharmacol 11(2):123–133

Yungaporta T (2009) p 26 https://researchonline.jcu.edu.au/10974/2/01thesis.pdf

C. F. Black is an indigenous intellectual explorer and creative practitioner. She is a descendant of the Yugambeh speakers and Associate Professor adjunct to the Griffith Centre for Coastal Management, Griffith University, Australia. She has published two books (Routledge) relating to indigenous legal theory and many other publications. Black is now combining her research into coastal sea rise with her artistic performances, as part of the Ensayos creative research collective who have the Summer Digital Residency (2020) at the New Museum of Contemporary Arts, New York. Ongoing research on sea rise awareness will be done in northern Australia and Scandinavia.